MORAL TRIBES

MORAL TRIBES

EMOTION, REASON, AND THE GAP
BETWEEN US AND THEM

Joshua Greene

THE PENGUIN PRESS

NEW YORK

2013

THE PENGUIN PRESS
Published by the Penguin Group
Penguin Group (USA) Inc., 375 Hudson Street,
New York, New York 10014, USA

USA · Canada · UK · Ireland · Australia
New Zealand · India · South Africa · China

penguin.com
A Penguin Random House Company

First published by The Penguin Press, a member of Penguin Group (USA) LLC, 2013

Illustration credits appear on pages 407–408.

Library of Congress Cataloging-in-Publication Data

Greene, Joshua David, 1974-
Moral tribes : emotion, reason, and the gap between us and them / Joshua D. Greene.
pages cm
Includes bibliographical references and index.
ISBN 978-1-59420-260-5
1. Ethics. 2. Emotions. 3. Civilization. I. Title.
BJ1031.G75 2013
170—dc23
2013007775

Printed in the United States of America
1   3   5   7   9   10   8   6   4   2

BOOK DESIGN BY AMANDA DEWEY

For Andrea

Man will become better when you show him what he is like.

—ANTON CHEKHOV

☺ *The philosophy of one century is the common sense of the next* ☺

—FORTUNE COOKIE, TIGER NOODLES, PRINCETON, NEW JERSEY

Contents

Introduction

The Tragedy of Commonsense Morality

To the east of a deep, dark forest, a tribe of herders raises sheep on a common pasture. Here the rule is simple: Each family gets the same number of sheep. Families send representatives to a council of elders, which governs the commons. Over the years, the council has made difficult decisions. One family, for example, took to breeding exceptionally large sheep, thus appropriating more of the commons for itself. After some heated debate, the council put a stop to this. Another family was caught poisoning its neighbors' sheep. For this the family was severely punished. Some said too severely. Others said not enough. Despite these challenges, the Eastern tribe has survived, and its families have prospered, some more than others.

To the west of the forest is another tribe whose herders also share a common pasture. There, however, the size of a family's flock is determined by the family's size. Here, too, there is a council of elders, which has made difficult decisions. One particularly fertile family had twelve children, far more than the rest. Some complained that they were taking up too much of the commons. A different family fell ill, losing five of their

six children in one year. Some thought it unfair to compound their tragedy by reducing their wealth by more than half. Despite these challenges,
the Western tribe has survived, and its families have prospered, some more
than others.

To the north of the forest is yet another tribe. Here there is no common pasture. Each family has its own plot of land, surrounded by a fence.
These plots vary greatly in size and fertility. This is partly because some
Northern herders are wiser and more industrious than others. Many such
herders have expanded their lands, using their surpluses to buy land from
their less prosperous neighbors. Some Northern herders are less prosperous than others simply because they are unlucky, having lost their flock,
or their children, to disease, despite their best efforts. Still other herders
are exceptionally lucky, possessing large, fertile plots of land, not because
they are especially wise or industrious but because they inherited them.
Here in the North, the council of elders doesn't do much. They simply
ensure that herders keep their promises and respect one another's property.
The vast differences in wealth among Northern families have been the
source of much strife. Each year, some Northerners die in winter for want
of food and warmth. Despite these challenges, the Northern tribe has survived. Most of its families have prospered, some much more than others.

To the south of the forest is a fourth tribe. They share not only their
pasture but their animals, too. Their council of elders is very busy. The
elders manage the tribe's herd, assign people to jobs, and monitor their
work. The fruits of this tribe's labor are shared equally among all its members. This is a source of much strife, as some tribe members are wiser and
more industrious than others. The council hears many complaints about
lazy workers. Most members, however, work hard. Some are moved to
work by community spirit, others by fear of their neighbors' reproach. Despite their challenges, the Southern tribe has survived. Its families are not,
on average, as prosperous as those in the North, but they do well enough,
and in the South no one has ever died in winter for want of food or warmth.

One summer, a great fire burned through the forest, reducing it to
ash. Then came heavy rains, and before long the land, once thick with
trees, was transformed into an expanse of gently rolling grassy hills,

perfect for grazing animals. The nearby tribes rushed in to claim the land. This was a source of much strife. The Southern tribe proclaimed that the new pastures belonged to all people and must be worked in common. They formed a new council to manage the new pastures and invited the other tribes to send representatives. The Northern herders scoffed at this suggestion. While the Southerners were making their big plans, Northern families built houses and stone walls and set their animals to graze. Many Easterners and Westerners did the same, though with less vigor. Some families sent representatives to the new council.

The four tribes fought bitterly, and many lives, both human and animal, were lost. Small quarrels turned into bloody feuds, which turned into deadly battles: A Southern sheep slipped into a Northerner's field. The Northerner returned it. Another Southern sheep did the same. The Northerner demanded a fee to return it. The Southerners refused to pay. The Northerner slaughtered the sheep. Southerners took three of the Northerner's sheep and slaughtered them. The Northerner took ten of the Southerners' sheep and slaughtered them. The Southerners burned down the Northerner's farmhouse, killing a child. Ten Northern families marched on the Southerners' meetinghouse and set it ablaze, killing dozens of Southerners, including many children. Back and forth they went with violence and vengeance, soaking the green hills with blood.

To make matters worse, tribes from distant lands arrived to settle the new pastures. One tribe claimed the new pastures as a gift to them from their god. The burning of the great forest and the greening of the hills had been prophesied in their holy book, they said. Another tribe claimed the new pastures as their ancestral homeland, from which they had been driven many generations ago, before there was a forest. Tribes arrived with rules and customs that seemed to outsiders rather strange, if not downright ridiculous: Black sheep must not sleep in the same enclosure as white sheep. Women must have their earlobes covered in public. Singing on Wednesdays is strictly forbidden. One man complained of a neighboring woman who, while tending her sheep, bared her earlobes in plain view of his impressionable sons. The woman refused to cover her earlobes, and this filled her pious neighbor with rage. A little girl told a little boy that the god to which his family prayed did not exist. The shocked boy reported this to his father, who complained to the girl's father. The father defended his daughter, praising her fierce intelligence, and refused to apologize. For this he was killed, as required by the laws of the tribe he had offended. And so began another bloody feud.

Despite their fighting, the herders of the new pastures are, in many ways, very similar. For the most part, they want the same things: healthy families, tasty and nutritious food, comfortable shelter, labor-saving tools, leisure time to spend with friends and family. All herders like listening to music and hearing stories about heroes and villains. What's more, even as they fight one another, their minds work in similar ways. What they perceive as unjust makes them angry and disgusted, and they are motivated to fight, both by self-interest and by a sense of justice. Herders fight not only for themselves but for their families, friends, and fellow tribe members. They fight with honor and would be ashamed to do otherwise. They guard their reputations fiercely, judge others by their deeds, and enjoy exchanging opinions.

Despite their differences, the tribes of the new pastures share some core values. In no tribe is it permissible to be completely selfish, and in no tribe are members expected to be completely selfless. Even in the South, where the herd is shared, workers are free at day's end to pursue their own interests. In no tribe are ordinary members allowed to lie, steal, or harm one another at will. (There are, however, some tribes in which certain privileged individuals are free to do as they please.)

The tribes of the new pastures are engaged in bitter, often bloody conflict, even though they are all, in their different ways, moral peoples. They fight not because they are fundamentally selfish but because they have incompatible visions of what a moral society should be. These are not merely scholarly disagreements, although their scholars have those, too. Rather, each tribe's philosophy is woven into its daily life. Each tribe has its own version of moral common sense. The tribes of the new pastures fight not because they are immoral but because they view life on the new pastures from very different moral perspectives. I call this the *Tragedy of Commonsense Morality.*

The Parable of the New Pastures is fictional, but the Tragedy of Commonsense Morality is real. It's the central tragedy of modern life, the deeper tragedy behind the moral problems that divide us. This book is about understanding and, ultimately, solving these problems. Unlike many authors of popular books, I make no promise of helping you solve your personal problems. What I'm offering you, I hope, is clarity—and with this clarity, the motivation and opportunity to join forces with like-minded others.

This book is an attempt to understand morality from the ground up. It's about understanding what morality is, how it got here, and how it's implemented in our brains. It's about understanding the deep structure of moral problems as well as the differences between the problems that our brains were designed to solve and the distinctively modern problems we face today. Finally, it's about taking this new understanding of morality and turning it into a universal moral philosophy that members of all human tribes can share.

This is an ambitious book. I started developing these ideas in my late teens, and they've taken me through two interwoven careers—as a philosopher and as a scientist. This book draws inspiration from great philosophers of the past. It also builds on my own research in the new field of moral cognition, which applies the methods of experimental psychology and cognitive neuroscience to illuminate the structure of moral thinking. Finally, this book draws on the work of hundreds of social scientists who've learned amazing things about how we make decisions and how our choices are shaped by culture and biology. This book is my attempt to put it all together, to turn this new scientific self-knowledge into a practical philosophy that can help us solve our biggest problems.

LIFE ON THE NEW PASTURES

Two issues dominated Barack Obama's first presidential term: healthcare and the economy. Both reflect the tension between the individualism of the

Northern herders and the collectivism of the Southern herders. The Patient Protection and Affordable Care Act, also known as Obamacare, established national health insurance in the United States. Liberals praised it, not as a perfect system but as a historic step in the right direction. The United States had finally joined the rest of the modern world in providing basic health-care to all its citizens. Conservatives—many of them—despise Obamacare, which they regard as a step toward ruinous socialism. The recent healthcare debate has been awash in misinformation,* but amid the lies and half-truths there can be found an honest philosophical disagreement.

At its core, this disagreement, like so many others, is about the tension between individual rights and the (real or alleged) greater good. Universal health insurance requires everyone to buy in, either through an individual purchase of health insurance or through taxes. Conservatives mounted a legal challenge to Obamacare, culminating in a landmark Supreme Court decision. The Supreme Court upheld Obamacare on the grounds that it's funded through a combination of voluntary purchases and taxes (which are both constitutional) rather than by the government's forcing people to buy something (which is arguably not constitutional). But the tax-versus-forced-purchase distinction is really just a legal technicality. The people who hate Obamacare don't hate it because they believe that it's funded by forced purchases rather than forced taxes; what they hate is the *forcing*. Obamacare might not be socialism, but it's certainly more collectivist than some people care for, restricting individual freedom in the name of the greater good.

During the 2012 Republican presidential primary, candidates de-nounced Obamacare as loudly and often as possible, calling it social-ism and vowing to repeal it. During one of the primary debates, journalist

---

This is the only footnote in this book, but the endnotes are packed with supporting material, in addition to source citations. Nowadays, many books leave endnotes unmarked in the main text. I don't want to clutter your view with hundreds of little numbers, but I want you to know when you may be missing something of interest, and how much you may be missing. I've therefore devised the following notation system, analogous to the chili pepper heat index used on Asian food menus: Asterisks are used to indicate additional material in the notes (= sentences; ** = para-graphs; *** = pages). Notes that simply list sources are unmarked in the main text. (For more on "awash in misin-formation," please see the endnotes for the introduction.)

Wolf Blitzer had the following exchange with Texas congressman Ron Paul.

> BLITZER: A healthy 30-year-old young man has a good job, makes a good living, but decides, you know what? I'm not going to spend $200 or $300 a month for health insurance because I'm healthy, I don't need it. But something terrible happens, all of a sudden he needs it. Who's going to pay if he goes into a coma, for example? Who pays for that?
>
> PAUL: Well, in a society that you accept welfarism and socialism, he expects the government to take care of him.
>
> BLITZER: Well, what do you want?
>
> PAUL: But what he should do is whatever he wants to do, and assume responsibility for himself. My advice to him would have a major medical policy, but not be forced—
>
> BLITZER: But he doesn't have that. He doesn't have it, and he needs intensive care for six months. Who pays?
>
> PAUL: That's what freedom is all about, taking your own risks. This whole idea that you have to prepare and take care of everybody—
>
> [applause]
>
> BLITZER: But Congressman, are you saying that society should just let him die?

As Paul prepared his hesitant answer, a chorus of voices from the crowd shouted, "Yeah! Let him die!" These are the Northern herders. Paul couldn't quite bring himself to agree—or disagree. He said that neighbors, friends, and churches should take care of such a man, implying, but not explicitly stating, that the government should let him die if no one else is willing or able to pay. As you might expect the more Southerly herders disagree.

(Note: In the Parable of the New Pastures, the Southern herders are extreme collectivists, communists, and are thus far to the left of contemporary mainstream liberals, despite frequent accusations to the contrary. Thus, as we discuss contemporary politics, I refer to contemporary liberals as "more Southerly" rather than "Southern." Contemporary U.S. conservatives, in contrast, resemble more closely their fictional Northern counterparts.)

Along with healthcare, the miserable state of the U.S. economy took center stage during President Obama's first term. When Obama took office in 2009, the economy was in free fall, thanks to a housing bubble that burst after a decade of inflated growth and a financial sector that placed enormous bets on housing prices. The government did several things in an attempt to stave off complete financial disaster. First, in late 2008, while President Bush was still in office, the federal government bailed out several of the investment banks at the heart of the crisis.* Later, the Obama administration bailed out the auto industry and extended aid to homeowners facing foreclosure. These measures were opposed, to varying degrees, by Northern herders who argued that the banks, the automakers, and the desperate homeowners should, like Ron Paul's hypothetical patient, be allowed to "die." Why, they asked, should American taxpayers have to pay for these people's poor judgment? The more Southerly herders didn't especially relish the thought of bailing out irresponsible decision makers, but they argued that these measures were necessary for the greater good, lest their bad choices sink the whole economy. During Obama's first year, congressional Democrats passed his $787 billion stimulus bill, the American Recovery and Reinvestment Act of 2009. This, too, was opposed by Northern herders who favored less government spending and more tax cuts. Better, they said, to put money into the pockets of individuals who can decide for themselves how to spend it.

Related to both healthcare and the economy is the broader issue of economic inequality, which came to the fore in 2011 with the Occupy Wall Street protests. From 1979 to 2007, the incomes of the wealthiest U.S. households skyrocketed, with the top 1 percent enjoying income gains of 275 percent, while the bulk of Americans gained around 40 percent. (The gains at the tippy top, the top 0.1 percent, were even larger, around 400 percent.) These trends inspired the Occupy slogan "We are the 99%," calling for economic reforms to restore a more egalitarian distribution of wealth and power.

The story of rising income inequality comes in two versions. According to the individualist Northern herders, the winners earned their winnings fair and square, and the losers have no right to complain. "Occupy a Desk!" read the sign of a Wall Street counterprotester. Presidential hopeful Herman Cain called the protesters "un-American," and the eventual Republican nominee, Mitt Romney, accused them of waging "class warfare."

In September 2012, the liberal magazine *Mother Jones* dropped one of the biggest bombshells in U.S. electoral history. They posted online a secret recording of Romney in which he described roughly half of the American population as willful government dependents who will never "take personal responsibility and care for their lives." According to Romney's infamous speech, the "47 percent" of the population that earns too little to pay income taxes (on top of payroll taxes) deserve no better than what they've got.

The more Southerly herders tell a different story. They say that the wealthy have rigged the system in their favor, noting that rich people like Mitt Romney pay taxes at a lower rate than many middle-class workers, thanks to lower tax rates on investment income, myriad tax loopholes, and overseas tax havens. And now, thanks to the Supreme Court's decision in *Citizens United v. Federal Election Commission*, which legalized unlimited campaign contributions to "independent" political groups, the rich can use their money to buy elections like never before. These more Southerly herders say that even in the absence of nefarious system rigging, maintaining a just society requires active redistribution of wealth. Otherwise the rich use their advantages to get richer and richer, passing on their advantages to their children, who then begin life with a big head start. Without redistribution of wealth, they say, our society will bifurcate into permanent classes of haves and have-nots.

During her first political campaign, Massachusetts senator Elizabeth Warren made a Southerly case for redistribution in a stump speech that went viral on YouTube:

There is nobody in this country who got rich on their own. Nobody. You built a factory out there—good for you. But I want to be clear. You moved your goods to market on roads the rest of us paid for. You hired workers the rest of us paid to educate. You were safe in your factory because of police forces and fire forces that the rest of us paid for. You didn't have to worry that marauding bands would come and seize everything at your factory . . . Now look. You built a factory and it turned into something terrific or a great idea—God bless! Keep a hunk of it. But part of the underlying social contract is you take a hunk of that and pay forward for the next kid who comes along.

Responding to these remarks, Ron Paul called Warren a socialist and said that the government can do nothing but "steal and rob people with a gun and forcibly transfer wealth from one person to another." Conservative commentator Rush Limbaugh went a step further, calling Warren a communist and "a parasite who hates her host."

Other tribal disagreements are less obviously related to the fundamental divide between individualism and collectivism. In the United States there is enormous disagreement over what, if anything, we ought to do about global warming. This may appear to be, at bottom, not a debate about values but a factual disagreement over whether global warming is a real threat and whether humans are causing it. But is this argument just about how to interpret the data? Those who believe in global warming are saying that all of us must make sacrifices (use less fuel, pay carbon taxes, and so on) to ensure our collective well-being. Individualists are, by nature, skeptical of such demands; collectivists, far less so. Our values may color our view of the facts.

Some of our troubles on the new pastures are not about individualism versus collectivism per se but about the boundaries of our respective collectives. Nearly all of us are collectivists to some extent. The only pure individualists are hermits. Consider, once again, Ron Paul's prescription for the man who neglected to buy health insurance. Paul didn't say that we should let the man die. He said that *friends, neighbors,* and *churches* should take care of him. What this suggests is that our tribal disagreements are not necessarily between individualist and collectivist tribes, but between tribes that are more versus less *tribal,* more versus less inclined to see the world in terms of Us versus Them, and thus more versus less open to collective enterprises that cross tribal lines, such as the U.S. federal government and the United Nations. For many conservatives, the circle of "Us" is just smaller.

Some tribal disagreements arise because tribes have values that are inherently *local*, particular to the tribe in question. Some tribes grant special authority to specific gods, leaders, texts, or practices—what one might call "proper nouns."* For example, many Muslims believe that no one—Muslim or otherwise—should be allowed to produce visual images of the prophet Muhammad. Some Jews believe that Jews are God's "chosen people" and that the Jews have a divine right to the land of Israel. Many American Christians believe that the Ten Commandments should be displayed in public buildings and that all Americans should pledge allegiance to "one nation under God." (And they're not talking about Vishnu.)

The moral practices of some tribes are (or appear to be) arbitrary, but, at least in the developed world, tribes generally refrain from imposing their most arbitrary rules on one another: Orthodox Jews don't expect non-Jews to forgo lobster and to circumcise their male children. Catholics don't expect non-Catholics to wear ash crosses on their foreheads on Ash Wednesday. The tribal differences that erupt into public controversy typically concern sex (e.g., gay marriage, gays in the military, the sex lives of public officials) and death at the margins of life (e.g., abortion, physician-assisted suicide, the use of embryonic stem cells in research). That such issues are moral issues is surely not arbitrary. Sex and death are the gas pedals and brakes of tribal growth. (Gay sex and abortion, for example, are both alternatives to reproduction.) What's less clear is why different tribes hold different views about sex, life, and death, and why some tribes are more willing than others to impose their views on outsiders.

This has been a whirlwind tour of the new pastures in the United States during the period in which I completed this book. If you're reading this book at a later time, or in another place, the specific issues will be different but the underlying tensions will likely be the same. Look around and you'll see Northern and Southern herders fighting over whether government should do more versus less; tribes that have smaller versus larger conceptions of "Us"; tribes engaged in bitter arguments over the morality of sex and death; and tribes demanding deference to their respective proper nouns.

TOWARD A GLOBAL MORAL PHILOSOPHY

If you were an alien biologist, dropping by Earth every ten thousand years or so to observe the progress of life on our planet, there might be a page in your field notebook like this:

Homo sapiens sapiens: big-brained, upright primates, vocal language, sometimes aggressive

VISIT#	POPULATION	NOTES
1	< 10 million	hunter-gatherer bands, some primitive tools
2	< 10 million	hunter-gatherer bands, some primitive tools
3	< 10 million	hunter-gatherer bands, some primitive tools
4	< 10 million	hunter-gatherer bands, some primitive tools
5	< 10 million	hunter-gatherer bands, some primitive tools
6	< 10 million	hunter-gatherer bands, some primitive tools
7	< 10 million	hunter-gatherer bands, some primitive tools
8	< 10 million	hunter-gatherer bands, some primitive tools
9	< 10 million	hunter-gatherer bands, some primitive tools
10	> 7 billion	global indust. economy, advanced technology w/ nuc. power, telecom., artificial intel., extraterrestrial travel, large-scale social/political institutions, democratic governance, advanced scientific inquiry, widespread literacy, and advanced art (See addendum)

For all but the past ten thousand years of our existence, it didn't look like we'd amount to much. Yet here we are, sitting in our climate-controlled, artificially illuminated homes, reading and writing books about ourselves. Our progress goes well beyond creature comforts. Contrary to popular lamentation, humans are getting better and better at getting along. Violence has declined over the course of human history, including recent history, and participation in modern market economies, far from turning us into selfish bean counters, has expanded the scope of human kindness.

Nevertheless, we've plenty of room for improvement. The twentieth century was the most peaceful on record (controlling for population growth), yet its wars and assorted political conflicts killed approximately 230 million people, laying down enough human bodies to circle the globe seven times. In this new century, the death toll continues to climb, albeit at a reduced rate. For example, the ongoing conflict in Darfur has killed, through violence or increased disease, about 300,000 people. A billion people—about one in seven humans—live in extreme poverty, with so few resources that mere survival is an ongoing struggle. More than twenty million people are forced into labor (i.e., slavery), many of them children and women forced into prostitution.

Even in the world's happier quarters, life is still systematically unfair to millions of people. When researchers in the United States sent out identical résumés to prospective employers, some with white-sounding names (e.g., Emily and Greg) and others with black-sounding names (e.g., Lakisha and Jamal), the white résumés generated 50 percent more calls from employers. Worst of all, we face two problems that may severely disrupt, or even reverse, our trend toward peace and prosperity: the degradation of our environment and the proliferation of weapons of mass destruction.

Amid such doom and gloom, the premise of this book is fundamentally optimistic: that we can improve our prospects for peace and prosperity by improving the way we think about moral problems. Over the past few centuries, new moral ideas have taken hold in human brains. Many people now believe that no human tribe ought to be privileged over any other, that all humans deserve to have certain basic goods and freedoms, and that violence should be used only as a last resort. (In other words, some tribes have become a lot less tribal.) We subscribe to these ideals more in principle than in practice, but the fact that we subscribe to them at all is something new under the sun. As historians tell us, we've made a lot of progress, not just technologically but morally.

Inverting the usual question about today's morals, Steven Pinker asks: What are we doing right? And how can we do better? What we lack, I think, is a coherent global moral philosophy, one that can resolve disagreements among competing moral tribes. The idea of a universal moral philosophy is not new. It's been a dream of moral thinkers since the Enlightenment. But it's never quite worked out. What we have instead are some shared values, some unshared values, some laws on which we agree, and a common *vocabulary* that we use to express the values we share as well as the values that divide us.

Understanding morality requires two things: First, we must understand the *structure of modern moral problems* and how they differ from the problems that our brains evolved to solve. We'll do this in part 1 of this book. Second, we must understand *the structure of our moral brains* and how different kinds of thinking are suited to solving different kinds of problems. That's part 2. Then, in part 3, we'll use our understanding of moral problems and moral thinking to introduce a solution, a candidate global moral philosophy. In part 4 we'll address some compelling arguments against this philosophy, and in part 5 we'll apply our philosophy to the real world. I'll now describe this plan in a bit more detail.

THE PLAN

In part 1 ("Moral Problems"), we'll distinguish between the two major kinds of moral problems. The first kind is more basic. It's the problem of Me versus Us: selfishness versus concern for others. This is the problem that our moral brains were designed to solve. The second kind of moral problem is distinctively modern. It's Us versus Them: our interests and values versus theirs. This is the Tragedy of Commonsense Morality, illustrated by this book's first organizing metaphor, the Parable of the New

Pastures. (Of course, Us versus Them is a very old problem. But historically it's been a tactical problem rather than a moral one.) This is the larger problem behind the moral controversies that divide us. In part 1, we'll see how the moral machinery in our brains solves the first problem (chapter 2) and creates the second problem (chapter 3).

In part 2 ("Morality Fast and Slow"), we'll dig deeper into the moral brain and introduce this book's second organizing metaphor: The moral brain is like a dual-mode camera with both automatic settings (such as "portrait" or "landscape") and a manual mode. Automatic settings are efficient but inflexible. Manual mode is flexible but inefficient. The moral brain's automatic settings are the *moral emotions* we'll meet in part 1, the gut-level instincts that enable cooperation within personal relationships and small groups. Manual mode, in contrast, is a general capacity for practical reasoning that can be used to solve moral problems, as well as other practical problems. In part 2, we'll see how moral thinking is shaped by both emotion and reason (chapter 4) and how this "dual-process" morality reflects the general structure of the human mind (chapter 5).

In part 3, we'll introduce our third and final organizing metaphor: Common Currency. Here we'll begin our search for a *metamorality*, a global moral philosophy that can adjudicate among competing tribal moralities, just as a tribe's morality adjudicates among the competing interests of its members. A metamorality's job is to make *trade-offs* among competing tribal values, and making trade-offs requires a *common currency*, a unified system for weighing values. In chapter 6, we'll introduce a candidate metamorality, a solution to the Tragedy of Commonsense Morality. In chapter 7, we'll consider other ways of establishing a common currency, and find them lacking. In chapter 8, we'll take a closer look at the metamorality introduced in chapter 6, a philosophy known (rather unfortunately) as *utilitarianism*. We'll see how utilitarianism is built out of values and reasoning processes that are universally accessible and, thus, how it gives us the common currency that we need.*

Over the years, philosophers have made some intuitively compelling arguments against utilitarianism. In part 4 ("Moral Convictions"), we'll reconsider these arguments in light of our new understanding of moral cognition. We'll see how utilitarianism becomes more attractive the better we understand our dual-process moral brains (chapters 9 and 10).

Finally, in part 5 ("Moral Solutions"), we return to the new pastures and the real-world moral problems that motivate this book. Having defended utilitarianism against its critics, it's time to apply it—and to give it a better name. A more apt name for utilitarianism is *deep pragmatism* (chapter 11). Utilitarianism is pragmatic in the good and familiar sense: flexible, realistic, and open to compromise. But it's also a *deep* philosophy, not just about expediency. Deep pragmatism is about making *principled* compromises. It's about resolving our differences by appeal to shared values—common currency.

We'll consider what it means, in practice, to be a deep pragmatist: When should we trust our automatic settings, our moral intuitions, and when should we shift into manual mode? And once we're in manual mode, how should we use our powers of reasoning? Here we have a choice: We can use our big brains to *rationalize* our intuitive moral convictions, or we can *transcend* the limitations of our tribal gut reactions. I'll make the case for transcendence, for getting beyond point-and-shoot morality, and for changing the way we think and talk about the problems that divide us. I'll close in chapter 12 with six simple, pragmatic rules for life on the new pastures.

PART I

Moral Problems

1.

The Tragedy of the Commons

As you may have noticed, the Parable of the New Pastures is a sequel. The original parable comes from Garrett Hardin, a worldly ecologist who in 1968 published a classic paper entitled "The Tragedy of the Commons." In Hardin's parable, a single group of herders shares a common pasture. The commons is large enough to support many animals, but not infinitely many. From time to time, each herder must decide whether to add another animal to her flock. What's a rational herder to do? By adding an animal to her herd, she receives a substantial benefit when she sells the animal at market. However, the cost of supporting that animal is shared by all who use the commons. Thus, the herder gains a lot, but pays only a little, by adding an additional animal to her herd. Therefore, she is best served by increasing the size of her herd indefinitely, so long as the commons remains available. Of course, every other herder has the same set of incentives. If each herder acts according to her self-interest, the commons will be completely eroded, and there will be nothing left for anyone.

Hardin's Tragedy of the Commons illustrates the *problem of cooperation*.

Cooperation is not always a problem. Sometimes cooperation is a foregone conclusion, and sometimes it's just impossible. In between these two extremes, things get interesting.

Suppose that two people, Art and Bud, are at sea in a rowboat, trying to stay ahead of a violent storm. Neither will survive unless both row as hard as possible. Here self-interest and collective interest (in this case, a collective of two) are in perfect harmony. For both Art and Bud, doing what's best for "Me" and what's best for "Us" is the same. In other cases, cooperation is impossible. Suppose, for example, that Art and Bud's boat is now sinking and that they've only one life vest, which can't be shared. Here there is no Us, just two different Me's.

When cooperation is easy or impossible, as in the two scenarios above, there's no social problem to be solved. Cooperation becomes a challenging but solvable problem when, as in Hardin's parable, individual interest and collective interest are neither perfectly aligned nor perfectly opposed. Once again, any one of Hardin's herders is better off adding more animals to her herd, but this leads to collective ruin, which is in no one's best interest. The problem of cooperation, then, is the problem of getting collective interest to triumph over individual interest, when possible. The problem of cooperation is the central problem of social existence.

Why should any creature be social? Why not just go it alone? The reason is that individuals can sometimes accomplish things together that they can't accomplish by themselves. This principle has guided the evolution of life on earth from the start. Approximately four billion years ago, molecules joined together to form cells. About two billion years later, cells joined together to form more complex cells. And then a billion years later, these more complex cells joined together to form multicellular organisms. These collectives evolved because the participating individuals could, by working together, spread their genetic material in new and more effective ways. Fast-forward another billion years to our world, which is full of social animals, from ants to wolves to humans. The same principle applies. Ant colonies and wolf packs can do things that no single ant or wolf can do, and we humans, by cooperating with one another, have become the earth's dominant species.

Most cooperation among humans is of the interesting kind, the kind in which self-interest and collective interest are partially aligned. In the first case involving Art and Bud above, we stipulated that their interests are perfectly aligned: Both must row as hard as possible or both are sunk. But cases like this are rare. In a more typical case, either Art or Bud could row a little less hard and their boat would still arrive. More generally, it's rare to find a cooperative enterprise in which individuals have no opportunity to favor themselves at the expense of the group. In other words, nearly all cooperative enterprises involve at least some tension between self-interest and collective interest, between Me and Us. And thus, nearly all cooperative enterprises are in danger of eroding, like the commons in Hardin's parable.

The tension between individual and collective interest exists in many situations that we don't ordinarily think of as cooperative. Suppose Art is traveling through the Wild West along an isolated mountain trail. Up ahead, he sees the silhouette of a lone traveler coming over the next ridge. Is he armed? Art doesn't know, but Art sure is, and he's a good shot. Eyeing the stranger over the barrel of his rifle, Art thinks he can take him out with a single bullet. Should he do it? From Art's selfish point of view, there's nothing to lose. If he kills the stranger, he doesn't have to worry about being robbed. Thus, it's in Art's self-interest to shoot the stranger.

Bud, who is also traveling through these parts, faces a similar choice while traversing a mountain range to retrieve a stash of gold. Bud encounters a sleeping stranger on the trail. He knows that he will likely encounter the stranger on his way back, at which point Bud will be carrying his gold. Will the stranger try to rob him? Bud doesn't know, but he knows that if he poisons the sleeping stranger's whiskey, he won't have to find out.

The logic of self-interest unfolds: Bud poisons Art's whiskey. A few hours later, Art shoots Bud dead. And then a few hours after that, Art downs his whiskey and dies. Had Art and Bud both cared a bit more about the well-being of strangers, they'd have both survived. Instead, like the herders in Hardin's parable, their self-interest got the best of them. The lesson: Even the most basic form of decency, nonaggression, is a form of cooperation, and not to be taken for granted—in our species or any other. Consider, for example, one of our two nearest living relatives, the chimpanzees. If male chimpanzees from different troops encounter one another on the trail, and one party has a clear numerical advantage over the other, it's a good bet that the larger party will kill the members of the smaller party, simply because they can. And why not? Who needs the competition? Peace is a cooperation problem.

Nearly all economic activity poses a cooperation problem as well. When you buy something from a store, you count on the storekeeper to give you what you've paid for (e.g., ground beef, not ground squirrel). Likewise, the storekeeper counts on you to hand over a real ten-dollar bill (not counterfeit) and to refrain from filling your pockets with additional merchandise. Of course, in our society we have laws and police officers to ensure that people hold up their end of a bargain. And that is precisely the point. Because nearly all economic activity involves the interesting kind of cooperation, the kind that pits individual interest against collective interest, we need additional machinery to make it work.

Beyond the marketplace, nearly all human relationships involve give-and-take, and all such relationships break down when one or both parties do too much taking and not enough giving. In fact, the tension between individual and collective interest arises not only between us but within us. As noted above, complex cells have been cooperating for about a billion years. Nevertheless, it is not uncommon for some of the cells in an animal's body to start pulling for themselves instead of for the team, a phenomenon known as cancer.

THE FUNCTION OF MORALITY

After Darwin, human morality became a scientific mystery. Natural selection could explain how intelligent, upright, linguistic, not so hairy, bipedal primates could evolve, but where did our morals come from? Darwin himself was absorbed by this question. Natural selection, it was thought, promotes ruthless self-interest. Individuals who grab up all the resources and destroy the competition will survive better, reproduce more often, and thus populate the world with their ruthlessly selfish offspring. How, then, could morality evolve in a world that Tennyson famously described as "red in tooth and claw"?

We now have an answer. Morality evolved as a solution to the problem of cooperation, as a way of averting the Tragedy of the Commons:

> Morality is a set of psychological adaptations that allow otherwise selfish individuals to reap the benefits of cooperation.

How does morality do this? We'll spend the next chapter answering this question in more detail, but here is the gist: The essence of morality is altruism, unselfishness, a willingness to pay a personal cost to benefit others. Selfish herders will keep adding animals to their herds until the individual costs outweigh the individual benefits, and this, as we saw, leads to ruin. Moral herders, however, may be willing to limit the sizes of their herds out of concern for others, even though such restraint imposes a net cost on oneself. Thus, a group of moral herders, through their willingness to put Us ahead of Me, can avert the Tragedy of the Commons and prosper.

Morality evolved to enable cooperation, but this conclusion comes with an important caveat. Biologically speaking, humans were designed for cooperation, *but only with some people.* Our moral brains evolved for cooperation *within groups*, and perhaps only within the context of personal relationships. Our moral brains did not evolve for cooperation *between groups* (at least not *all* groups). How do we know this? Why couldn't morality have evolved to promote cooperation in a more general way? Because universal cooperation is inconsistent with the principles governing evolution by natural selection. I wish it were otherwise, but there's no escaping this conclusion, as I will now explain. (I hasten to add that this *does not* mean that we are doomed to be less than universally cooperative. More on this shortly.)

Evolution is an inherently competitive process: The faster lion catches more prey than other lions, produces more offspring than other lions, and thus raises the proportion of fast lions in the next generation. This couldn't happen if there were no competition for resources. If lion food existed in unlimited supply, the faster lions would have no advantage over the slower ones, and the next generation of lions would be, on average, no faster than the last generation. No competition, no evolution by natural selection.

For the same reason, cooperative tendencies cannot evolve (biologically) unless they confer a competitive advantage on the cooperators. Imagine, for example, two groups of herders, one cooperative and one not. The cooperative herders limit the sizes of their individual herds, and thus preserve their commons, which allows them to maintain a sustainable food supply. The members of the uncooperative group follow the logic of self-interest, adding more and more animals to their respective herds. Consequently, they erode their commons, leaving themselves with very little food. As a result, the first group, thanks to their cooperative tendencies, can take over. They can wait for the uncooperative herders to starve, or, if they are more enterprising, they can wage a lopsided war of the well fed against the hungry. Once the cooperative group has taken over, they can raise even more animals, feed more children, and thus increase the proportion of cooperators in the next generation. Cooperation evolves, not because it's "nice" but because it confers a survival advantage.

As with the evolution of faster carnivores, competition is essential for the evolution of cooperation. Suppose that both groups of herders live on magical pastures capable of supporting infinitely many animals. Under these magical conditions, the uncooperative group has no disadvantage. Selfish herders can go on adding animals to their respective herds, and their herds will simply grow and grow. Cooperation evolves only if individuals who are prone to cooperation outcompete individuals who are not (or who are less so). Thus, if morality is a set of adaptations for cooperation, we today are moral beings only because our morally minded ancestors outcompeted their less morally minded neighbors. And thus, insofar as morality is a biological adaptation, it evolved not only as a device for putting Us ahead of Me, but as a device for putting Us ahead of Them. (And note that in saying this I am not assuming that morality evolved by group selection.*) This has profound implications.

The idea that morality evolved as a device for intergroup competition may sound strange for at least two reasons. First, much of morality appears to be unrelated to intergroup competition. What, for example, does being pro-choice or pro-life on the issue of abortion have to do with intergroup competition? Likewise for people's moral opinions about gay marriage, capital punishment, not eating certain foods, and so on. As we'll see in the chapters that follow, moral thinking can be related to intergroup competition in ways that are indirect and not at all obvious. We'll put this issue aside for now.

The second strange thing about morality as a device for beating Them is that it makes morality sound *amoral* or even *immoral*. But how could this be? The paradox is resolved when we realize that morality can do things that it did not evolve (biologically) to do. As moral beings, we may have values that are opposed to the forces that gave rise to morality. To borrow Wittgenstein's famous metaphor, morality can climb the ladder of evolution and then kick it away.

As an analogy, consider the invention of birth control. We evolved big, complex brains that allow us to invent technological solutions to complex problems. In general, our technical problem-solving skills help us produce and support more offspring, but in the case of birth control, we've taken our big brains and used them to *limit* our offspring, thus thwarting nature's "intentions."* In the same way, we can take morality in new directions that nature never "intended." We can, for example, donate money to faraway strangers without expecting anything in return. From a biological point of view, this is just a backfiring glitch, much like the invention of birth control. But from our point of view, as moral beings who can kick away the evolutionary ladder, it may be exactly what we want. Morality is more than what it evolved to be.

METAMORALITY

Two moral tragedies threaten human well-being. The original tragedy is the Tragedy of the Commons. This is a tragedy of selfishness, a failure of individuals to put Us ahead of Me. Morality is nature's solution to this problem. The new tragedy, the modern tragedy, is the Tragedy of Commonsense Morality, the problem of life on the new pastures. Here morality is undoubtedly part of the solution, but it's also part of the problem. In the modern tragedy, the very same moral thinking that enables cooperation within groups undermines cooperation between groups. Within each tribe, the herders of the new pastures are bound together by their moral ideals. But the tribes themselves are divided by their moral ideals. This is unfortunate, but it should come as no surprise, given the conclusion of the last section: Morality did not evolve to promote universal cooperation. On the contrary, it evolved as a device for successful intergroup competition. In other words, *morality evolved to avert the Tragedy of the Commons, but it did not evolve to avert the Tragedy of Commonsense Morality.*

What, then, are we modern herders to do? That's the question I'm trying to answer in this book. How can we adapt our moral thinking to the circumstances of the modern world? Is there a kind of moral thinking that can help us live peacefully and happily together?

Morality is nature's solution to the problem of cooperation within groups, enabling individuals with competing interests to live together and prosper. What we in the modern world need, then, is something like morality but one level up. We need a kind of thinking that enables groups with *conflicting moralities* to live together and prosper. In other words, we need a *metamorality*. We need a moral system that can resolve disagreements among groups with different moral ideals, just as ordinary, first-order morality resolves disagreements among individuals with different selfish interests.

The idea of a metamorality is not wholly new. On the contrary, identifying universal moral principles has been a dream of moral philosophy since the Enlightenment. The problem, I think, is that we've been looking for universal moral principles that *feel right*, and there may be no such thing. What feels right may be what works at the lower level (within a group) but not at the higher level (between groups). In other words, commonsense morality may be enough to avert the Tragedy of the Commons, but it might be unable to handle the Tragedy of Commonsense Morality.

Herders on the new pastures who want to live peacefully and happily may need to think in new and uncomfortable ways.

To find the metamorality that we're looking for, we must first understand basic morality, the kind that evolved to avert the Tragedy of the Commons.

2.

Moral Machinery

I've said that morality is a device for enabling cooperation, for averting the Tragedy of the Commons. In fact, morality is a collection of devices, a suite of psychological capacities and dispositions that together promote and stabilize cooperative behavior. In this chapter, we'll see how these devices actually work at the psychological level, how they are implemented in our moral brains. What we really want to know, of course, is why we fight. We want to understand why our moral machinery breaks down on the new pastures. But to understand how our moral machinery fails us (the subject of the next chapter), we first need to know how it works when everything is functioning properly.

Hardin's Parable of the Commons describes a multiperson cooperation problem. In this chapter, we'll simplify things by focusing on another famous parable describing a two-person cooperation problem. This parable, known as the Prisoner's Dilemma, is about two criminals who are trying to stay out of prison. Despite the criminal context, the abstract principles behind the Prisoner's Dilemma explain why our moral brains are the way they are.

THE MAGIC CORNER

For our version of the Prisoner's Dilemma, we'll bring back our friends Art and Bud, this time as a bank-robbing duo. Following an otherwise successful heist, they are picked up by the police and brought in for questioning. The police know that Art and Bud are guilty, but they're short on hard evidence. They do, however, have enough evidence to convict both Art and Bud of a lesser crime, tax evasion, which would put them away for two years each. Still, what the police really want is two convictions for bank robbery—eight years each, minimum. To get their conviction, the police need a confession. They separate the two suspects and go to work.

Art and Bud each face the same choice: Confess or keep quiet. If Art confesses and Bud doesn't, Art gets a light sentence, just one year, and Bud gets ten years. The reverse happens if Bud confesses and Art doesn't. If they both confess, they both get eight years. And if they both keep quiet, they both get two years. This set of contingencies is laid out in the payoff matrix shown in figure 2.1.

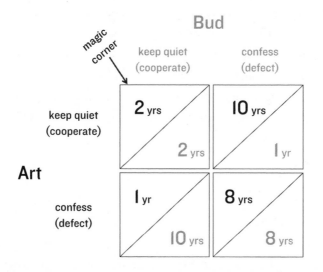

Figure 2.1. Payoff matrix in a classic Prisoner's Dilemma. Collectively the two players are better off keeping quiet (cooperating), but individually each player is better off confessing (defecting).

The four boxes in the matrix describe the four possible outcomes. Art's choice determines the row; Bud's choice determines the column. If Art confesses and Bud doesn't, they end up in the lower-left corner—good for Art, bad for Bud. If Bud confesses and Art doesn't, they end up in the upper-right corner—good for Bud, bad for Art. If both confess, they end up in the lower-right corner, which is pretty bad for both of them. And if they both keep quiet, they end up in the upper-left corner, the magic corner, the one that is pretty good for both of them and that minimizes their joint prison time.

So what will Art and Bud do? You might expect them both to keep quiet, putting themselves in the magic corner. However, if Art and Bud are selfish, and if all else is equal, that won't happen. Both will confess, putting them in the lower-right corner, maximizing their joint prison time with eight years each. This "tragic" outcome is analogous to the tragic outcome in Hardin's parable, and follows the same logic. Work through the payoff matrix in figure 2.1 and you'll see that Art is better off confessing, no matter what Bud does, and vice versa. If they're selfish and rational, both will confess. Great for the police, tragic for them.

The Prisoner's Dilemma, like the Tragedy of the Commons, involves a tension between individual interest and collective interest. Individually, Art and Bud are better off confessing, but collectively they are better off keeping quiet. Our question now is this: What would it take to get Art and Bud into that magic corner? How can they defeat their selfish inclinations and reap the benefits of cooperation? And how can we humans do this more generally? Wheel in the moral machinery.

FAMILY VALUES

In a famous episode related in the Talmud, Rabbi Hillel was approached by a skeptical man who vowed to convert to Judaism, on one condition: The great rabbi had to teach him the entire Torah in the time that he could stand on one foot. Rabbi Hillel replied, "That which is hateful to you, do not do to your neighbor. That is the whole Torah. The rest is commentary. Go and study it."

This, of course, is a version of the "Golden Rule," affirmed in one form or another by every major religion and every recognizably moral philosophy. It is also, not coincidentally, the most straightforward solution to Art and Bud's cooperation problem. Spending ten years in prison is "hateful" to both Art and Bud, and thus, if they take Rabbi Hillel's advice, they'll both keep quiet and find the magic corner. (Of course, if they *really* took Rabbi Hillel's advice, they wouldn't be robbing banks in the first place, but that's a separate problem.)

But why would Art and Bud care about doing "hateful" things to each other? Perhaps Art and Bud are brothers. That would explain it, but this only pushes our question back further: Why do brothers care about each other? Brotherly love (and familial love more generally) is explained by the well-known theory of kin selection,* which takes a gene's-eye view of behavior. Genetically related individuals share genes (by definition), and therefore, when an individual does something to enhance the survival of a genetic relative, that individual is, in part, doing something that enhances the survival of his or her own genes. Or, to take a gene's-eye view, genes that promote beneficence toward kin are enhancing their own survival, helping equally good copies of themselves inside the bodies of others.

In many species, what counts as caring in the biological sense—conferring a benefit on another individual at a cost to oneself—does not involve caring in the psychological sense. Ants, for example, confer benefits on their genetic relatives, but, so far as we can tell, ants are not motivated by tender feelings. Among humans, of course, caring behavior is motivated by feelings, including the powerful emotional bonds that connect us to our close relatives. Thus, familial love is more than just a warm and fuzzy thing. It's a strategic biological device, a piece of moral machinery that enables genetically related individuals to reap the benefits of cooperation.

TIT FOR TAT

Familial love helps genetic relatives find the magic corner, but what about people who aren't related? They, too, can find the magic corner by giving one another the right incentives.

Suppose that Art and Bud care not at all for each other but happen to work very well together, so well that their ability to rob banks together far outstrips their ability to rob banks separately or with other partners. If their most recent bank robbery were guaranteed to be their last, then they would, for reasons outlined above, both have reason to rat each other out. But what if they have a bright bank-robbing future ahead of them? Bright, that is, so long as they can resist the temptation to talk to the police. If the Prisoner's Dilemma occurs not as an isolated episode but as part of a series of such episodes, then the logic of the game changes. Sure, Art can get himself a quick one-year turn in prison by sticking Bud with a long sentence. But by doing this, he might throw away a glorious future with Bud at his side—not worth it to avoid one measly year in prison. Thus, if Art and Bud take the long view, they'll keep quiet—not because they care about each other but because they are useful to each other, because they have a productive future that depends on their present cooperation. This kind of conditional cooperation—"I scratch your back because you scratch mine"—is known as *reciprocity*, or reciprocal altruism.*

In the early 1980s, Robert Axelrod and William Hamilton published a classic paper reporting on the results of a Prisoner's Dilemma tournament. The competitors were not people but algorithms, computer programs implementing different strategies for playing the Prisoner's Dilemma. The two simplest strategies are to always cooperate (always keep quiet) or never cooperate (always confess). (Not cooperating is typically called "defecting.") Axelrod and Hamilton asked their colleagues to submit programs to compete in their tournament. Many of the programs were quite complicated, but the winning program, submitted by Anatol Rapoport, employed a strategy almost as simple as "always cooperate" and "never cooperate." The program, known as Tit for Tat, started out by cooperating (keeping quiet) and then, after that, doing whatever its partner had done on the previous round. If the other program had cooperated last time, then it would cooperate this time. If not, not. Hence, "Tit for Tat." In more recent tournaments, other programs have edged out Tit for Tat, but these other programs are all variations on the Tit for Tat theme. Reciprocity works very well.

In humans, the logic of reciprocity could be implemented through conscious reasoning: "Bud ratted on me last time. This makes it more likely that he will rat on me this time. Therefore, I will not attempt to cooperate with him this time." And Bud, of course, could anticipate Art's reasoning with some reasoning of his own: "If I rat on Art this time, Art will conclude that I am unlikely to cooperate in the future. But I have more to gain from future cooperation with Art than from ratting on him now. Therefore, I will cooperate now." This kind of explicit strategic thinking can get Art and Bud into the magic corner, but it's often unnecessary. That's because we humans have feelings that do the thinking for us. Suppose Bud rats on Art. Art could reason his way to the conclusion that he should dump Bud. But the same effect might be achieved, and achieved more reliably, if Art were automatically disposed to respond to Bud's ratting with *anger, disgust,* or *contempt.** Likewise, Bud might understand intuitively that if he rats on Art, Art will harbor such ill will toward him, with detrimental effects for Bud's professional future. Bud might shudder at the thought of betraying Art. Positive emotions can also support cooperation through reciprocity. By cooperating with Art, Bud might expect Art's *gratitude,* and with it an increased willingness to cooperate with Bud in the future.

Our primate relatives appear to engage in conditional cooperation, and to the extent that they do, they do it with feelings rather than explicit strategic reasoning. A classic study of food-sharing behavior in chimpanzees found that adult chimps are more likely to share food with chimps who have recently groomed them, and chimps are also more likely to protest aggressively against individuals who come looking for food if they've not recently provided grooming services. Studies such as these suggest that our ability to reciprocally scratch each other's backs depends, at least in part, on emotional dispositions that we inherited from our primate ancestors.

Reactive emotions, properly configured, can incentivize cooperative behavior, but they can ruin a cooperative relationship if they're applied too vigorously. Suppose that Bud, in a moment of uncharacteristic weakness, confesses to the police, leaving Art in the lurch. Many years later, Art and Bud have the bank-robbing opportunity of a lifetime. If Art's still angry at Bud, he'll miss out. Forgiveness pays. (Think, for example, of aging rockers burying the hatchet for the lucrative reunion tour.) Consistent with this, computer simulations show that conditional cooperators who are a bit forgiving do better than individuals who hold grudges indefinitely, so long as they live in a world in which things don't always go as planned. Chimps seems to follow this logic. De Waal and Roosmalen analyzed records of hundreds of postconflict episodes and found that it's common for chimps to kiss and embrace after a fight. This suggests that our capacity for forgiveness, which tempers our negative reactive emotions, has deep biological origins, following the logic of reciprocity in an uncertain world.

BESTIES

Art might keep quiet because he fears Bud's wrath, and with it the dissolution of their lucrative partnership. But after years of robbing banks together, you might expect them to operate differently. Art and Bud can motivate their cooperative behavior by thinking explicitly about the long-term costs and benefits. But it would also work, and may work better, if they had feelings that made them abide by this logic intuitively. More specifically, it would be useful for bank robbers like Art and Bud to have automated psychological programs that make them care about individuals with whom they have cooperative futures.

How might such a program work? More specifically, how would such a program identify the individuals with whom one has promising cooperative futures? The best guide to the future is the past. If one has cooperated extensively with an individual in the past, that's a sign of more to come.

Thus, cooperation may be efficiently automated by a psychological program that makes one care about one's historical cooperation partners. Such a program might be called *friendship*.

It may seem strange to conceive of friendship as principally about cooperation rather than, say, hanging out and having fun, but appearances can be misleading. First, nature's purposes need not be revealed in our experience. Sex, for example, is primarily about making babies, but that's not necessarily what motivates people to do the deed. Likewise, friendship may ultimately be about things that are far from our minds when we're being friendly. Indeed, if you're constantly thinking about the material advantages of your friendship, that's a sign that you're not really a friend. Second, if the idea of friendship as a cooperation device seems strange, that may be because of the unusually good times in which we live. In the feast-and-famine world of our hunter-gatherer ancestors, having friends who were willing to have you over for dinner wasn't just a nicety but a matter of life and death. The world of our ancestors was also a lot more violent. In our world, few friends can say that they've saved one another's lives, but that might not have been true in the past. Third, it's important to remember that a lot of cooperation doesn't feel like "cooperation." Friends are friends not only because of what they do together but also because of what they *don't do* separately. Your friends don't steal your stuff, make snide remarks about you, or try to bed your significant other. These everyday acts of nonaggression are inconspicuous forms of cooperation, as when Art and Bud pass each other on the trail without incident. Thus, the cooperation device that we call "friendship" begins with benign familiarity and takes off from there.

MINIMAL DECENCY

Suppose you're Art, looking for a bank-robbing partner. You hear about this guy named Bud who's quick on the draw and who can drive a getaway car like nobody's business. The only problem with Bud is that he'll put a bullet through your brain in a second flat if it's to his advantage. Art's no

Boy Scout himself, but given the uncertain nature of bank robbing, it's just not worth it for him to partner up with a psychopath like Bud. Lesson: For two strangers to cooperate, it helps if they have at least a minimal regard for each other's well-being.

As noted above, male chimpanzees are predisposed to killing strangers when they can do so with little risk. At times, humans, too, may regard strangers as nothing more than threats to be eliminated, or as sources of protein. (Cannibals of the South Pacific have been known to refer to edible outsiders as "long pig.") Nevertheless, humans can adopt less malicious attitudes toward strangers, and in the modern world they typically do. During the American Civil War, the Union Army's top brass was dismayed by frequent reports of Union soldiers found dead on the battlefield with their guns fully loaded. Many soldiers couldn't bring themselves to shoot at strangers, even ones who were trying to kill them. From this experience, the U.S. military concluded that soldiers need to have their reluctance to kill trained out of them—the birth of modern military training.

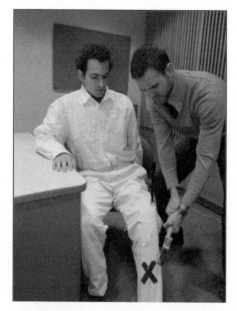

Figure 2.2. People simulating violent actions exhibit robust physiological responses, despite knowing that their actions are harmless. Underneath the X is a false leg.

Recently, Fiery Cushman, Wendy Mendes, and their colleagues conducted a laboratory study of the human aversion to violence. They monitored people's vital signs while having them simulate a variety of violent actions, such as smashing someone's leg with a hammer (see figure 2.2).

The people who participated in this experiment knew perfectly well that these actions were harmless. Nevertheless, simply pretending to do these nasty things caused their peripheral blood vessels to constrict dramatically, literally giving them "cold feet."

What's more, the researchers found that this vasoconstriction effect is specific to performing the pseudo-violent action *oneself*. It doesn't occur to the same extent when one watches another person perform a pseudo-violent action or when one performs a kinetically similar action (such as hammering a nail) that is not pseudo-violent. Many of the people in this experiment performed the pseudo-violent actions as halfheartedly as possible, for example, giving their supposed victims' legs a perfunctory tap with the hammer. One person simply refused to play along. Of course, humans can be extremely violent, and often with apparently little cause. But, as aggressive as we are, our aggression is nothing compared with what it could be. Under ordinary circumstances, we shudder at the thought of behaving violently toward innocent people, even total strangers, and this is most likely a crucial feature of our moral brains. (Try to imagine our world without it.)

Our basic decency extends beyond nonaggression to positive acts of kindness. Sadly, we are not anywhere near as kind as we could be, but we are often willing to do nice things for others, even strangers, without expecting anything in return. In a classic study from the 1960s, Stanley Milgram and his colleagues left "lost" letters in public places and found that most of them were eventually returned, in many cases even when the letters had been left without postage. We leave tips at restaurants that we do not intend to visit again, and some of us make anonymous donations to charity. Decades of research in social and developmental psychology confirm what most of us suspect, but what some researchers have questioned: When we help people, it's often because we *feel bad* for them and want to relieve their suffering. Indeed, feeling bad for someone can make one more likely to cooperate with that person in a prisoner's dilemma, played with money instead of years in prison. (We'll spend a lot of time discussing laboratory experiments in which people play cooperation games with money.) Such feelings are generally referred to as *empathy*, an emotional state in which one experiences the feelings of others as one's own.* In recent years, cognitive neuroscientists have studied the neural bases of empathy and found that this definition is quite apt: Watching another person experience pain, for example, engages the same emotion-related neural circuits that are engaged when one experiences pain oneself, and the brains of people who report having high levels of empathy toward others exhibit this effect more strongly.

The neural circuits that support empathetic responses to strangers may derive originally from circuits that evolved for maternal care. Oxytocin is a neurotransmitter and hormone that plays an important role in maternal care in many mammalian species. Genes that increase the human brain's sensitivity to oxytocin are correlated with higher levels of empathy, and spraying oxytocin into people's noses (from where it can enter the brain) makes people more likely to initiate cooperation in a version of the Prisoner's Dilemma.

Our capacity to care about others, including unrelated individuals, is almost certainly an elaboration of traits we inherited from our primate ancestors. For decades, primatologists have reported on incidents in which apes and monkeys behaved with apparent compassion. The pioneering primatologist Nadezhda Ladygina-Kohts raised a young chimpanzee named Joni in her Moscow home. Joni liked to play on the roof of her house and often refused to come down. Over time, Kohts found that the best way to get Joni to come down was to appeal to his sympathy. She would pretend to cry, and Joni would immediately rush to her side, look about suspiciously for an offender, and comfort her by gently touching her face. Chimpanzees sometimes appear to help one another as well. Jaki, a seven-year-old chimp living at the Arnhem Zoo, in the Netherlands, observed an older caregiver named Krom trying unsuccessfully to retrieve a tire that had been filled with water. After Krom gave up in frustration, Jaki went over to the tire, removed the other tires that were blocking it, and carried it over to Krom, being careful not to spill any water.

Such anecdotes are fascinating, and may well reflect deep truths about our primate cousins, but if we're feeling skeptical, we can explain them away. More recently, however, primatologists have conducted controlled laboratory experiments that strengthen the case for genuine caring in nonhuman primates. In a series of experiments, Felix Warneken, Michael Tomasello, and their colleagues have demonstrated that chimpanzees will help both other chimps and humans spontaneously, and without expectation of a reward. In one experiment, chimps spontaneously volunteered to help a human experimenter by retrieving an out-of-reach object for him. In another experiment, chimps performed similar good deeds for an unfamiliar human, even when doing so required climbing over obstacles. In yet another experiment, chimps actively chose to release a chain, thus granting another chimp access to food while gaining nothing for themselves. It seems that neighborliness goes even further down our evolutionary tree. Recent studies by Venkat Lakshminarayanan and Laurie Santos show that capuchin monkeys, given a choice between rewarding themselves *only* and rewarding themselves *and a neighbor*, typically choose to do the neighborly thing, even when the neighbor's reward is bigger. There is even evidence of empathy in rats, who will forgo an immediate reward in order to free another rat from a restraining device.

In sum, we are a caring species, albeit in a limited way, and we probably inherited at least some of our caring capacity from our primate ancestors, if not our more distant ancestors. We care most of all about our relatives and friends, but we also care about acquaintances and strangers. Under ordinary circumstances, we're highly reluctant to harm strangers, so much so that even pretending to do so causes our veins to constrict. We're also willing to help strangers, expecting nothing in return, so long as it's not too costly. Because we care about one another, because our individual payoffs are not the only ones that matter to us, we can more easily get ourselves into the magic corner.

THREATS AND PROMISES

Art and Bud can find the magic corner if they care about each other or if they have a productive future together. But what if they're just strangers with no future? Suppose that Art and Bud have a once-in-a-lifetime bank-robbing opportunity. They've never worked together before and never will again. The police will surely bring them in and try to turn them against each other. Can Art and Bud hold it together?

Perhaps they can make a pact beforehand to keep quiet. Making this kind of promise is easy. The hard part is *keeping* it. The problem is that making promises does not, by itself, change the payoff matrix. When it comes time to confess or keep quiet, Art is still going to be better off if he confesses, and likewise for Bud. If they don't care about each other, and they've no cooperative future to protect, then they'll both confess—promise or no promise.

What they need is some way to enforce their contract. To that end, Art might say this to Bud: "If you rat on me, as soon as I get out I'll hunt you down and kill you." Unfortunately, this threat-based strategy has the same problem as the more good-natured promising strategy considered above. Suppose Art threatens, and suppose Bud goes ahead and rats on him anyway. Ten years later, they're both out of prison, and it's time for Art to make good on his threat. Why should Art bother? Trying to kill someone is risky, and here it brings no benefit. If Bud knows from the start that Art won't bother, then Art's threat is idle. Bud will ignore Art's threat and confess. And, of course, Art will do the same if the tables are turned. No cooperation.

Thus, merely threatening doesn't work, for much the same reason that merely promising doesn't work. But a threat can work if it's set up properly. Suppose that Art has a high-tech programmable robotic hit man. Art can program his robot to kill Bud if Bud rats on him. Critically, we'll assume that Art's robot works perfectly and that once the robot has been programmed, it cannot be stopped, not even by Art. If Bud knows that he will be killed if he rats on Art, then he won't rat. Art's threat is a bit crazy, because Art will be held responsible for whatever his robot does, and so he would never want to see his threat carried out. If, for some reason, Bud were to ignore Art's robotic threat and rat on Art, Art would try his best to shut down his own robot. (To no avail, of course.) Nevertheless, by committing to this crazy threat up front, he can secure Bud's cooperation, so long as Bud is informed and rational. And, of course, Bud can secure Art's cooperation in the same way. (You may recognize this strategy as MAD—mutually assured destruction.)

Alas, we humans have yet to invent programmable robotic hit men, but according to the economist Robert Frank, our brains have emotional machinery that performs the same function.* Suppose that Art is a real hothead. If Bud rats on him, Art will be so enraged that nothing will stop him from killing Bud, even if he has to wait ten years, and even if he has to chase Bud to the ends of the earth. If Bud knows of Art's vengeful nature, then Bud has a strong incentive not to rat on him. Thus, by being vengeful, and being known for it, Art can be his own robotic hit man, incentivizing others to cooperate with him through his high-flying, credible threats. Of course, being vengeful can be very costly. Art could lose everything if he does, in fact, devote his life to exacting vengeance on Bud. Still, if all goes well, Art will never actually need to go after Bud, because people like Bud won't dare cross him. Thus, the emotions that fuel vengeful behaviors are, or can be, a kind of rational irrationality. They serve our interests by publicly committing us to doing things that are not in our own interest.

We humans are not the only ones with a taste for vengeance. Keith Jensen and his colleagues conducted an experiment in which chimpanzees could prevent other chimps from getting food. They found that if Chimp A steals food from Chimp B, Chimp B is more likely to pull a rope that will cause a table to collapse out from under Chimp A's food, placing it out of reach. Field studies suggest that chimps do much the same in the wild.

We have negative social emotions that give others incentives to cooperate with us, but cooperation might also be enabled by nobler feelings. If Art and Bud are mere scoundrels, their promises are useless, because, as noted above, scoundrels have no reason to keep their promises, and everyone knows it. But what if Art and Bud are *honorable* thieves? Art may not care about Bud, but he might care very much about keeping his word. Art might be the kind of guy who, upon breaking his word, would be so displeased with himself that he would immediately hurl himself into the nearest lava pit. Now, *that's* a guy you can work with. Just as vengeful anger makes one's threats to others credible, being prone to powerful feelings of *guilt* and *shame* enables credible threats to *oneself.* As one might expect, breaking a promise—or even thinking about breaking a promise—elicits increased activity in emotion-related brain regions.

Above we talked about familial love and friendship as forms of cooperative caring. Such feelings can also be strategic straitjackets, much like
vengeful anger, rationally committing us to irrational behavior. In this
case, however, the emotional straitjackets are worn not by people who have
no future together, but by people who might have even better futures with
others. Suppose the police offer Art a really sweet deal: If he rats on Bud,
they'll not only let Art off the hook—they'll also give him a job as an
in-house bank-robbing expert. In other words, the police invite Art to
partner with them instead of Bud. Art cares about Bud, and that is fitting,
as they share a cooperative past and the prospect of a cooperative future.
But the police are offering Art something that Bud can only dream of—an
exciting, respectable career with good, steady pay. Art, it seems, has an
even better cooperative future with the police. If Art's friendly feelings for
Bud are strong, but merely proportional to the cooperative opportunities
Bud offers, then Art will ditch Bud and join the police. Trading up is great
for Art, but if Art is *known* for his willingness to trade up, that's not so
great for him. Bud, among others, might not work with a guy who's willing to ditch his partner as soon as a better gig comes along.

Enter the virtue of *loyalty*. If Art values his bank-robbing partners over
and above their "market value" (the value of the cooperative opportunities
they offer), that makes Art a more attractive partner. As Steven Pinker
observes, the logic of loyalty is particularly clear in the domain of romantic relationships: You're a great catch, but there is bound to be someone out
there who's got everything you've got plus a little more. Knowing that your
partner might someday meet such a person, you'd be reassured by the
knowledge that your partner isn't going to leave you as soon as something
better comes along. This would make you much more willing to settle
down with your partner and start a family—a high-stakes cooperative
endeavor if ever there was one. It's wonderful that your partner fully ap-
preciates your many marketable qualities, but that may not be enough to
keep you together. What you really want is for your partner to have a deep,
unshakable desire to be with you and you alone. In short, you want your
partner to *love* you, to want you not only for your wonderful qualities but
just because you're you. Only love provides the kind of loyalty you need in

order to take the parenting plunge. Thus, love appears to be more than just an intense form of caring. It's a highly specialized piece of psychological machinery, an emotional straitjacket that enables cooperative parenting by assuring our parenting partners that they won't be abandoned.

There is yet another kind of loyalty that can grease the wheels of cooperation. Just as personal loyalty makes one a more attractive friend or lover, a disposition to respect authority can make one a more attractive foot soldier within a larger cooperative enterprise. Indeed, if you're a general or a CEO, whom do you want in your organization? Someone who's going to do what he thinks is best, even when you say otherwise, or someone who will reliably follow your orders? Likewise, do you want someone who will jump ship as soon as a more attractive vessel floats by or someone who's willing to follow your ship to the seafloor? Good foot soldiers have the virtues of loyalty and *humility*. They know their place and dare not abandon it.

This sense of place can be motivated by both positive and negative emotions. In nearly all primates, lower-ranking individuals are negatively disposed toward higher-ranking individuals, regarding them primarily with fear. But humans sometimes regard their leaders with powerful feelings of admiration. We can be inspired by leaders we've never met and devoted to organizations with no fixed membership, such as nations, churches, corporations, and schools. Jonathan Haidt has argued that this capacity for devotion to leaders, organizations, and more abstract ideals might have evolved to facilitate cooperation in large groups, just as romantic love evolved to facilitate cooperative parenting. This capacity may depend on our ability to experience *awe*—to be moved by, and devoted to, things larger than ourselves and our familiar social circles.

WATCHFUL EYES AND DISCERNING MINDS

Art and Bud can find the magic corner if they care about each other, if they have a cooperative future together, or if they're emotionally committed to carrying out their threats or keeping their promises. But what if they have none of these advantages?

Art has vowed to kill Bud if he rats, but Bud, now alone in his cell, is nonetheless tempted to talk. He's tempted because he knows that Art is rational. Unlike some hotheads, Art isn't going to go after Bud just because he's angry. Still, Bud had better think twice. Art might kill Bud for ratting—not because Art is irrationally vengeful, but because *others are watching*. The bank-robbing community wants to know: Are Art's threats credible? If Art kills Bud for ratting, he can answer that question affirmatively, and that's good business. Bud better keep quiet.

Thus, people with reputations to maintain can more easily cooperate: Reputations give people reasons to follow through on their threats, which makes their threats credible, which gives those who are threatened an incentive to cooperate. Reputations can also enhance cooperation in a more direct way. If Bud develops a reputation for being a rat, then others won't want to rob banks with him, and he will lose out in the end. Thus, reputation can enhance cooperation in two ways: by giving people incentives to demonstrate their cooperativeness and by giving people incentives to demonstrate their intolerance of noncooperativeness. Our moral brains appear to be designed for both strategies.

Kevin Haley and Daniel Fessler brought people into their lab and gave half of them ten dollars each. The lucky recipients were then given the option to share some, all, or none of their money with one of the less fortunate participants. This is known as a "Dictator Game," because the person choosing has complete control over the money. All of this was done anonymously over a set of networked computers, ensuring that none of the participants knew anything about who gave what to whom. In this experiment, the critical manipulation was very subtle. For half the dictators, the computer desktop "wallpaper" in the background displayed a pair of eyes, as shown in figure 2.3. The other dictators, the ones in the control condition, just saw a standard background with the lab's logo.

Only about half (55 percent) of the people who saw the standard background gave something to the other player, while an overwhelming majority (88 percent) of the people who saw the eyes chose to give. In a subsequent

Figure 2.3. The eyespots used in Haley and Fessler's experiment. People who saw these watchful eyes were more generous toward strangers.

field study using an "honor box" for buying drinks, a picture of eyes made people pay more than twice as much for their milk.

We all know that people behave better when they think they're being watched, when they are feeling *self-conscious*. What's surprising here is that an irrelevant low-level cue, a picture of eyes, can put people on their best behavior. I say "irrelevant" because no one would ever consciously choose to respond to this cue: "I'll now pay for my milk because there's a picture of someone's eyes taped to the cupboard." This is, instead, the work of an automated program, an efficient piece of moral machinery.

If watchful eyes have great power over us, it's probably because watchful eyes are nearly always connected to running mouths. According to anthropologist Robin Dunbar, humans spend about 65 percent of their conversation time talking about the good and bad deeds of other humans— that is, *gossiping*. He argues that we devote an enormous amount of time to gossip because in humans gossip is a critical mechanism for social control—that is, for enforcing cooperation. Indeed, the prospect of having "everyone" know what you've done gives one a very strong incentive not to do it in the first place. What's more, it's not just that people *can* gossip. Gossip seems to happen *automatically*. For many people, *not* gossiping requires great effort.

In a world full of watchful eyes and loose mouths, people are bound to get caught doing uncooperative things. For one who does get caught, the worst-case scenario can be pretty bad: No one wants to have anything to do with you for the rest of your life. How might one avoid such a fate? It would help if there were some way to convince "everyone" that, in the future, you will be a better cooperator. You could say that you're sorry, but that, by itself, is not very convincing. Anyone can say, "I'm sorry." It would be much more convincing if your face were to involuntarily turn an unusual color—say, bright red—providing a credible signal that you are genuinely displeased with your own behavior. Indeed, it seems that *embarrassment* was designed to play precisely this kind of signaling role, restoring one's social standing by signaling a genuine desire to behave differently in the future. This signal seems to work. Research shows that people like transgressors better if, after committing a transgression, the transgressor appears to be embarrassed.

Of course, the mere fact that "everyone" knows that a transgressor has transgressed makes no difference by itself. What matters is that people treat people differently depending on what they've seen or heard. Thus, our sensitivity to watchful eyes and perked-up ears makes sense only if the minds behind those eyes and ears are *judgmental*—poised to treat us differently depending on what they see and hear. It's no news that we humans are judgmental. What is news, however, is that we're judgmental as *babies*, as demonstrated by one of the last decade's most remarkable psychological experiments.

Kiley Hamlin, Karen Wynn, and Paul Bloom presented six- and ten-month-old infants with movement sequences involving googly-eyed geometric figures that can go up and down a hilly landscape, as shown in figure 2.4.

In the sequence depicted on the left, the circle tries to climb up the hill but can't quite make it to the top on its own. Then along comes the *helpful* triangle, who comes up from below and pushes the circle up to the top. In

Figure 2.4. Preverbal infants like the little triangle who helps the circle get up the hill and dislike the square who pushes the circle down.

the sequence depicted on the right, the circle once again tries and fails to make it to the top on its own. Then along comes the *hindering* square, who comes down from the top of the hill and pushes the circle back down to the bottom. The infants watched each of these sequences several times, until they got bored. Then, in the critical test phase, the experimenters presented the infants with a tray carrying, on one side, a toy resembling the helpful triangle and, on the other side, a toy resembling the hindering square.* Fourteen out of sixteen ten-month-olds and all twelve six-month-olds reached out for the helpful toy—an amazingly robust result.

Next, the researchers repeated the experiment with a new group, making the circle look like an inanimate object rather than an agent with goals. They did this by removing the circle's googly eyes and by preventing the infants from seeing the circle engaged in self-propelled motion, a sign of being alive and having intentions. In this version, then, the triangle and square aren't helping or hindering anyone. They're just pushing a circle up or down the hill. This time, as predicted, the infants showed no preference for the triangle over the square (i.e., for the pusher-upper over the pusher-downer). This shows that the infants' preferences are specifically *social*. It's helping, not pushing up, that they like, and it's hindering, not pushing down, that they dislike.

Thus, at the age of six months, long before they can walk or talk, human infants are making value judgments about actions and agents, reaching out to individuals who show signs of being cooperative (caring about others) and passing over individuals who do the opposite. Because these children are so young, their behavior is clearly not produced by conscious reasoning: "That square did not treat the red circle well. This suggests that the square is unlikely to treat me well. Therefore, I will avoid the square." Instead, these judgments are produced by automated programs, ones that are sensitive to low-level cues—certain types of movement and the presence of things resembling eyes. And given how early this machinery comes online, it is almost certainly part of our genetic inheritance.

MEMBERS ONLY

Two prisoners who care about each other, or who have a cooperative future together, can get into the magic corner. And two strangers can get into the magic corner with the right kinds of threats, or if they have reputations to maintain. But can strangers cooperate without threats and without reputations?

Suppose there's a large group of bank robbers called the League of Tight-Lipped Bank Robbers. As their organization's name suggests, these thieves abide by a strict code of silence when confronted by the authorities. The league is large enough that most members do not know one another, either personally or by reputation. In other words, most league members are complete strangers. League membership promises great benefits. Despite being strangers, league members can rob banks together, knowing that their partners won't betray them. Problem solved?

More like problem stipulated away. To suppose that such a league exists is essentially to stipulate the existence of a cooperative group. The challenge is to get such a group started and to prevent it from falling apart. To be a league member is, essentially, to promise not to rat on other league members in return for the same favor. And, as noted above, mere promises are useless in a selfish world. Why should league members keep their promises if there's no cost for breaking them? Perhaps the offending members will be punished by the league. That might work, but this only pushes our question back further: Who are these league officials, and what's in it for them to punish strangers for ratting on other strangers? We'll consider the role of punishing authorities shortly, but for now, we'll make things easy and assume that the league members were born with tight lips. As long as they work exclusively with one another, the league members are fine. The challenge, for them, is to avoid being exploited by loose-lipped outsiders. A rogue bank robber would gladly cozy up to a group of tight-lipped bank robbers, doing job after job with league members, each time sending his unwitting partner up the river when the cops start asking questions.

League members could avoid rogue bank robbers if only the baddies were preceded by bad reputations. But here we're assuming that's not possible. In the absence of information about untrustworthy outsiders, league members could actively provide information about trustworthy insiders. League members could carry little ID cards, emblazoned with the tight-lipped seal. This would allow them to find one another and avoid working with nonmembers. Such an ID system can work, so long as outsiders can't make fake IDs. The league, to function well, needs a reliable indicator of membership.

This is a common problem. All cooperative groups must protect themselves from exploitation. This requires the ability to distinguish Us from Them, and the tendency to favor Us over Them. While there are some rare individuals who treat strangers like family, there are no human societies in which this is the norm, and for good reason. Such a society would be an open-access resource pump, waiting to shower its treasures upon any strangers who arrive at its doorstep, as if those strangers were long-lost brothers and sisters. Consistent with this, anthropologist Donald Brown, in his survey of human cultural differences and similarities, identifies in-group bias and ethnocentrism as universal.

Each of us occupies the center of a set of concentric social circles. Immediately surrounding us are our closest relatives and friends, bounded by a larger circle of more distant relatives and acquaintances. Beyond our circles of kith and kin are strangers to whom we are related through our memberships in groups of various kinds and sizes: village, clan, tribe, ethnic group; neighborhood, city, state, region, country; church, denomination, religion. In addition to these nested groups, we organize ourselves by political affiliation, the schools we attended, social class, the sports teams we root for, and other likes and dislikes. Social space is complex and multidimensional, but at least one thing is clear from both common sense and boatloads of social scientific research: We humans pay exquisitely close attention to where people reside in our egocentric social universes, and we tend to favor people who are closer to us. Call this tendency *tribalism*, which is sometimes known as *parochial altruism*.

It's easy to identify the people in our innermost social circles (family, friends, acquaintances) as members of our cooperative groups, but humans cooperate in much larger groups, both actively (e.g., building a bridge, fighting a war) and more passively (e.g., through nonaggression). However, to cooperate with strangers, we need some means of distinguishing the strangers with whom we can cooperate from those who might exploit us. In other words, we need the ability to display and read social ID badges and to adjust our behavior based on what we've read.

The Hebrew Bible tells the story of the Gileadites, who defeated the Ephraimites around 1200 BC, driving them from their homes and across the Jordan River. Following the battle, many surviving Ephraimites attempted to return home, seeking passage from the Gileadites who guarded the river crossings. To ferret out Ephraimite refugees, the Gileadite guards employed a simple test: They asked travelers seeking passage to pronounce the Hebrew word *shibboleth*. (The word refers to the grain-bearing part of a plant.) The ancient Ephraimite dialect had no *sh* sound, making it hard for Ephraimites to pronounce this word. According to the Bible, forty-two thousand Ephraimites were killed because they couldn't say *sh*.

Today the word *shibboleth* refers to any reliable marker of cultural group membership. Research by Katherine Kinzler and her colleagues indicates that humans are predisposed from an early age to use the original shibboleths—linguistic cues—as markers of group identity and as a basis for social preference. In a series of experiments using English and French children, they showed that six-month-old infants prefer to look at speakers who lack foreign accents, that ten-month-old infants prefer to accept toys from native language speakers, and that five-year-old children prefer to be friends with children who lack foreign accents. It seems that human brains, even before they can generate speech, use language to distinguish trustworthy Us from untrustworthy Them.

Shibboleths illustrate a more general point about tribalism, which is that arbitrary differences can serve a nonarbitrary function. Gileadites' pronunciation doesn't matter per se. What matters is that Gileadites pronounce things *differently* from Ephraimites. In the same way, arbitrary cultural practices may play a vital role in supporting cooperation. How people dress, wash, eat, work, dance, sing, joke, court, have sex, et cetera— all the rules that govern daily life—can serve the nonarbitrary function of making strangers seem strange, thus separating Us from Them.

In the modern world, one of the most salient delimiters of Us versus Them is race. In recent years, psychologists have studied racial attitudes using the Implicit Association Test (IAT), which measures associations between concepts by measuring how quickly people can sort items into different conceptual categories. (Try it yourself.*) In a typical IAT, one performs two interleaved sorting tasks on a computer. For example, one might sort words as they appear on the screen based on whether they refer to good things or bad things. One might press the left-hand button for good words (e.g., "love") and the right-hand button for bad words (e.g., "hate"). At the same time, one might sort pictures of faces based on their race, pressing the left-hand button for white faces and the right-hand button for black faces. The IAT measures the speed with which one makes these categorizations and the changes in one's speed depending on how categories are paired with buttons. For example, you might use the right-hand button for both good words and white faces, while using the left-hand button for both bad words and black faces. Alternatively, you might use the right-hand button for both bad words and white faces, while using the left-hand button for both good words and black faces. If you're quicker when you use the same button for "bad" and black, that indicates that you have an implicit association between "bad" and "black." And so on for other conceptual pairings. The IAT shows that most whites have an implicit preference for whites over blacks, more readily pairing good words with white faces and bad words with black faces, et cetera. These IAT scores are reflected in brain activity: White people who strongly associate black faces with "bad" exhibit stronger neural responses to black faces in the brain region associated with heightened vigilance (the amygdala). A version of the IAT developed for children shows that children as young as six years old have the same kind of race-based biases as adults. And, rather amazingly, an IAT developed for monkeys shows that they, too, exhibit implicit preferences for in-group members, associating good things like fruit with in-group members and bad things like spiders with out-group members.

Sadly, racial bias is not just a laboratory phenomena. As noted earlier, economists have shown that résumés with white-sounding names (Emily, Greg) generate many more calls from prospective employers than identical résumés with black-sounding names (Lakisha, Jamal). Even more chilling, studies of U.S. court records show that in death penalty cases involving white victims, black defendants are more likely to receive the death penalty than white defendants, and this is especially true for black defendants with stereotypically black facial features. Race has profound political implications as well. Economist Seth Stephens-Davidowitz produced a U.S. map of the frequency of Google searches including the words "nigger" or "niggers." Regions high in "nigger" searches (mostly aimed at finding racial jokes) yielded significantly fewer votes for Barack Obama in the 2008 U.S. presidential election than votes for John Kerry in 2004. This racial animus appears to have given Obama's opponent a 3 to 5 percent advantage, the equivalent of a home-state advantage nationwide, which is enough to swing most presidential elections.

Given the strength and pervasiveness of racial bias, you might think that we are "hardwired" for racial discrimination. But if you think about it, this makes little sense. In the world of our hunter-gatherer ancestors, one was unlikely to encounter someone whom, today, we would classify as a member of a different race. On the contrary, the "Them" on the other side of the hill would likely be physically indistinguishable from "Us." This suggests that race, far from being an innate trigger, is just something that we happen to use today as a marker of group membership. From an evolutionary perspective, one would expect the human mind's social sorting system, if it has one, to be more flexible, sorting people based on culturally acquired characteristics, such as language and clothing, rather than genetically inherited physical features.

With this in mind, Robert Kurzban and his colleagues conducted an experiment in which they pitted people's sensitivity to race against their sensitivity to cultural markers of group membership. They had people watch an argument unfold between members of two mixed-race basketball teams. The participants saw pictures of various players paired with partisan statements such as "You were the ones that started the fight." The researchers then gave their participants a surprise memory test, asking them to pair pictures of people with the things those people said. By looking at the kinds of mistakes people made on the test, the experimenters could see how the participants were categorizing the players. If people in this experiment are highly sensitive to race, then they should rarely misattribute a white person's statement to a black person, or vice versa. Likewise, if people are highly sensitive to team membership, they should rarely misattribute one player's statement to a player on the other team. Kurzban and his colleagues found that, in the absence of salient markers of team membership, people paid a lot of attention to race and not a lot of attention to team membership. That is, people were relatively unlikely to misattribute statements across racial lines and more likely to misattribute statements across team lines. However, when the players wore colored T-shirts indicating team membership, everything reversed. Suddenly, race mattered much less, and team membership mattered much more.

Kurzban and his colleagues made an additional prediction based on their evolutionary theory. As explained above, they expected race-based categorization to be mutable, because race is not a deep evolutionary category. The same logic does *not* apply to gender (male vs. female). Our hunter-gatherer ancestors did routinely encounter males and females, and males and females differ in biologically important ways. This suggests that gender-based categorization, as compared with race-based categorization, should be harder to change, and that's exactly what they found. No matter who was black or white, and no matter who was wearing which T-shirts, their participants were unlikely to confuse what women said with what men said.

This experiment shows that we readily characterize people based on arbitrary markers of group membership, but it doesn't, by itself, tell us anything about how we use these social categories. Classic studies by

Henri Tajfel and his colleagues show how easily social categories become the basis for social preferences. Tajfel brought people into his lab and divided them into two groups based on incidental differences. For example, in one version he pretended to sort people into two groups based on whether they had tended to overestimate or underestimate in a prior estimation task. (The division was actually random.) He then had his subjects anonymously allocate money to different individuals participating in the study. He found that people tended to favor in-group members, even though the groups had no past and no future and were based on nothing of any importance. Indeed, Tajfel found that people favored in-group members even when group assignments were explicitly made randomly. Favoring in-group members was not just a strategy for breaking ties. People often gave less money to in-group members, rather than giving more money to out-group members.

Tribalism has recently been linked to specific neural systems. As noted above, oxytocin is a neurotransmitter and hormone that is involved in maternal care across mammalian species, one that is associated with increased empathy and trust in humans. It turns out that oxytocin, sometimes described as the "cuddle chemical," is more discriminating than had previously been thought. Recent experiments by Carsten De Dreu and colleagues show that intranasal administration of oxytocin makes men more cooperative with in-group members, but not out-group members, especially when fear of out-group members is high. Oxytocin also increases in-group favoritism as measured by IATs and shows modest signs of increasing out-group antipathy. Finally, oxytocin affects people's responses to moral dilemmas that pit in-group members against out-group members, making them less comfortable with sacrificing in-group members, but not out-group members.

In sum, our brains are wired for tribalism. We intuitively divide the world into Us and Them, and favor Us over Them. We begin as infants, using linguistic cues, which historically have been reliable markers of group membership. In the modern world, we discriminate based on race (among other things), but race is not a deep, innate psychological category. Rather, it's just one among many possible markers for group membership.

As Tajfel's experiments show, we readily sort people into Us and Them based on the most arbitrary of criteria. This sounds crazy, and in many ways it is. But it's what one might expect from a species that survives by cooperating in large groups—large enough that members cannot identify one another without the help of culturally acquired identity badges.

Before moving on, I hasten to add that being wired for tribalism does not mean being *hardwired* for tribalism. Brains can be rewired through experience and active learning. What's more, our brains include many different circuits that compete for control of behavior, some of which are more modifiable than others. More on this in later chapters.

INTERESTED PARTIES

Art and Bud can coax each other into the magic corner with credible threats. A powerful third party can also perform the same service. For example, Art and Bud might be members of a crime syndicate, with a crime boss who makes them both an offer they can't refuse: "If you rat on your partner, I'll kill you." In keeping with the old joke, I called this threat an "offer," but genuine offers can achieve the same outcome: "Keep quiet and I'll make it worth your while."

Enforced cooperation is surely one of the driving forces of history: Chiefs and kings and emperors have used their increasingly large carrots and sticks to enforce productive cooperation (and skim the proceeds off the top). According to the seventeenth-century English philosopher Thomas Hobbes, this is a good thing. He praised the king for being a peace-keeping Leviathan, the earthly god who lifts us out of our natural state, in which life is "nasty, brutish, and short."

Leviathans need not be *earthly* gods. Among believers, a supernatural authority is an ideal guarantor of cooperation, because supernatural beings can be omniscient and omnipotent, guaranteeing maximal rewards for cooperativeness and maximal punishments for uncooperativeness. As David Sloan Wilson has argued, religion may be a device that evolved through cultural evolution to enable cooperation in large groups. The idea that respect for God and being a good cooperator are related is not new, of course. Believers have long been, and continue to be, wary of people who are not "God-fearing."

From an evolutionary perspective, enforced cooperation makes sense because it requires nothing more than straightforward self-interest from anyone involved: The cooperative underlings receive rewards and avoid punishments from the boss, and the boss—if it's an earthly boss—benefits by exploiting a more productive group of underlings. Still, one might wonder whether rewards and punishments from interested third parties can stabilize cooperation without a Leviathan. This is an important question, because ethnographic studies indicate that pre-agricultural societies are rather egalitarian, with no Leviathan telling everyone what to do.

Consider, once again, the League of Tight-Lipped Bank Robbers. Above we envisioned the league as a collection of naturally tight-lipped individuals whose challenge is to prevent exploitation from the outside. But the league might also contain individuals who are perfectly capable of choosing to be uncooperative. Here the challenge is to keep the league's members in line, and to do so without a powerful boss. Can league members police themselves?

If the league were small, uncooperative behavior could be punished through direct retaliation, à la Tit for Tat: If Art rats on Bud, Bud can punish Art so that next time Art will be more cooperative. This is called *direct reciprocity*, because Bud receives a direct benefit from punishing Art. But if the league is large—and we're assuming that it is—it's not worth it for Bud to punish Art because it's unlikely that there will be a next time. Still, if somehow the members of the league were willing to punish one another for ratting, even when they had nothing to gain from it personally, that would be a great boon to cooperation. In a league full of willing punishers, there would be very little ratting, and also very little punishment, because the odds of getting punished for ratting would be very high.*

This general willingness to punish uncooperative behavior is a kind of *indirect reciprocity*. It's indirect because league members pay a direct cost when they punish but receive an indirect benefit from the punishing that other members do. If this reminds you of the original Tragedy of the

Commons, it should. Indirect reciprocity of this sort is itself a form of cooperation and a form of altruism, putting group interest ahead of self-interest. Thus, such punishment is often called "altruistic punishment." This term can be misleading, however, because an altruistic punisher need not be thinking about the good of the group. An altruistic punisher may simply enjoy sticking it to people who've done him—or others—wrong. To make things more clear, I'll call this kind of costly punishment *pro-social* punishment.

Are we pro-social punishers? Test yourself: Suppose that in some foreign city, thousands of miles away, there is a serial rapist-murderer who has terrorized dozens of women and girls. He will continue to do this until he is caught. Would you anonymously pay $25 to ensure that this rapist-murderer is brought to justice? If not $25, how about a dollar? If your answer is yes, congratulations—you're a pro-social punisher. You are willing to pay a personal cost to ensure that others cooperate. (Recall, once again, that nonaggression is a form of cooperation.)

Numerous lab experiments confirm that people are, indeed, pro-social punishers. The most famous such experiment was conducted by Ernst Fehr and Simon Gächter, using what's called the "Public Goods Game," a multiperson prisoner's dilemma that is analogous to the Tragedy of the Commons. Players are each given an allotment of money and then divided into groups. In each round of play, each player can contribute a sum of money to a common pool. The money that goes into the pool gets multiplied by the experimenter and then distributed evenly across the players. All of this is done anonymously.

Here the collectively rational behavior is for each player to put all of her money into the common pool. This maximizes the amount of money that gets multiplied by the experimenter, and thus maximizes the group's total take. For example, suppose that four players start with $10 each and together put all $40 into the common pool. The experimenter doubles the pool, turning it into $80, and then gives each player $20. That's a tidy profit. Still, the individually rational thing to do (if one is selfish) is to put nothing in—that is, to "free ride" off the contributions of the other players. Free riders get to keep all of their original allotment of money plus a share of what's in the common pool. In this case, a lone free rider amid three cooperators nets $25, while the others net only $5 each. Free riding in the Public Goods Game is analogous to ratting in the Prisoner's Dilemma and to growing one's herd *ad libitum* in the Tragedy of the Commons.

In a typical repeated Public Goods Game, most people start out cooperating, putting at least some money into the common pool. But then some people free ride, contributing little or nothing to the common pool. The cooperators see that they are being exploited and reduce or eliminate their contributions. As contributions decline from one round to the next, more players say "To hell with this," and contributions drop to almost nothing. Tragic.

However, when cooperators have the opportunity to punish free riders, things often change. Here punishment means "costly punishment"— paying a sum of money to reduce the payoff of another player. For example, one might pay a dollar to reduce a free rider's payoff by four dollars, following a round of play. This is the economic equivalent of bapping someone on the head. When experimenters introduce the opportunity to punish, contributions typically go up. Critically, this happens even when the punishers have nothing to gain from punishing and everyone knows it. What's more, the boost in cooperation from the introduction of punishment is often immediate, before anyone actually does any punishing. This means that would-be free riders anticipate that they will be punished for free riding, even by people who gain nothing (materially) by punishing them.

There is a vigorous debate about *why* we are pro-social punishers. Some say that pro-social punishment is just a by-product of an evolved tendency toward direct reciprocity and reputation management: We punish people with whom we have no cooperative futures because our brains automatically assume that everyone is a cooperation partner and that someone might always be watching. For life in a small hunter-gatherer band, these are not unreasonable assumptions. Others think that pro-social punishment evolved through biological or cultural selection at the level of groups: Pro-social punishment is good for one's group, and by punishing pro-socially, one helps one's group outcompete other groups. This is a fascinating debate, but we need not take a stand on it. What matters for our purposes is that pro-social punishment happens and that it fits a now familiar psychological profile.

As you might expect, pro-social punishment is driven by emotions. Fehr and Gächter asked their participants how they would feel about a free rider, upon meeting him or her outside the lab. Most people indicated that they would feel angry, and that they would expect others to feel anger toward *them* if the roles were reversed. Call this distinctively moral kind of anger *righteous indignation.*

Nowhere is our concern for how others treat others more apparent than in our intense engagement with fiction. Were we purely selfish, we wouldn't pay good money to hear a made-up story about a ragtag group of orphans who use their street smarts and quirky talents to outfox a criminal gang. We find stories about imaginary heroes and villains engrossing because they engage our social emotions, the ones that guide our reactions to real-life cooperators and rogues. We are not disinterested parties.

MORAL MACHINERY

From simple cells to supersocial animals like us, the story of life on Earth is the story of increasingly complex cooperation. Cooperation is why we're here, and yet, at the same time, maintaining cooperation is our greatest challenge. Morality is the human brain's answer to this challenge. (For a lively and extensive discussion of this idea, see Jonathan Haidt's book *The Righteous Mind.*)

As Art and Bud have taught us, there are several complementary strategies for getting otherwise selfish individuals into the magic, cooperative corner. The strategies that enable cooperation in the Prisoner's Dilemma are general strategies that apply to any social dilemma, any situation in which there's a tension between Me and Us. For example, Art and Bud's cooperative strategies readily translate into strategies for averting the Tragedy of the Commons: Herders who care about each other will want to limit the sizes of their herds. A Leviathan herder can ensure that herders play by the rules. Herders can maintain cooperation by keeping track of cheaters within the group and by barring exploitative outsiders. And so on. More important, these strategies translate into solutions to real-world problems: What goes for greedy herders goes for tax evaders, shady businesspeople, illegal polluters, aggressors, "frenemies," and so on. (More on this in the next chapter.)

For each cooperative strategy, our moral brains have a corresponding set of emotional dispositions that execute the strategy. Let's review:

Concern for others: Two prisoners can find the magic corner if they place some value on each other's payoffs in addition to their own. Corresponding to this strategy, humans have *empathy*. More generally, we have emotions that make us *care* about what happens to others, especially family, friends, and lovers. Our emotions also make us reluctant to directly and intentionally harm others and (to a lesser extent) to allow others to be harmed. I've called this *minimal decency*.

Direct reciprocity: Two prisoners can find the magic corner if they know that being uncooperative now will deny them the benefits of their future cooperation. Corresponding to this strategy, we humans have—and are *known* to have—negative reactive emotions such as *anger* and *disgust*, which motivate us to punish or avoid uncooperative individuals. At the same time, these emotional dispositions are moderated by a tendency toward *forgiveness*, an adaptive strategy for a world in which mistakes happen. We also give one another positive incentives to cooperate through our *gratitude*.

Commitments to threats and promises: Two prisoners can find the magic corner if they are committed to punishing each other's uncooperative behavior. Corresponding to this strategy, humans are often *vengeful*. Many of us have—and are known for having—emotional dispositions that commit us to punishing uncooperative behavior, even when the costs outweigh the benefits. Likewise, two prisoners can find the magic corner if they are committed to punishing *themselves* for being uncooperative. Corresponding to this strategy, humans are sometimes *honorable* and known for their dispositions toward *shame* and *guilt*, self-punishing emotions. Humans can also exhibit the related virtue of *loyalty*, including the loyalty that comes with love. Loyalty to higher authorities also involves the virtue of *humility* and the capacity for *awe*.

Reputation: Two prisoners can find the magic corner if they know that being uncooperative now will deny them the benefits of future cooperation with *others* who are in the know. Corresponding to this strategy, we humans are *judgmental*, even as infants. We attend to how people treat others and adjust our behavior toward them accordingly. What's more, we amplify the influence of our judgments through our irrepressible tendency to produce and consume *gossip*. Consistent with this, we are highly sensitive to the watchful eyes of others, which make us *self-conscious*. When our self-consciousness fails us and we get caught transgressing, we get visibly *embarrassed*, signaling that we will not transgress in the future.

Assortment: Two prisoners can find the magic corner by belonging to a cooperative group, provided that group members can reliably identify one another. Corresponding to this strategy, humans are *tribalistic*—highly sensitive to signals of group membership and intuitively disposed to favor in-group members (including strangers) over out-group members.

Indirect reciprocity: Two prisoners can find the magic corner if there are others who will punish them for not cooperating (or reward them for cooperating). Corresponding to this strategy, humans are *prosocial punishers* whose *righteous indignation* inclines them to punish noncooperators, despite having nothing to gain. Likewise, people expect others to feel righteously indignant toward noncooperators.

Empathy, familial love, anger, social disgust, friendship, minimal decency, gratitude, vengefulness, romantic love, honor, shame, guilt, loyalty, humility, awe, judgmentalism, gossip, self-consciousness, embarrassment, tribalism, and righteous indignation: These are all familiar features of human nature,* and all socially competent humans have a working understanding of what they are and what they do. Nevertheless, until recently we lacked an understanding of how all of these apparently disparate features of human psychology fit together and why they exist. All of this psychological machinery is perfectly designed to promote cooperation among otherwise selfish individuals, implementing strategies that can be formalized in abstract mathematical terms and illustrated by imprisoned bank robbers. There's currently no way to prove that all of this psychological machinery evolved, either biologically or culturally, to promote cooperation, but if it didn't, it's a hell of a coincidence.

According to this view of human morality, cooperation is typically intuitive. We need not reason through the logic of cooperation in order to cooperate. Instead, we have feelings that do this thinking for us. To test this idea, David Rand, Martin Nowak, and I conducted a series of studies. First, we reanalyzed the data from several published experiments that used Prisoner's Dilemma games and Public Goods Games. More specifically, we looked at people's decision times. Over and over, we found the same pattern. The faster people decided, the more they cooperated, consistent with the idea that cooperation is intuitive (see figure 2.5).

Later we conducted our own Public Goods Games, in which we forced some people to decide quickly (less than ten seconds) and forced others to decide slowly (more than ten seconds). As predicted, forcing people to decide faster made them more cooperative and forcing people to slow down made them less cooperative (more likely to free ride). In other experiments, we asked people, before playing the Public Goods Game, to write about a time in which their intuitions served them well, or about a time in which careful reasoning led them astray. Reflecting on the advantages of intuitive thinking (or the disadvantages of careful reflection) made people more cooperative. Likewise, reflecting on the advantages of careful reasoning (or the disadvantages of intuitive thinking) made people less cooperative. These studies underscore this chapter's main point: Built

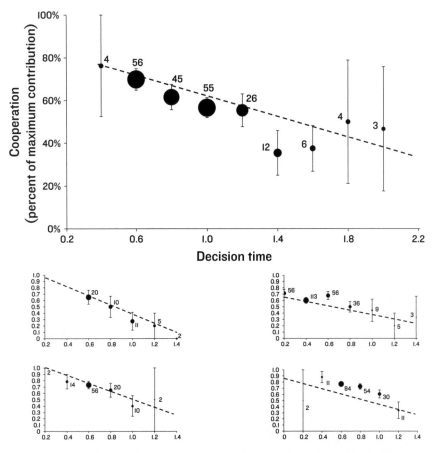

Figure 2.5. Decision-time data from five cooperation experiments: Faster deciders are more likely to put collective interest above individual interest, indicating that being cooperative is (at least in some contexts) more intuitive than being selfish.**

into our moral brains are automated psychological programs that enable and facilitate cooperation.

(Note: From these studies, you might conclude that intuition is the source of all things good and that careful reasoning is the enemy of morality. This would be a mistake. Indeed, it's the mistake that this book was written to correct. What these studies show is that our social instincts are great at averting the Tragedy of the Commons. As noted earlier, this is not the only tragedy. More on this shortly.)

I've called the psychological machinery described in this chapter "moral machinery," but equating "moral" with "cooperation-enhancing" may seem strange, for at least two reasons. First, there are recognizably moral phenomena that appear to be unrelated to cooperation. For example, in some cultures, eating certain foods or engaging in certain consensual sexual acts is considered immoral. How do these prohibitions help people cooperate?

To be clear, in calling our brains' psychological tools for cooperation "moral machinery," I'm not saying that this machinery is used exclusively to promote cooperation. Rather, I'm saying that we have this machinery in our heads because of the role it plays in promoting cooperation. This doesn't mean that moral machinery can't be used for other things. Your nose, for example, can be used to hold up your glasses, but noses did not evolve for this purpose. Likewise, righteous indignation directed toward gays, for example, may do nothing to promote cooperation, and yet our capacity for righteous indignation may nonetheless exist because of the role it plays in promoting cooperation. That said, some moral practices that appear to be unrelated to cooperation may in fact be related. For example, the Hindu prohibition against eating cows may increase the food supply by making cows long-term sources of milk rather than short-term sources of meat. The Protestant work ethic, which combines high productivity with limited consumption, makes more resources available to the community. Even prohibitions against masturbation—a private act if ever there was one—may serve a social function: A cooperative institution such as a church may increase its power by maintaining a monopoly on the blessing of marriages, while blocking alternative routes to sexual gratification.

Second, some of what I've called "moral machinery" may appear to be *amoral*, if not downright *immoral*. Caring about others is certainly moral, and one might say the same about the selfless enforcement of cooperative rules. But what about direct reciprocity? Our tendency to avoid or punish people who've failed to cooperate with us may promote cooperation, but this doesn't seem particularly moral. On the contrary, this looks like simple self-interest. And what about our all-too-human taste for vengeance?

This, too, may promote cooperation, and yet it strikes many of us as far from morally admirable.

Indeed. In calling this psychological machinery "moral," I am not endorsing it, at least not all of it. On the contrary, as we'll see shortly, I believe that our moral machinery gets us into a lot of unnecessary trouble. Nevertheless, from a purely descriptive, scientific perspective, it's important to understand that these features of our psychology, many of which are not especially admirable, are parts of an organic whole—a suite of psychological adaptations that evolved to enable cooperation. Moreover, it's important to understand that this psychological machinery is the earthly source of everything that *is* undeniably moral. In other words, not everything that evolved to promote cooperation is rightly praised as "moral," but nothing that is rightly praised as "moral" would exist on earth were our brains not designed for cooperation.

Why, then, are our brains designed for cooperation? It could be because God designed them that way. Or it could just be an accident of nature. But we are no longer left with a stark choice between divine will and chance. We have cooperative brains, it seems, because cooperation provides material benefits, biological resources that enable our genes to make more copies of themselves. Out of evolutionary dirt grows the flower of human goodness.

3.

Strife on the
New Pastures

The herders of the new pastures have brains packed full of moral machinery designed for cooperation, and yet their lives are marred by intertribal violence. And even in their more peaceful moments, the tribes of the new pastures have deep disagreements about how humans ought to live. Why is this? In the last chapter, we toured the moral machinery that makes us cooperative, the psychological programs that enable us to find the magic corner and avert the Tragedy of the Commons. In this chapter we'll take a second look at our moral machinery, this time to understand why it so often fails us in the modern world. Why do our moral brains, which are so good at averting the Tragedy of the Commons, so often fail to avert the Tragedy of Commonsense Morality?

THE PSYCHOLOGY OF CONFLICT

The psychological obstacles to intertribal cooperation come in two general flavors. First, there is plain old selfishness at the group level, also known as *tribalism*. Humans nearly always put Us ahead of Them. Second, beyond tribalism, groups have genuine differences in values, disagreements concerning the proper *terms* of cooperation. The disagreement between the Northern and Southern herders, for example, goes beyond mere tribal selfishness. The individualistic Northerners genuinely believe that it's wrong to force wise and industrious herders to support herders who are foolish or lazy. Likewise, the collectivistic Southerners genuinely believe that it's wrong to allow members of their group, especially victims of unfortunate circumstance, to starve when others have plenty. Southern and Northern herders can find plenty to fight over without being selfish.

These two flavors of tribal conflict naturally blend. That is, groups can have *selfish* reasons for favoring some moral values over others, a phenomenon I call *biased fairness*. The Northerners are extremely individualistic; the Southerners are extremely collectivist. What about the more moderate Easterners and Westerners? Suppose that the Eastern pastures are more fertile than those to the west, making the Easterners comparatively rich and the Westerners comparatively poor. Faced with the prospect of subsidizing their poorer cousins, one might expect Easterners to tilt toward individualism and away from collectivism. One might expect the Westerners to do the opposite. As they tilt in opposite moral directions, the Easterners and Westerners may not see themselves as in any way biased. Indeed, this tilting process might occur over generations, such that no individual ever changes her mind about how societies ought to be organized.

Some genuine moral disagreements are essentially matters of emphasis. It's not that the Southerners are blind to the injustice of taking from the wise and industrious and giving to the foolish and lazy. Indeed, the most common complaint among Southerners is of foolish or lazy freeriding neighbors. Nevertheless, the collectivist Southerners think it unconscionably cruel to let members of their group, even foolish or lazy ones, die in times of plenty. Likewise, it's not that prosperous Northerners have no sympathy for those who are less fortunate, even the foolish and lazy. Prosperous Northerners often make charitable donations to help such people. Nevertheless, they object to being *forced* to help the foolish and lazy, and to *legitimizing* foolishness and laziness by granting the foolish and lazy the *right* to be helped. This, they say, undermines the whole society and is even worse than letting some people die.

Other moral differences between groups are not matters of emphasis. Some groups have moral values that outsiders simply do not share, at least not in their particulars. From the outside, these values seem arbitrary and strange, but from the inside they make perfect sense and are often sacrosanct. Take, for example, the tribe that thinks it abominable for women to bare their earlobes in public. Other tribes see nothing wrong with bare female earlobes, and they see no reason to accommodate this prohibition if it inconveniences them. Likewise, some tribes grant moral and political authority to particular individuals, institutions, texts, and deities. For example, one tribe's holy book states that black and white sheep may not be housed together. This principle is affirmed by the tribe's Supreme Leader, who speaks for the God of all Gods and whose word is infallible. Here, too, the disagreement is not a matter of degree: Members of other tribes grant no authority whatsoever to this book, this god, and this leader.

Disagreements over which people, deities, and texts are authoritative also lead to disagreements over matters of earthly fact. According to one tribe's holy book, the new pastures are this tribe's ancestral homeland, from which they were driven long ago. Other tribes dismiss this as a self-serving fiction. "Where is the evidence?" they ask. "Right here in the holy book!" the believers reply. Beliefs such as these are *local*, meaning that they are inextricably bound up with commitments to specific people, texts, and deities—entities referred to by proper nouns. A less neutral—and, some would say, more apt—term for these beliefs is *parochial*. However, the believers rarely, if ever, see such beliefs as parochial, or even local. From the believers' point of view, these beliefs reflect knowledge of a universal moral order that, for whatever reason, members of other tribes have failed to appreciate.

Thus, the tribes of the new pastures fight in part because each tribe selfishly favors Us over Them and in part because different tribes see the world through different moral lenses. In the sections that follow, we'll examine the psychology and sociology of moral conflict.

TRIBALISM

The most straightforward cause of strife on the new pastures is tribalism, the (often unapologetic) favoring of in-group members over out-group members. This is going to be a very short section, because there's little doubt that humans have tribalistic tendencies that promote conflict. Insofar as there is a debate about our tribalistic tendencies, it's not about whether we have them, but about *why*. In my view, the evidence strongly suggests that we have innate tribalistic tendencies. Once again, anthropological reports indicate that in-group favoritism and ethnocentrism are human universals. Young children identify and favor in-group members based on linguistic cues. Reaction-time tests (IATs) reveal widespread negative associations with out-group members in adults, children, and even monkeys. People readily favor in-group members over out-group members, even when the groups are arbitrarily defined and temporary. People readily replace racial classification schemes with alternative coalitional classification schemes, but they don't do the same for classification by gender, as predicted by evolutionary accounts of human coalitional psychology. And there is a neurotransmitter, oxytocin, that makes people selectively favor in-group members. Finally, all biological accounts of the evolution of cooperation with non-kin involve favoring one's cooperation partners (most or all of whom belong to one's group) over others. Indeed, some mathematical models indicate that altruism within groups could not have evolved without hostility between groups.

In short, we appear to be tribalistic by nature, and, in any case, we are certainly tribalistic. This is bound to cause problems—though by no means *insurmountable* problems—when human groups attempt to live together.

COOPERATION, ON WHAT TERMS?

Tribalism makes it hard for groups to get along, but group-level selfishness is not the only obstacle. Cross-cultural studies reveal that different human groups have strikingly different ideas about the appropriate terms of cooperation, about what people should and should not expect from one another.

In a landmark set of experiments, Joseph Henrich and colleagues teamed up with anthropologists studying small-scale societies around the world, including ones in Africa, South America, Indonesia, and Papua New Guinea. They had members of these societies play three economic games, all designed to measure people's willingness to cooperate and their expectations about the willingness of others. Two of these games, the Dictator Game and the Public Goods Game, we encountered in the last chapter. The third game is called the "Ultimatum Game."

In the Ultimatum Game, one player, the proposer, makes a proposal about how to divide a sum of money between herself and the responder. The responder can either accept or reject the proposal. If the responder accepts the proposer's offer (e.g., "I get six; you get four"), then the money is divided as proposed. If the responder rejects the offer, then no one gets anything. As usual, all of this is done anonymously. The Ultimatum Game essentially measures people's sense of fairness in dividing up resources. High offers reflect a willingness to share, either because the proposer sees it as the fair thing to do or because the proposer expects that the responder will see it that way. Low offers reflect a sense of individual entitlement, and an expectation that others will respect it. To reject an offer is to say, "Your offer is unfair, and I'm willing to pay to say so."

Henrich and colleagues found that typical Ultimatum Game offers varied widely from society to society. At one end of the spectrum we have the Machiguenga of Peru, who offered on average 25 percent of the money to the responder. Consistent with this, only one Machiguenga responder out of twenty-five rejected an offer. The Machiguenga offered little and expected little from one another. This is very different from what we see in the United States and other Western industrialized nations, where the average offer is about 44 percent, the most common offer is 50 percent, and offers below 20 percent are rejected roughly half the time. Some small-scale societies look more or less like Western societies. For example, a group of resettled villagers in Zimbabwe offered 45 percent on average and rejected about half of the low offers. In contrast to all of the above, the Aché of Paraguay and the Lamelara of Indonesia gave average offers *over* 50 percent and accepted all offers. The Au of Papua New Guinea often made offers greater than 50 percent, but Au responders rejected these extra-generous offers as often as they rejected low offers. The moral machinery in our brains function differently in different places.

The Public Goods Game is, once again, the laboratory version of the Tragedy of the Commons. Individuals can contribute to a common pool, which gets multiplied by the experimenter and then divided evenly among all players. Individuals maximize their payoffs by not contributing (free riding), but groups maximize their payoffs by fully contributing. Typical Public Goods Games in the West (using college students) generate average contributions between 40 percent and 60 percent, with most players contributing either everything or nothing. (Interestingly, the cooperative behavior of Americans is highly sensitive to contextual cues. For example, cooperation in the Prisoner's Dilemma varies dramatically depending on whether the game is labeled the "Wall Street Game" or the "Community Game.") Among the Machiguenga, in contrast, players contributed only 22 percent on average, and not a single player fully contributed. The Tsimané of Bolivia, unlike Westerners, clustered in the middle, with very few individuals giving nothing and very few giving everything. Here, too, we see wide variation from place to place.

The Dictator Game is not really much of a game, because the "proposer" has complete control. Once again, in the Dictator Game one person is given a sum of money along with the option to give all, some, or none of it to another person. In the Dictator Game, Western college students typically offer 50 percent or nothing, consistent with their behavior in Public Goods Games. (And here, too, Americans' behavior varies dramatically with context. Americans tend to give nothing in the Dictator Game when the choice is reframed by a third option to take money *away* from a stranger.) Likewise, the Tsimané show a consistent cultural pattern, offering 32 percent on average and always offering something. Among the Orma of Kenya, the most common offer was 50 percent. Among the Hadza of Tanzania, the most common offer was 10 percent. As you might expect, the societies in which people are most cooperative are also the societies in which people are most willing to punish people who are not cooperative. (One might wonder what the Dictator Game has to do with cooperation, given that this one-sided "game" involves no *co*-operation.**)

Why do people from different cultures play these games so differently? As you might expect, the way people play these games reflects the way they live. Henrich and colleagues characterized these societies in two ways. First, they gave each society a "payoffs to cooperation" ranking, indicating the extent to which people in that society benefit from cooperation. For example, Machiguenga families make their livings independently, while the Lamelara of Indonesia hunt whales in large parties of a dozen or more. Consistent with their economic lifestyles, the Lamelara offer about twice as much as the Machiguenga in the Ultimatum Game. The experimenters also ranked these societies in terms of their "market integration," that is, the extent to which they rely on market exchange in their daily lives (e.g., buying food vs. producing it oneself). As noted in chapter 1, participating in a market economy is a form of large-scale cooperation. Henrich and colleagues found that payoffs to cooperation and market integration explain more than two thirds of the variation across these cultures. A more recent study shows that, accross societies, market integration is an excellent predictor of altruism in the Dictator Game. At the same time, many factors that you might expect to be important predictors of cooperative behavior—things like an individual's sex, age, and relative wealth, or the amount of money at stake—have little predictive power.

These experimental findings dovetail with cultural practices in more specific ways. Consider, for example, the Au and Gnau of Papua New Guinea, who frequently offer more than 50 percent in the Ultimatum Game and frequently reject such hypergenerous offers. It turns out that these groups have gift-giving cultures in which accepting a large gift obligates one to repay the gift and subordinates the recipient to the giver. The Aché of Paraguay were among the most generous Ultimatum Game players, with nearly all players offering more than 40 percent. This group is highly collectivist. Successful Aché hunters typically leave their kill by the edge of

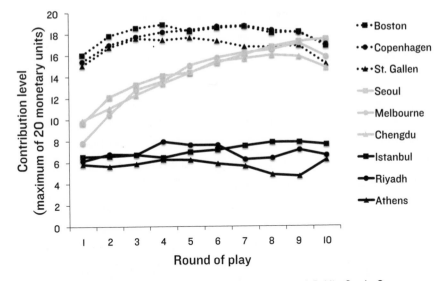

Figure 3.I. People from different cities played a series of Public Goods Games. Cooperation levels and trajectories varied widely.*

the camp and report having been unsuccessful; others then find the kill and share it equally among all in the camp. Members of the Orma of Kenya, another highly collectivist group, spontaneously dubbed the Public Goods Game the *harambee* game, referring to their practice of working together on collective projects such as building schools and roads. The Orma contributed 58 percent of their allotments in the Public Goods Game.

More recently, Benedikt Herrmann and colleagues have examined cooperation and punishment in a set of large-scale societies, and the results are equally striking. People in cities around the world played repeated Public Goods Games in which players could punish free riders. Figure 3.1 shows some of the results.

On the x-axis, we have the round of play (first round, second round . . .), and on the y-axis we have the mean contribution. First, you'll notice that, right from the start, people in different cities contribute at very different levels, with people in Athens, Riyadh, and Istanbul contributing a little over 25 percent of their allotments on average and people in Boston, Copenhagen, and St. Gallen contributing more than 75 percent. Second, there are roughly three different patterns concerning how the games play out over time. In places like Copenhagen, contributions start out high and stay high, because most people are willing to cooperate initially and because people will pay to punish the few who aren't. (Even in places like Copenhagen, however, cooperation unravels over time if there's no opportunity to punish.) And then there are places like Seoul, where contributions start out moderately high and then rise up to very high levels as free riders get reined in by punishment. Finally, there are place like Athens, Riyadh, and Istanbul, where contributions start low and stay low. This last set of results is surprising: Given that cooperators in these places can punish the free riders, why doesn't cooperation ramp up over time the way it does in Seoul?

It turns out that in places like Athens, Riyadh, and Istanbul, there is an opposing social force. In this version of the Public Goods Game, cooperators can punish free riders, but free riders can also punish cooperators, a phenomenon known as "antisocial punishment." In places like Athens, people who didn't contribute to the common pool often paid to punish those who did. Why would anyone do that? In part, it's about revenge. Free riders resent being punished by cooperators and strike back. But it can't be about revenge only because, in some places, low contributors will punish cooperators on the first round! It's as if they are saying, "To hell with you do-gooders! Don't even *think* about trying to make me play your little game!" As shown in figure 3.2, the prevalence of antisocial punishment is an excellent predictor of a group's failure to cooperate.

Thus, in some places, the forces that might otherwise sustain cooperation in the Public Goods Game—altruism and pro-social punishment— are overwhelmed by antisocial punishment. Here, too, the way people play seems to reflect the culture on the ground. The experimenters examined the responses of thousands of people from each city to questions on the World Values Survey. They found that antisocial punishment is high in places where people report having lax attitudes toward things like tax evasion and dodging fares for public transportation. Likewise, my colleagues and I found in our experiments using Public Goods Games (pp. 62–63)

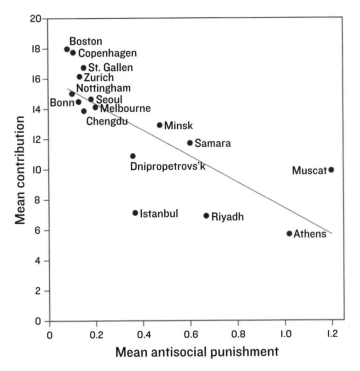

Figure 3.2. Cooperation levels in repeated Public Good Games played in cities around the world are negatively correlated with levels of "antisocial punishment," punishing people for cooperating.

that the people who cooperated intuitively were the ones who reported trusting their daily interaction partners. In a sad convergence between lab and field, the European economy, during much of the writing of this book, was in crisis, primarily because Greece (see the bottom-right corner of figure 3.2) had become financially insolvent. Greece's troubles threatened to tear apart the European Union as the leaders of nations such as Denmark (see the upper-left corner) debated whether, and on what terms, to bail out Greece in the name of the greater good.

(Before moving on, let me say that my intention here—in this chapter and elsewhere—is not to pick on Greeks or the members of any other nation or tribe. Instead, my hope is that we can learn from the successes and failures of different social systems—systems for which few individuals bear any significant responsibility. To learn these lessons, however, we must be willing to say things that could be interpreted as insulting and that may sound distressingly similar to the things people say when they are airing their prejudices.***)

HONOR VERSUS HARMONY

In the early 1990s, Dov Cohen and Richard Nisbett conducted a series of studies examining cultural differences among Americans. Male students at the University of Michigan were brought into the lab for an experiment on "limited response time conditions on certain facets of human judgment." The students arrived one at a time, filled out some forms, and were instructed to take their completed forms to a table at the end of a long, narrow hallway. On the way to the table, each student would pass a man—a confederate of the experimenters—who was standing in the hallway, working at a file cabinet. As each student returned down the hall and passed the man for the second time, the man would slam the file cabinet drawer shut, bump into the student with his shoulder, and call the student an "asshole."

How the students responded to this insult depended on where they were from. On average, students from the southern United States responded to this insult with more anger and less amusement than students from northern states, as measured by independent observers positioned in the hallway. Not only that, but these two groups of students also exhibited different physiological responses. The experimenters collected saliva samples before and after the insult and found that southerners who were insulted exhibited greater increases in their levels of cortisol (a hormone associated with stress, anxiety, and arousal) than both northerners who were insulted and southerners who were not insulted. Likewise, the insulted southerners exhibited greater increases in their testosterone levels following the insult.

Later in the experiment, the students read and responded to the following vignette:

It had only been about twenty minutes since they had arrived at the party when Jill pulled Steve aside, obviously bothered about something.

"What's wrong?" asked Steve.

"It's Larry. I mean, he knows that you and I are engaged, but he's already made two passes at me tonight."

Jill walked back into the crowd, and Steve decided to keep his eye on Larry. Sure enough, within five minutes Larry was reaching over and trying to kiss Jill.

The students were asked to complete the ending to this story. Seventy-five percent of the southerners who were insulted by the confederate (that is, the experimenter's covert accomplice—not a member of the Southern Confederacy) completed this story in a way that involved either violence or the threat of violence, while only 20 percent of the noninsulted southerners did. The northerners, by contrast, didn't respond to the scenario differently depending on whether they were insulted.

Cohen, Nisbett, and their colleagues wanted to see whether these insults would affect real behavior. To find out, they arranged for their experimental subjects to play a game of "chicken." After being insulted (or not), the students encountered a second confederate, a six-foot-three-inch, 250-pound man walking quickly down the hall. The middle of the hall was lined with tables and thus too narrow for both the subject and the burly confederate to pass through at once. Someone had to give way. The big confederate walked down the hall on a collision course with the subject, giving way only at the very last moment. The experimenters measured the distance at which the subjects gave way to the big man. The insulted southerners, on average, gave way to the approaching confederate when he was 37 inches away, while the noninsulted southerners gave way, on average, at 108 inches. The insult had no effect on the northerners' "chicken point." At the same time, the southerners were also more *polite* than the northerners when *not* insulted. The noninsulted northerners typically gave way at about 75 inches.

Why did the southerners and northerners react so differently to being insulted (or not insulted)? Cohen and Nisbett predicted these results, based on the idea that the American South, like some other regions of the world, has a strong "culture of honor." Like Henrich and his colleagues, they began their analysis with economics. The economy of the South was initially based on the herding of livestock, and many of the original settlers of the South came from regions on the fringes of Britain that are dominated by a herding economy. Herders are especially vulnerable to opportunistic aggression, because their wealth is so portable. (It's easier to steal sheep than cornfields.) The threat of aggression is amplified when there is no reliable law enforcement, as was historically the case in both the highlands of Britain and parts of the American South. Herders need to stand their ground, lest they lose everything. Not only that, as explained in the last chapter, herders need to *make it known* that they are willing to stand their ground, lest they encourage others to test their mettle. A herder who is known as a pushover will spend much time and energy defending his lot, and may succumb in the end. An insult is a test of one's honor, and herders who take even small insults in stride run the risk of advertising weakness. On the flip side, a herder who is hotheaded, and known for being hotheaded, has a strategic advantage.

The southern students who walked down the narrow hallway at the University of Michigan were not herders, but they had grown up in a culture that takes traditional notions of honor very seriously. The southern culture of honor has had profound social effects. Homicide rates are higher in the South than in the North, but this is only because the South has a higher rate of homicides that are argument- or conflict-related. In surveys, southerners are not more likely than other Americans to approve of violence in general, but they are more likely to approve of violence committed in defense of one's home or in response to an affront to one's wife. Likewise, they are more likely to stigmatize men who do not respond violently to personal affronts.

The southern culture of honor appears to have had a profound effect on American foreign policy. According to historian David Hackett Fischer, the South has "strongly supported every American war no matter what it was about or who it was against," a pattern that he attributes to "Southern ideas of honor and the warrior ethic." For example, the South was highly enthusiastic about helping the British fight the French in 1798, and then equally enthusiastic about helping the French fight the British in 1812. Though regional political allegiances have shifted dramatically in the United States, with the Republican and Democratic parties reversing their strongholds, southern support for war has remained consistent and has often cut across party lines. For example, southern Democrats who strongly opposed F.D.R.'s New Deal legislation nevertheless supported his military engagement in World War II. Likewise, Harry S. Truman and Lyndon Johnson enjoyed far more southern support for their anti-Soviet foreign policies than for their domestic initiatives.

While the culture of honor that pervades the American South emphasizes self-reliance and individual autonomy, the more collectivist cultures of East Asia emphasize interdependence and group harmony. Nisbett and others argue that collectivism, like the South's culture of honor, is a cultural adaptation to economic circumstances, in this case an economic system based on cooperative agriculture. With this in mind, Kaiping Peng, John Doris, Stephen Stich, and Shaun Nichols presented both American and Chinese subjects with a classic moral dilemma known as the Magistrates and the Mob case:

> An unidentified member of an ethnic group is known to be responsible for a murder that occurred in a town. . . . Because the town has a history of severe ethnic conflict and rioting, the town's Police Chief and Judge know that if they do not immediately identify and punish a culprit, the townspeople will start anti-ethnic rioting that will cause great damage to property owned by members of the ethnic group, and a considerable number of serious injuries and deaths in the ethnic population. . . . The Police Chief and Judge are faced with a dilemma. They can falsely accuse, convict, and imprison Mr. Smith, an innocent member of the ethnic group, in order to prevent the riots. Or they can continue hunting for the guilty man, thereby

allowing the anti-ethnic riots to occur, and do the best they can to combat the riots until the guilty man is apprehended. . . . [T]he Police Chief and Judge decide to falsely accuse, convict, and imprison Mr. Smith, the innocent member of the ethnic group, in order to prevent the riots. They do so, thereby preventing the riots and preventing a considerable number of ethnic group deaths and serious injuries.

Most Americans find the idea of deliberately convicting an innocent person appalling, regardless of the benefits. Philosophers are known for their willingness to hear all sides of an argument, but the renowned philosopher Elizabeth Anscombe said that she would draw the line at someone who was willing to defend the magistrates. "I do not want to argue with him; he shows a corrupt mind."* Peng and colleagues predicted that Chinese people, as members of a collectivist culture emphasizing group harmony over individual rights, would be more comfortable with harming one person to save others. They were right. Chinese people were less likely to condemn the imprisoning of an innocent person to stave off the riot and less likely to say that the police chief and judge should be punished for their decisions. Interestingly, Chinese subjects were more likely to hold the would-be rioters responsible for the scapegoating.

Given the sensitive nature of these topics,* I'd like to ward off a few misconceptions before moving on. First, the experimental results presented above, like nearly all experimental results in psychology, concern differences in group *averages*. The research suggests that, *on average*, southerners are more likely than northerners to endorse violence in defense of honor, but those are just averages. There are docile southerners, hotheaded northerners, and everything in between in both cultural groups. The same goes for Chinese people and their collectivist tendencies. Second, my point here is not to praise or condemn these cultural tendencies. On the contrary, as I'll explain later, I believe that such tendencies should be evaluated based on how well they function in context, and these tendencies may function very well in their natural contexts. As explained in the previous chapter, punishment can play a key role in sustaining cooperation. Consistent with this, the southern culture of honor is not indiscriminately violent. Rather, it's a culture that is particularly vigilant about punishing certain kinds of uncooperative behavior. And, as noted above, it's a culture that emphasizes politeness and respect under ordinary conditions. I've suggested that the southern culture of honor has had an important influence on U.S. foreign policy, but I didn't say whether this influence is for better or for worse. That's because I don't know the answer. It's possible that the South's strong support for certain wars has been essential for keeping the United States and other nations free. Likewise, I'm not here passing judgment on Chinese collectivism. In fact, later in this book I'll defend a moral philosophy that some regard as too collectivist.

Thus, my point here is not that the southern culture of honor is good or bad, or that Chinese collectivism is good or bad, but rather that these cultural differences are further examples of moral diversity, reflecting the diversity of human social circumstances. These are both cooperative cultures, but they cooperate on different terms. Chinese collectivism emphasizes active cooperation and the necessity of individual sacrifice for the greater good. Southern honor culture emphasizes passive cooperation (respect for the property and prerogative of others) and endorses aggressive responses to inappropriate aggression, or the threat thereof.

LOCAL MORALITY

In September 2005, the Danish newspaper *Jyllands-Posten* published a series of cartoons depicting and satirizing the Prophet Muhammad. This was a deliberately provocative gesture in defiance of Muslim law, which explicitly prohibits the visual depiction of Muhammad. The newspaper published the cartoons as a contribution to an ongoing debate over self-censorship among journalists, artists, and other intellectuals, many of whom were reluctant to criticize Islam for fear of violent retribution. Such fears were not unfounded. The previous year, a lecturer at the University of Copenhagen had been assaulted by five people who opposed his reading of the Koran to non-Muslims during a lecture.

The cartoons were indeed provocative. Danish Muslim organizations held protests in Denmark. Newspapers around the world covered the controversy and reprinted the cartoons. This led to violent protests around the Muslim world, with more than a hundred deaths, due primarily to police firing on protesters. Crowds set fire to the Danish embassies in Syria, Lebanon, and Iran. Several of the cartoonists went into hiding because of death threats. Haji Yaqoob Qureishi, an Indian state minister, offered a reward of roughly US$11 million to anyone who would behead "the Danish cartoonist" responsible for the drawings. Muslim boycotts of Danish goods cost Danish business approximately $170 million during the five-month period following these events. More recently, in 2012, a YouTube video portraying Muhammad in a highly unflattering manner sparked protests around the world, many of which turned violent.

These conflicts are not simply a matter of different groups' emphasizing different values. The aggrieved Muslims were deeply opposed to the cartoons, but the Danish journalists had no problem with them at all. (Indeed, their lack of a gut reaction to the cartoons may explain why they so severely underestimated the magnitude of the Muslim world's response.) And insofar as non-Muslims opposed the publication of the cartoons, it was out of respect for Muslim values, not because they had objections of their own. In other words, the prohibition against depicting Muhammad is a *local* moral phenomenon. By this I mean, once again, that it is inextricably bound up with the authority of certain entities named by proper nouns, such as Muhammad, the Koran, and Allah.

The Danish cartoon conflict illustrates two familiar points that are nevertheless worth making explicit. First, religious moral values and local moral values are intimately related. More specifically, local moral values are nearly always religious values, although many religious values—arguably the most central ones—are not local. For example, as noted in the last chapter, all major religions affirm some version of the Golden Rule as a central principle and, along with it, general (though not exceptionless) prohibitions against killing, lying, stealing, et cetera. If it's local morality, it's probably religious, but if it's religious morality, it's not necessarily local.

Second, the cartoon controversy reminds us that local moral values are, and have long been, a major source of conflict. Indeed, compared with other conflicts involving local religious values, the Danish cartoon affair is a minor dustup. The ongoing Israeli-Palestinian conflict, arguably the world's most divisive political dispute, is bedeviled by competing claims to specific parcels of land, grounded in the authority of various proper-noun entities. Likewise, the ongoing conflicts within Sudan and between Pakistan and India run along religious lines. Local values and their associated proper nouns play central roles in many domestic controversies, such as those over prayer in public schools in the United States and the banning of Muslim women's traditional facial coverings from public spaces in France. Likewise, many controversial issues, such as abortion and gay rights, which can be discussed in purely secular terms, are nevertheless deeply intertwined with local religious moral values.

In short, serious conflicts between groups arise not only because they have competing interests, and not only because they emphasize shared values differently, but also because they have distinctive local values, typically grounded in religion. As noted above, many of the most widely held moral values, such as a commitment to the Golden Rule, are actively promoted by the world's religions. Thus, religion can be a source of both moral division and moral unity.

BIASED FAIRNESS

In 1995, a *U.S. News & World Report* survey posed the following question to readers: "If someone sues you and you win the case, should he pay your legal costs?" Eighty-five percent of respondents said yes. Others got this question: "If you sue someone and lose the case, should you pay his costs?" This time, only 44 percent said yes. As this turnabout illustrates, one's sense of fairness is easily tainted by self-interest. This is biased fairness, rather than simple bias, because people are genuinely motivated to be fair. Suppose the magazine had posed both versions of the question simultaneously. Few respondents would have said, "The loser should pay if I'm the winner, but the winner should pay if I'm the loser." We genuinely want to be fair, but in most disputes there is a range of options that might be seen as fair, and we tend to favor the ones that suit us best. Many experiments have documented this tendency in the lab. The title of a Dutch paper nicely summarizes the drift of these findings: "Performance-based pay is fair, particularly when I perform better."

A series of negotiation experiments by Linda Babcock, George Loewenstein, and colleagues illuminates the underlying psychology of biased fairness. In some of these experiments, pairs of people negotiated over a settlement for a motorcyclist who had been hit by a car. The details of the hypothetical case were based on a real case that had been tried by a judge in Texas. At the start of the experiment, the subjects were randomly assigned to their roles as plaintiff and defendant. Before negotiating, they separately read twenty-seven pages of material about the case, including witness testimony, maps, police reports, and the testimonies of the real defendant and plaintiff. After reading this material, they were asked to guess what the real judge had awarded the plaintiff, and they did this knowing which side they would be on. They were given a financial incentive to guess accurately, and their guesses were not revealed to the opponents, lest they weaken their bargaining positions. Following the subsequent negotiation, the subjects were paid real money in proportion to the size of the settlement, with the plaintiff subject getting more money for a larger settlement and the defendant subject getting more money for a smaller one. The settlement could be anywhere from $0 to $100,000. The pairs negotiated for thirty minutes. Both subjects lost money in "court costs" as the clock ticked, and failure to agree after thirty minutes resulted in an additional financial penalty for both negotiators.

On average, the plaintiffs' guesses about the judge's award were about $15,000 higher than those of the defendants, and the bigger the discrepancy between the two guesses, the worse the negotiation went. In other words, the subjects' perceptions of reality were distorted by self-interest. What's more, these distortions played a big role in the negotiation. Pairs with relatively small discrepancies failed to agree only 3 percent of the time, while the negotiating pairs with relatively large discrepancies failed to agree 30 percent of the time. In a different version of the experiment, the negotiators didn't know which side they would be on until after they made their guesses about the judge's settlement. This dropped the overall percentage of negotiators who failed to agree from 28 percent to 6 percent.

These experiments reveal that people are biased negotiators, but, more important, they reveal that their biases are *unconscious*. Plaintiffs guessed high about the judge's award, and defendants guessed low, but they weren't consciously inflating or deflating their guesses. (Once again, they had financial incentives to guess accurately.) Rather, it seems that knowing which side of a dispute you're on unconsciously changes your thinking about what's fair. It changes the way you process the information. In a related experiment, the researchers found that people were better able to remember pretrial material that supported their side. These unconsciously biased perceptions of fairness make it harder for otherwise reasonable people to reach agreements, often to the detriment of both sides.

To test their ideas about biased fairness in the real world, the research team examined historical records of negotiations over public school teachers' salaries in Pennsylvania. In such negotiations, teachers' unions and school boards typically base their cases on the salaries paid in other, comparable school districts. However, which districts count as "comparable" is an open question. The researchers hypothesized that bargaining impasses over teachers' salaries would be exacerbated by biased selections of comparable school districts. The researchers polled school board and union presidents, asking them to identify comparable nearby districts. As predicted, the average salary paid in the districts listed as comparable by union presidents was considerably higher than the average salary in districts listed by school board presidents. The researchers then examined the school district's records and found, as predicted, that districts in which union presidents and school board presidents gave highly discrepant lists of comparable districts were about 50 percent more likely to experience teacher strikes.

In the original Tragedy of the Commons described by Hardin, all of the herders are in symmetrical positions. As a result, there is only one plausibly fair solution: Divide the commons equally among all the herders. But in the real world, interested parties are almost never in perfectly symmetrical positions. Indeed, even in Hardin's stylized parable, it's hard to flesh out the details without raising sticky issues: Should each family get the same number of animals, or should the number vary with family size? And so on. As long as people's starting points are asymmetrical, people will be tempted, unconsciously if not consciously, to tailor their conceptions of fairness to suit their interests.

An experiment conducted by Kimberly Wade-Benzoni, Ann Tenbrunsel, and Max Bazerman illustrates the problem of biased fairness in the context of an environmental commons problem. Subjects in their experiment negotiated as stakeholders in the fish stocks off the coast of the northeastern United States, where overfishing has become a serious economic and environmental problem. In the control condition, the negotiators represented different companies, but they, like the original herders of the commons, were all in more or less the same situation. In the critical experimental condition, the various negotiators had different payoff structures. For example, some had a more long-term stake in the fish stocks, while others had a more short-term stake, though all had an interest in achieving a sustainable policy. In the control condition, in which the negotiators occupied symmetrical economic positions, 64 percent of the negotiating groups agreed on sustainable solutions. But when the negotiators' positions were asymmetrical, only 10 percent did. Thus, when everyone has competing selfish interests, but those interests are symmetrical, people can fairly easily put their selfish interests aside and find a mutually agreeable solution. But when people's selfish interests come in different forms, people gravitate toward different conceptions of what's fair, and agreement becomes much harder.

Ironically, our tendency toward biased fairness is sufficiently strong that, in some situations, we may be better off if everyone thinks selfishly rather than morally. Fieke Harinck and colleagues at the University of Amsterdam had pairs of strangers negotiate over penalties for four hypothetical criminal cases, modeled after real-life cases. Each pair of negotiators negotiated over all four cases simultaneously. One member of each pair was randomly assigned to the role of defense lawyer and, as such, attempted to get lighter penalties for the defendants. The other negotiator played the role of district attorney and, as such, attempted to get stiffer penalties.

There were, in each criminal case, five possible penalties for the defendant, ranging from a light fine to a long jail sentence. Each negotiator received a confidential document telling her how good or bad each outcome was from her point of view as defense lawyer/district attorney. For two of the criminal cases, the outcome values were arranged to make it a "zero sum" game. That is, a gain for one player necessarily involved an equally large loss for the other player. For the other two cases, however, the outcome values were arranged to allow for "win-win" solutions. In these cases, a gain for one side would still involve a loss for the other side, but the two cases were weighted differently for each negotiator. This meant that each player could make concessions on the case that mattered less to him and, in return, gain concessions on the case that mattered more. In other words, the experiment was set up so that both sides could come out ahead if both sides were willing to make concessions. Unbeknownst to the negotiators, each outcome had a pre-assigned point value, corresponding to the goodness/badness of the outcome for that negotiator. By adding up the points earned by both members of each negotiating pair, the experimenters could measure how well each pair did at finding hidden "win-win" solutions.

All of this is part of the standard setup for a negotiation experiment. The twist, in this case, was in how the negotiators were told to *think* about the negotiation. Some pairs were told to think about the negotiation in purely *selfish* terms, to try to get lighter/stiffer penalties because doing so would advance their careers and help them get promoted. Other pairs of negotiators were told to think about the negotiation in *moral* terms; here, the defense lawyers were told to pursue lighter penalties because lighter penalties are, in these cases, more *just*. Likewise, the district attorneys were told to pursue stiffer penalties because stiffer penalties would be more just.

So who did better, the selfish careerists or the seekers of justice? The surprising answer is that the selfish careerists did better. Bear in mind that the selfish careerists did not succeed by trampling over the seekers of justice. The selfish careerists were negotiating with *each other*. What Harinck and colleagues found was that two people told to negotiate selfishly were, on average, better at finding win-win solutions than two people told to seek justice. Why is this?

Once again, in this set of negotiations, the key to mutual success is for both negotiators to make concessions on the issues that are less important to them, in order make greater gains on the issues that are more important to them. As a selfish negotiator, you're willing to make these concessions because they result in a net gain. Moreover, you understand that your opponent, who is also selfish, will make only those concessions that result in a net gain for her. Thus, two selfish and rational negotiators who see that their positions are symmetrical will be willing to make the concessions necessary to enlarge the pie, and then split the pie evenly. However, if negotiators are seeking justice, rather than merely looking out for their bottom lines, then other, more ambiguous, considerations come into play, and with them the opportunity for biased fairness. Maybe your clients *really deserve* lighter penalties. Or maybe the defendants you're prosecuting *really deserve* stiffer penalties. There is a range of plausible views about what's truly fair in these cases, and you can choose among them to suit your interests. By contrast, if it's just a matter of getting the best deal you can from someone who's just trying to get the best deal for himself, there's a lot less wiggle room, and a lot less opportunity for biased fairness to create an impasse. When you're thinking of the negotiation as a mutually self-serving endeavor, it's harder to convince yourself that there is a relevant asymmetry between you and your negotiation partner. Two selfish negotiators with no illusions about their selfishness have nowhere to hide. This surprising result doesn't imply that we should forsake all moral thinking in favor of pure selfishness, but it does highlight one of the dangers in moral thinking. Biased fairness is sufficiently destructive that, in some cases, we're better off putting morality aside and simply trying to get a good deal.

In some cases, we ourselves may have little sense of what's fair, but we can bias our judgments by adopting the opinions of trusted members of our tribes. This is especially likely to happen in the domain of public policy, in which it's often impossible for ordinary citizens to master the details necessary to make a truly informed decision. An experiment by Geoffrey Cohen nicely illustrates this tribalistic brand of biased fairness. Cohen presented self-described conservative and liberal Americans with two different welfare policy proposals: one offering generous welfare benefits, more generous than any existing welfare program, and one offering meager welfare benefits, more meager than any existing welfare program. As you might expect, the liberals tended to like the generous program more than the conservatives did, and vice versa. In the next part of the experiment, using a new set of liberal and conservative subjects, Cohen presented the same proposals, but this time he labeled the proposals as coming either from Democrats or from Republicans. As you would expect, support from Democrats made programs more appealing to liberals and support from Republicans made programs more appealing to conservatives. More surprising, however, was the strength of this partisan bias. In this experiment, the effect of partisan support completely obliterated all effects of policy content. Liberals liked extreme conservative policies in liberal clothing better than they liked extreme liberal policies in conservative clothing. The conservatives did the same thing, valuing conservative endorsement well above conservative substance. And, as you should expect by now, most subjects denied that their judgments were affected by the partisan packaging. It's all unconscious.*

BIASED PERCEPTION

Our judgments about what's fair or unfair depend critically on our understanding of the relevant facts. Take, for example, people's views about the United States' 2003 invasion of Iraq. Many who opposed the invasion were baffled by Americans who supported it. "Why Iraq?" they asked. "Osama bin Laden attacked us, not Saddam Hussein!" What the baffled didn't know, or failed to fully appreciate, was that a majority of Americans at the time believed that Saddam Hussein had been personally involved in the attacks. And much of the rest of the world is under a different set of misconceptions about the 9/11 attacks. According to a 2008 World Public Opinion poll, a majority of people in Jordan, Egypt, and the Palestinian territories believe that someone other than Al Qaeda (typically the U.S. or Israeli government) was behind the 9/11 attacks.

Why is it so hard for people to get the facts straight? One explanation is simple self-serving bias. When the facts are at all ambiguous, people favor the version of the facts that best suits their interests. In a classic social psychology experiment, students from two colleges watched footage from a football game between their schools. The officials made a number of controversial calls, and the students were asked to evaluate them for accuracy. As you might expect, students from both schools attributed more errors to the officials when the officials made calls against the students' own teams. In another classic study, people with strong opinions about capital punishment were presented with mixed evidence concerning capital punishment's efficacy in deterring crime. One might think that mixed evidence would encourage more moderate views, but it appears to do the opposite. People found the evidence supporting their original views more compelling than the counterevidence, and as a result, both opponents and proponents of capital punishment became more confident in their views after considering the mixed evidence. In a later study, some of the same researchers had Arabs and Israelis watch news coverage of the 1982 Beirut massacre. The two groups saw the same coverage, yet both concluded that it was biased in favor of the other side, a phenomenon the researchers dubbed the "hostile media effect." In a more recent study by Dan Kahan and colleagues, people watched footage of protesters. Their task was to judge whether the protesters were merely exercising their right to free speech or whether they crossed the line by engaging in illegal behavior, such as obstructing and threatening pedestrians. Some subjects were told that the protesters were protesting abortion outside an abortion clinic. Others were told that they were protesting the military's "don't ask, don't tell" policy for gays outside a campus recruitment facility. Subjects who were unsympathetic to the protesters' cause, whether it was anti-abortion or pro–gay rights, were more likely to say that the protesters crossed the legal line.

In these cases, people capitalize on ambiguity to form self-serving beliefs. But sometimes our biases cause us to form beliefs that are, or appear to be, at odds with self-interest. Consider people's beliefs about climate change. There is an overwhelming consensus among experts that the earth's climate is changing due to human activities and that we ought to take strong measures to slow or, if possible, halt this trend. Nevertheless, many people, particularly in the United States, are skeptical of the evidence concerning climate change. Here the failure to get the facts right is not just a matter of simple bias. Some individuals—such as CEOs of corporations that emit large quantities of carbon—may have an interest in denying the reality of climate change, but most people, including most climate skeptics, do not. Nevertheless, according to a 2010 Gallup poll of Americans, only 31 percent of Republicans believe that the effects of global warming are already occurring, and 66 percent say that the seriousness of global warming is exaggerated in the news. You would think that, in the case of climate change, both Democrats and Republicans would be strongly motivated to get things right. After all, Republicans' stake in the earth's future habitability is no smaller than that of Democrats. (Save, perhaps, those who expect to be raptured away in the near future.) Why, then, do so many American political conservatives deny the facts on climate change, in apparent defiance of their own interests? One possibility is ideological: Conservatives are, in general, skeptical about the necessity of collective efforts to solve collective problems. That's likely to be an important part of this story, but it doesn't explain why conservatives are less concerned about global warming than they were only a few years ago (as I'll explain shortly) and why American conservatives are less concerned about global warming than conservatives elsewhere.

According to Kahan and colleagues, the key is to recognize that the problem of getting the facts right is actually a commons problem of its own, involving a tension between individual and collective self-interest. It is definitely in our collective self-interest to face the facts on climate change and act accordingly. But for some of us, as individuals, the payoff matrix is more complicated. Suppose you live in a community in which people are skeptical about climate change—and skeptical about people who *aren't* skeptical. Are you better off as a believer or a skeptic? What you, as a single ordinary citizen, think about climate change is very unlikely to have an effect on the earth's climate. But what you think about climate change is rather likely to have an effect on how you get along with the people around you. If you're a climate change believer living among climate change skeptics, when the topic comes up, your choices are to (a) keep suspiciously quiet, (b) lie about your views, or (c) say what you really think and risk being ostracized. None of these alternatives is particularly appealing, and the costs are palpable, as opposed to the distant possibility that you, by biting your tongue at Herb's barbecue, are going to alter the course of human history. Thus, says Kahan, many people's skepticism about climate change is actually quite rational if you view those people as not trying to manage the earth's physical environment but trying to manage their own social environments. It's individual rationality triumphing over collective rationality—unconsciously, of course.

Kahan's analysis of the problem makes some testable predictions. The conventional wisdom is that today's rank-and-file climate skeptics are simply ignorant and, perhaps, not very good at thinking critically in general. According to this view, people who are more scientifically informed in general (high in "scientific literacy") and generally better at processing quantitative information (high in "numeracy") will be more likely to believe that climate change and its associated risks are real. Kahan's prediction, by contrast, is that one's views on climate change have more to do with one's cultural outlook—one's tribal allegiances—than with one's scientific literacy and numeracy. Contrary to the conventional wisdom, Kahan's theory predicts that people who are more scientifically literate, rather than gravitating toward the truth, will simply be more adept at defending their tribe's position, whatever it happens to be.

To test these hypotheses, Kahan and colleagues administered math and science tests to a large, representative sample of U.S. adults. These subjects also filled out questionnaires designed to measure their cultural worldviews along two dimensions: the hierarchy-egalitarianism dimension and the individualism-communitarianism dimension. Hierarchical individualists are comfortable with having selected high-status individuals make decisions for the society and are wary of collective action aimed at interfering with their authority. Egalitarian communitarians, by contrast, favor less regimented forms of social organization and support collective action to protect the interests of ordinary individuals. For our present purposes, the important thing to know is that hierarchical individualists tend to be skeptical about climate change, while egalitarian communitarians tend to believe that climate change poses a serious threat requiring collective action.

Finally, the researchers asked those surveyed for their views on climate change. Contrary to conventional liberal wisdom, the researchers found that scientific literacy and numeracy were associated with slight *decreases* in the perceived risk of climate change. But the real story emerges when you sort people into tribes. The egalitarian communitarians, as expected, reported perceiving great risk in climate change, but within that group there was no correlation between scientific literacy/numeracy and perceived risk. Likewise, the hierarchical individualists were, as expected, skeptical about the risks of climate change, and within this group, those who were more scientifically literate/numerate were somewhat more skeptical about the risks of climate change. (This is why higher scientific literacy/numeracy was, overall, associated with more skepticism about climate change. The overall effect was driven by the effect within the hierarchical-individualist group.) But overall, scientific literacy and numeracy were not very good predictors of people's beliefs about the risks of climate change. Instead, their beliefs were well predicted by their general cultural outlooks—by their tribal memberships (see figure 3.3).

Now, to be clear, you should *not* conclude from these results that everyone's views on climate change are based simply on who their friends happen to be, and that therefore we have no good reason to worry about climate change. (Although you probably *will* draw this conclusion if that's what you want to believe!) If you want to know how to treat psoriasis, you don't ask your friend Jane, who is one standard deviation more scientifically literate than the average U.S. adult. Rather, you consult a

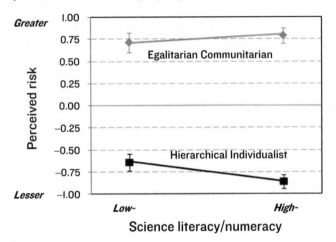

Figure 3.3. Scientific literacy and numeracy have little to do with ordinary people's views about the risks of climate change. Instead, people tend to adopt the beliefs of their respective tribes.

dermatologist—an expert. The people polled in this study were not climate science experts. They were ordinary U.S. adults whose scientific literacy and numeracy scores formed a bell curve around the mean for U.S. adults. While there is widespread disagreement about climate change among nonexperts, there is, once again, overwhelming consensus among experts that climate change is real and that the risks are serious. The lesson, then, is not that it's all "relative" or that there's no way to cut through the cultural clatter and find out the truth about climate change. Nor is the lesson that ordinary people are, in general, hopeless slaves to tribal prejudice. On the contrary, when it comes to most issues, people of all tribes are perfectly happy to accept the advice of experts. (People's views about treating psoriasis are *not* well predicted by their tribal allegiances.) Instead, the lesson is that false beliefs, once they've become culturally entrenched—once they've become tribal badges of honor—are very difficult to change, and changing them is no longer simply a matter of educating people.

In 1998, Republicans and Democrats were equally likely to believe that climate change is already under way. Since then, the scientific case for climate change has only gotten stronger, but the views of Republicans and Democrats have diverged sharply, to the point that in 2010, Democrats were twice as likely as Republicans to believe in the reality of climate change. This didn't happen because Republican scientific literacy and numeracy dropped over that decade. Nor did it happen because Democrats got dramatically more scientifically savvy. Rather, the two parties diverged on climate change because this issue got politicized, forcing some people to choose between being informed by experts and being good members of their tribe.

Note, however, that on some other issues it's the liberals whose views are at odds with expert consensus. Kahan and colleagues have found, for example, that liberals (egalitarian communitarians) are more likely to disagree with the experts on whether deep geologic isolation is a safe means of disposing of nuclear waste. No tribe has a monopoly on cultural bias.

BIASED ESCALATION

Our tribal allegiances can make us disagree about the facts. Other biases may be built into the way we perceive the world. Sukhwinder Shergill and colleagues at University College London conducted a simple experiment designed to test a hypothesis about the role of biased perception in the escalation of conflict. People came into the lab in pairs. The first person's finger was hooked up to a little squeezing machine, which applied a modest force of 0.25 newtons. The first person was then instructed to press down on the second person's finger using exactly the same amount of force that the machine had applied to his finger. Critically, the second person was not aware of this instruction. The force of the push was measured by a force transducer placed in between the fingers of the pusher and the pushee. The two people then reversed roles, with the second person attempting to push down on the first person's finger with no more and no less force than his own finger had received. The two people

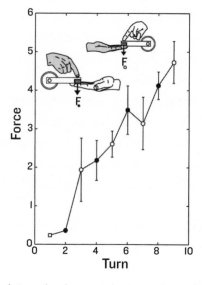

Figure 3.4. With each turn, the players apply more and more force to each other's fingers, even though they attempt to keep the level of force constant.

traded finger pushes back and forth, each attempting to match the force applied by the other in the previous round. In every pair of pushers tested the use of force quickly escalated until the two people were pushing down on each other about twenty times as hard as in the original push (see figure 3.4).

Why did this happen? Oddly enough, the escalation of force in this experiment seems to be related to your inability to tickle yourself. When you perform an action, your brain automatically anticipates the sensory consequences of the action and uses that information to damp down the sensory effects of the action. As a result, self-produced sensations are less salient than sensations produced by another person. (I know what you're thinking, and, yes, that's correct.) As a result, the anticipated force that you feel on your own finger when you're doing the pushing seems less intense than the unanticipated force that you receive when someone else is pushing on you. In other words, when we deliver a blow, we know it's coming, and it doesn't feel as forceful when we experience its impact. But when we receive a blow from someone else, our brains don't get the same kind of intracranial advance warning, and as a result it feels more intense.

Whether this experiment provides an explanation for the escalation of violence in the real world, or just a metaphor, is an open question, but the underlying mechanisms may indeed be parallel. It is plausible, if not inevitable, that we are more aware of the pain we suffer at the hands of others than of the pain that others suffer by our hands. Our society's nervous systems—the media, word of mouth—are far more likely to broadcast messages about our own group's painful experiences than about the painful experiences of others. As a result, our moral biases may, in some cases, be built into the systems that we use to perceive events in the world.

This cognitive principle, in addition to explaining why we underestimate the impact of the harm we cause, also explains why we overestimate the impact of the good we do. Eugene Caruso and colleagues asked the authors of four-author journal articles to separately estimate the amount of credit due to each of the four authors. On average, the sum of the credit that authors claimed for themselves added up to 140 percent. We're fully aware of the contributions we make, because we make them, but we're only partially aware of the contributions of others.

LIFE AND STRIFE ON THE NEW PASTURES

Like the fictional tribes of the new pastures, we in the modern world are attempting to live together despite our different values, beliefs, and interests. By historical standards, modern life is, on the whole, extremely good. As Steven Pinker explains in *The Better Angels of Our Natures*, human violence has declined dramatically over recent millennia, centuries, and decades—a trend Pinker attributes to profound, culturally driven changes in how we think, feel, and organize our societies. These changes include shifts toward democratic governance, states with legal monopolies on the use of force, entertainment that fosters empathy, legal rights for the vulnerable, science as a source of verifiable knowledge, and mutually advantageous commerce. These trends are reflected in some of the surprising findings mentioned above. Once again, people living in more market-integrated societies, rather than being hopelessly greedy, tend to be more altruistic toward strangers and more adept at cooperating with them.

Thus, from an aerial view of human history, the problems of life on the new pastures are 90 percent solved. Nevertheless, the view from the ground, which foregrounds the unsolved 10 percent, looks very different. Despite our immense and underappreciated progress, the problems we face are still enormous. The avoidable suffering produced by modern tragedies is no less intense than the suffering suffered in ages past, and still too vast for our limited minds to truly comprehend.

In this book's introduction, I highlighted some of our biggest problems:

Poverty: More than a billion people live in extreme poverty, such that mere survival is a struggle. Problems associated with poverty include hunger, malnutrition, lack of access to drinking water, poor sanitation, exposure to toxic pollutants, general lack of access to healthcare, lack of economic opportunity, and political oppression, especially for women.

Violent conflict: Ongoing conflicts in places like Darfur kill thousands every year, and hundreds of thousands of conflict refugees live in horrifying conditions.

Terrorism/weapons of mass destruction: Although violence between states is declining, weapons of mass destruction may enable small groups to inflict damage on a scale that historically could be inflicted only by powerful states. Of course, states in possession of such weapons may do extreme damage as well.

Global warming/environmental degradation: The damage we're doing to our environment threatens to reverse our trend toward peace and prosperity.

These are global problems. Within peaceful states, we have domestic problems that, though small by global and historical standards, profoundly affect millions of people and are, for many of them, matters of life and death. As discussed in this book's introduction, we in the United States have ongoing debates over taxes, healthcare, immigration, affirmative action, abortion, end-of-life issues, stem cell research, capital punishment, gay rights, the teaching of evolution in public schools, gun control, animal rights, environmental regulation, and the regulation of the financial industry. My hope is that we can use a better understanding of moral psychology to make progress on these problems.

In this chapter, we've considered six psychological tendencies that exacerbate intertribal conflict. First, human tribes are tribalistic, favoring Us over Them. Second, tribes have genuine disagreements about how societies should be organized, emphasizing, to different extents, the rights of individuals versus the greater good of the group. Tribal values also differ along other dimensions, such as the role of honor in prescribing responses to threats. Third, tribes have distinctive moral commitments, typically religious ones, whereby moral authority is vested in local individuals, texts, traditions, and deities that other groups don't recognize as authoritative. Fourth, tribes, like the individuals within them, are prone to biased fairness, allowing group-level self-interest to distort their sense of justice. Fifth, tribal beliefs are easily biased. Biased beliefs arise from simple self-interest, but also from more complex social dynamics. Once a belief becomes a cultural identity badge, it can perpetuate itself, even as it undermines the tribe's interests. Finally, the way we process information about social events can cause us to underestimate the harm we cause others, leading to the escalation of conflict.

Some of our biggest moral problems are clear examples of the Tragedy of Commonsense Morality—clashes between tribes that are moral, but *differently* moral. Perhaps the best example is the problem of global warming, which the philosopher Stephen Gardiner has called a "perfect moral storm." First, the problem of global warming is bedeviled by biased fairness. Consider, for example, the moral dimensions of the "cap-and-trade" approach to limiting carbon emissions. Cap-and-trade plans set a limit on global carbon emissions and then grant a limited number of emission credits to nations, which may use them or sell them to others. Critically, cap-and-trade plans require some initial distribution of carbon emission credits, and there is stark disagreement about what constitutes a fair distribution. (Sound familiar?) One approach is to use historical emission levels as a starting point. Another is to grant each nation emission rights in proportion to its population, thus granting each human a standard-size carbon credit. And there are, of course, any number of arrangements in between. As you might expect, developed nations tend to think that credits based on historical emissions are fair, while less developed nations favor distributions based on population. The fairness problem is not specific to cap-and-trade plans, of course. Proposals involving taxes on carbon emissions—the most widely favored alternative to cap-and-trade—require agreements about who gets taxed for carbon emissions and at what rates. There is currently no global agreement on carbon emissions, largely because many in the United States—the world's second-largest carbon emitter, both in absolute and per capita terms—regard the 1997 Kyoto Protocol for reducing global carbon emissions as unfair. George W. Bush expressed a popular American sentiment against the Kyoto plan while campaigning for the presidency in 2000: "I'll tell you one thing I'm not going to do is I'm not going to let the United States carry the burden for cleaning up the world's air, like the Kyoto treaty would have done." Whether it's the United States or the rest of the world, at least one side's sense of fairness is biased.

The problem of defining fairness arises in nearly all international conflicts. Is it fair for Israel to occupy the West Bank, given what Palestinians have done? Is it fair for Palestinians to kill Israeli civilians, given what Israelis have done? How much force is proportional, and how much is excessive? Is it fair that only certain nations are allowed to have nuclear weapons? Is it fair to impose painful economic sanctions on innocent people as a means of containing autocratic rulers? Is it fair to cause thousands of deaths in order to turn an autocracy into a democracy? Experts on negotiation and international relations have long lamented the problem of biased fairness in conflict resolution. Roger Fisher explains in his book *Basic Negotiating Strategy*:

An attempt to point out to an adversary that he ought to make a decision where the "oughtness" is based on *our* ideas of fairness, history, principle, or morality is at best a diversion from the immediate task at hand; at worst it is destructive of the result we want. . . .

Officials think of themselves as acting in morally legitimate ways. In order to get them to change their minds, we have to appeal to *their* sense of right and wrong. But this is the opposite of what most governments do. First they appeal to their own people's sense of right and wrong, attempting to whip up support by demonizing the opposition, and this may work. But then the opposition becomes much harder to deal with, . . . less willing to listen to what we have to say.

Historian Arthur Schlesinger, writing in the early 1970s, foreshadows the results of Harinck's negotiation study, in which the pursuit of biased fairness leaves everyone worse off.

Laying down the moral law to sinning brethren from our seat of judgment no doubt pleases our own sense of moral rectitude. But it fosters dangerous misconceptions about the nature of foreign policy. . . . For the man who converts conflicts of interest and circumstance into conflicts of good and evil necessarily invests himself with moral superiority. Those who see foreign affairs as made up of questions of right and wrong begin by supposing they know better than other people what is right for them. The more passionately they believe they are right, the more likely they are to reject expediency and accommodation and seek the final victory of their principles. Little has been more pernicious in international politics than excessive righteousness.

As noted earlier, the problem of defining fairness arises in nearly every domestic political issue as well, from economic inequality to how we treat the unborn.

———

How can we solve the problems of the new pastures? So far, we've talked about the structure of moral problems and the psychology behind them. Like all animals, we have selfish impulses. But more than any other animal, we also have social impulses, automated moral machinery that pushes us into the magic corner, solving the problem of Me versus Us. Unfortunately, as we've learned in this chapter, this moral machinery (along with good old-fashioned selfishness and bias) re-creates the fundamental moral problem at a higher level, at the level of groups—Us versus Them. Based on what we've seen so far, the problems of the new pastures might be hopeless: Our social impulses take us out of the frying pan of personal conflict and into the fire of tribal conflict. But fortunately, the human brain is more than a bundle of selfish and social impulses. We can *think*. To see moral thinking at work, and to appreciate the contrast between moral thinking and moral feeling, there's no better place to start than with philosophical dilemmas that pit "the heart" against "the head."

PART II

Morality
Fast and Slow

4.

Trolleyology

This is where my main line of research comes into our story. But first, I'll say a bit about how I got into this business and why I think it matters.

In the eighth grade, I joined my school's debate team. I did Lincoln–Douglas debate, in which two debaters argue opposite sides of a "resolution." The resolutions were determined by a national committee and changed every couple of months. I recently discovered the resolutions of my geeky youth online. Here are the ones from my sophomore year of high school:

- Resolved: that the United States ought to value global concerns above its own national concerns.
- Resolved: all United States citizens ought to perform a period of national service.
- Resolved: communities in the United States ought to have the right to suppress pornography.

- Resolved: development of natural resources ought to be valued above protection of the environment.
- Resolved: individual obedience to law plays a greater role in maintaining ethical public service than does individual obedience to conscience.

Although I didn't realize it at the time, these are all commons problems, questions about the terms of cooperation, both for individuals within a society and for nations among nations.

Filing into empty high school classrooms on evenings and weekends, donning ill-fitting Reagan-Thatcher power suits, my fellow debaters and I reenacted the philosophical battles of the new pastures, passionately defending the policies to which we were randomly assigned. I quickly developed a standard debating strategy. At the start of the debate, each debater presents a "value premise," the value that one regards as preeminent. For example, if you were arguing against the suppression of pornography, you might make your value premise "freedom." If you were arguing for obedience to law over obedience to conscience, you might make your value premise "security." Next you'd argue for the preeminence of your value premise. For example, if your value premise is "security," you might recite a few choice words from Thomas Hobbes and argue that security comes first because security is necessary for realizing other values. Then, with your value premise in place, you would argue that your side best serves the preeminent value.

I didn't like the standard value premises ("freedom," "security"), because it seemed to me that, no matter what your most favored value is, there could always be other considerations that take precedence. Sure, freedom is important, but is it everything? Sure, security is important, but is it everything? How can there be one preeminent value? Then I discovered utilitarianism, the philosophy pioneered by Jeremy Bentham and John Stuart Mill, British philosophers of the eighteenth and nineteenth centuries.*

Utilitarianism is a great idea with an awful name. It is, in my opinion, the most underrated and misunderstood idea in all of moral and political philosophy. In parts 3 through 5, we'll talk about why utilitarianism is so wise, misunderstood, and underappreciated. But for now we'll start with a simple understanding of utilitarianism that's good enough to get us into its psychology. By the end of this chapter, your feelings about utilitarianism will likely be mixed—or worse—but that's okay. I'll work on winning you over in parts 3 through 5.

So what's the big idea? Utilitarianism says that we should do whatever will produce the best overall consequences for all concerned. (Strictly speaking, what I'm describing here is *consequentialism*, a broader philosophical category that includes utilitarianism. More on this in chapter 6.) In other words, we should do whatever promotes the greater good. For example, if option A will kill six people and save four people, while option B will kill four people and save six people, and if all other consequences are equal, then we should choose option B over option A. This idea may strike you as so painfully obvious that it hardly deserves to be called an "idea," let alone a "great idea." But as we'll see shortly, it's not at all obvious that this is a good way to think about moral problems *in general*. And, as we'll see in parts 3 through 5, applying this principle in the messy real world is not at all simple, and very different from what people tend to envision when they imagine utilitarianism in practice.

As a debater, I liked utilitarianism because it gave me a value premise that itself allowed for the balancing of values: Does freedom outweigh security? Or does security outweigh freedom? Utilitarianism has a sensible answer: Neither value takes absolute precedence. We need to *balance* the values of freedom and security, and the best balance is the one that produces the best overall consequences.

Pleased with this general strategy, I made utilitarianism my value premise in every debate, no matter what side I was on. I began each round with my standard utilitarian spiel, peppered with authoritative quotations from Mill and friends. From there it was just a matter of cherry-picking the evidence to show that, indeed, the greater good was served by the side to which I'd been assigned.

This strategy worked pretty well. I used utilitarianism not only as my own value premise but also as a weapon for attacking the values of my opponents. Whatever my opponent proffered as the preeminent value, I would pit that value against the greater good, and in the most dramatic possible way. I would begin in cross-examination, the part of the debate in which opponents question each other directly, rather than making speeches. For example, if my opponent was pushing free speech, I'd whip out my stock counterexample:

> ME: You say that free speech is the most important value in this debate, is that right?
>
> MY UNWITTING OPPONENT: Yes.
>
> ME: And therefore, no other value can take precedence over free speech, is that right?
>
> MY UNWITTING OPPONENT: Yes.
>
> ME: So . . . Suppose that someone, just for fun, shouts "Fire!" in a crowded theater, and this causes people to rush to the exits, and some people get trampled and die. Is the right to shout "Fire!" more important than the right not to be trampled to death?

Touché! "Free speech" is an easy mark, but for most value premises, I could trot out or cook up a counterexample like the good old "crowded theater" case.

The eighteenth-century German philosopher Immanuel Kant—regarded by many, especially critics of utilitarianism, as the greatest moral philosopher of all time—was also popular among debaters. Opponents would sometimes appeal to Kant's "categorical imperative" for their value premises, arguing that "the ends don't justify the means." I would then ask questions like this: "Suppose a malfunctioning elevator is about to crush someone. To stop it, you need to press a button that you can't reach. But you can push someone into the button. Is it okay to use someone as a means to pressing the button, if it will save someone's life?"

I liked my utilitarian strategy not only because it was effective but because I believed in it. Of course, as noted above, I'd have to argue that the greater good was on my side, no matter what side I was on. But that was part of the game, and this seemed better to me than proffering a different flawed philosophy for every side of every debate.

One time, at a tournament in Jacksonville, Florida, I went up against a particularly sharp debater from Miami. I gave my standard utilitarian spiel, and then she cross-examined me:

MY OPPONENT: You say that we should do whatever will produce the greatest good, is that right?

UNWITTING ME: Yes.

MY OPPONENT: So . . . Suppose that there are five people who are all about to die, due to failed organs of various kinds. One has liver damage, one has damaged kidneys, et cetera.

UNWITTING ME: Uh-huh.

MY OPPONENT: And suppose that a utilitarian doctor can save them by kidnapping one person, anesthetizing him, removing his various organs, and distributing them to the other five people. That would seem to produce the greatest good. Do you think that would be right?

I was stunned. I don't remember how I responded. Perhaps I made a utilitarian appeal to realism, arguing that, in fact, rogue organ transplantation would not serve the greater good because people would live in fear, because of the potential for abuse, and so on. But whatever I said, it wasn't enough. I lost the round. Worse than that, I lost my winning strategy. And even worse than that, I was threatened with losing my burgeoning moral worldview. (And when you're a teenage guy without a girlfriend,* losing your burgeoning moral worldview can seem like a big loss!)

I quit debate midway through my senior year of high school, right after I got accepted to college. My parents were upset. My debate coach called me a traitor. But I was sick of arguing for sport. If I was going to make philosophical arguments, I wanted to make arguments that I could believe in, and at that point I didn't know what I believed.

I n the fall of 1992, I arrived at the University of Pennsylvania as a first-year student in the undergraduate program of the Wharton School of Business. It took me about a month to realize that business was not for me, but during that year I encountered some professors and ideas that have stayed with me ever since. During my first term, I took microeconomics, which introduced me to game theory. Game theory is the study of strategic decision problems like the Prisoner's Dilemma and the Tragedy of the Commons. I loved the abstract elegance of game theory. I loved the idea that seemingly unrelated social problems—from global warming to nuclear proliferation to the perpetual mess in the common kitchen of my dorm—shared an underlying mathematic structure, and that by understanding the mathematical essence of these problems we could solve them.

That year, I took my first psychology class with a brilliant teacher and scientist named Paul Rozin. It was a small seminar, not the standard auditorium-size intro class. Rozin asked questions, argued with us, and led us through demonstrations. In one of these demonstrations, we calculated the speed at which the human nervous system sends electrical signals. We used a method invented by the German physicist and physician Hermann von Helmholtz, one of the founders of experimental psychology in the nineteenth century.

First, Rozin had us hold hands in a chain. The first person squeezed the next person's hand, who then squeezed the next person's hand, and so on. How long would it take for a squeeze to propagate through the chain? We did this several times. Rozin stood by with a stopwatch, recording the time for each trial. He then averaged the results. Next, we did the same thing, but this time holding the next person's ankle instead of the next person's hand. As soon as you felt the squeeze on your left ankle, you used your right hand to squeeze the next person's left ankle, and so on down the line. Again, Rozin took repeated measurements and averaged them. This time, it took a bit longer, on average, for the squeeze to propagate through the chain. We then measured the distances between our hands and our brains, and then the distances between our ankles and our brains. The difference between these two distances was the extra distance that the squeeze signal had to travel when it went hand-to-ankle instead of hand-to-hand. Averaging over many trials, we estimated the time it took for the signal to travel that extra distance, and thus estimated the speed of the signal. Our estimate matched almost exactly the textbook answer that Rozin had, unbeknownst to us, written down in advance.

Demonstrations like this one turned me on to the power of scientific thinking. More specifically, they showed me how clever methods can turn the mysteries of the human mind into answerable questions. Part of what amazed me about this method was that it could have been used thousands of years ago, though no one thought of it until the nineteenth century. (With enough people and enough repetition, you don't need a stopwatch.) Here the power lay not in advanced technology but in a style of thinking that combined sharp reasoning with creativity.

Rozin also introduced me to sociobiology, the study of social behavior, especially human social behavior, from an evolutionary perspective. (This field has since split into several fields, including evolutionary psychology, which aims to understand how evolution has shaped the human mind.) Sociobiologists had an explanation for why—pardon my crude collegiate idiom—girls, but not guys, could get laid anytime they wanted. The explanation comes from Robert Trivers's theory of parental investment: Females have to put an enormous amount of work into producing viable offspring—nine months of pregnancy and years of breastfeeding, at a bare minimum. Males, by contrast, need only deposit a cheap wad of sperm. (Of course, males who invest more in their offspring are more likely to have successful offspring, but the *minimum* investment for males is low.) Thus, says Trivers, females will be more selective in their choices of mates. A female who takes sperm from the first willing donor is unlikely to have offspring as fit as those who are choosier. Males, in contrast, pay no cost for offering their genetic material more freely. (This reminded me of a song that was popular at the time, by the Red Hot Chili Peppers, entitled "Give It Away.") Many people dismissed—and continue to dismiss—this evolutionary theory as a pseudo-scientific rationalization for traditional gender roles, but I was impressed. I was impressed not because I was a fan of traditional gender roles—on the contrary, I'd have been very pleased if the females in my environment had become less selective—but because Trivers's theory, in addition to accounting for the observable social facts, made some nonobvious predictions that turned out to be correct. According to Trivers, what matters is not male versus female per se, but low versus high parental investment. If you could find a species in which the males were the high investors, then it should be the males, rather than the females, who would be more selective. And, sure enough, in some birds and fish, it's the males that guard and nourish developing offspring, and it's the males who are more choosy.

In college, I was in charge of my own finances for the first time. With this new freedom came a new sense of responsibility. I spent much of my monthly allotment on small luxuries—CDs from the music store, sampling interesting food in Center City Philadelphia. But how could I justify this? I imagined a desperately poor woman pleading with me: "Please may I have ten dollars? No money, no food. My child will die." Could I look her in the eye and say no? "Sorry, but I need another John Coltrane CD. Your child will just have to die." I knew that I could never do that. (Although I walked by homeless people all the time.) At the same time, however, I saw where this argument led. The world has no shortage of desperate people who need my money more than I need a bigger music collection. Where does my obligation end? (More on this problem in part 3.)

I found a psychology professor named Jonathan Baron who, according to his bio, was interested in psychology, economics, and ethics. He seemed like the guy to talk to. I made an appointment. He explained that philosophers had been arguing about this problem ever since Peter Singer first posed it, a few years before I was born. It was a problem that bothered him, too.

Baron and I were a great match, and we started doing research together. It was only later that I realized how unlikely it was that I, at that time, with my unusual preoccupations, had stumbled upon the world's only card-carrying utilitarian moral psychologist. For all I knew then, there was one at every university. Baron and I worked on the problem of "insensitivity to quantity" in environmental decision making. If you ask people, "How much would you pay to clean up two polluted rivers?" you'll get an answer. If you ask a different group of people, "How much would you pay to clean up twenty polluted rivers?" you'll get, on average, about the same answer. You might think—you might hope—that cleaning up twenty polluted rivers would seem to people about ten times better than cleaning up two. But often enough, the numbers just don't matter. Two rivers, twenty rivers: It all sounds the same.* Baron and I were trying to figure out why people were "insensitive to quantity." Our research didn't solve the problem, but we ruled out some promising theories, which was a kind of progress. This research project gave me my first scientific publication, but more important than that, working with Baron introduced me to the study of "heuristics and biases," the mental shortcuts (heuristics) that people use to make decisions and the irrational mistakes (biases) that result from heuristic thinking.

After losing interest in business, I transferred to Harvard and became a philosophy major. My first term there, I took a course called "Thinking About Thinking," team-taught by three legendary professors: the philosopher Robert Nozick, the evolutionary biologist Stephen Jay Gould, and the law professor Alan Dershowitz. (The course was nicknamed "Egos on Egos.") On the syllabus was a paper by the philosopher Judith Jarvis Thomson, entitled "The Trolley Problem."

THE TROLLEY PROBLEM

This paper, it turned out, was the source of the organ transplant dilemma that had blindsided me in high school. Thomson's ingenious paper worked through a series of moral dilemmas, all variations on the same theme of sacrificing one life to save five. In some cases, trading one life for five seemed clearly wrong, as in the *transplant* dilemma. Another such dilemma from Thomson's paper is the *footbridge* dilemma, which I present here in slightly modified form:

> A runaway trolley is headed for five railway workmen who will be killed if it proceeds on its present course. You are standing on a

footbridge spanning the tracks, in between the oncoming trolley and the five people. Next to you is a railway workman wearing a large backpack. The only way to save the five people is to push this man off the footbridge and onto the tracks below. The man will die as a result, but his body and backpack will stop the trolley from reaching the others. (You can't jump yourself because you, without a back-pack, are not big enough to stop the trolley, and there's no time to put one on.) Is it morally acceptable to save the five people by push-ing this stranger to his death? (See figure 4.2.)

Most people say that it's wrong to push the man off the footbridge in order to save the other five. This, however, is not the utilitarian answer, at least if we accept the premises of the dilemma. Pushing promotes the greater good, and yet it still seems wrong.

There are many ways to try to wiggle out of this problem. The most tempting wiggling strategy is to question the footbridge dilemma's as-sumptions: Will pushing the man really save the five? What if the five can be saved some other way? What if this act of pushing will be seen by others who will subsequently lose their respect for human life and end up killing other people? What if, as a result of this killing, millions of people will live in fear of enterprising utilitarians? These are perfectly reasonable

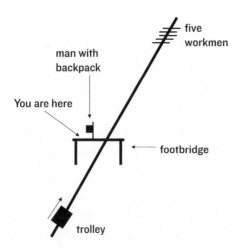

Figure 4.1. The *footbridge* dilemma.

questions, but raising them doesn't solve the utilitarian's problem. It may be true that, under more realistic assumptions, there would be good utilitarian reasons not to push. This is an important point, and one that I'll emphasize later. But for now we'll put these doubts aside and take seriously the idea that pushing the man off the footbridge is wrong, even if it promotes the greater good.

Why might this be wrong? The most common complaint about utilitarianism is that it undervalues people's *rights*, that it allows us to do things to people that are fundamentally wrong, independent of the consequences. Above I mentioned Kant's categorical imperative, which he famously summarized thus:

> Act so that you treat humanity, whether in your own person or that of another, always as an end and never as a means only.

A rough translation: Don't *use* people. It's hard to think of a more dramatic example of using someone than using someone as a human trolley-stopper.

One nice thing about the *footbridge* dilemma is that it has some interesting siblings. In one alternative version, which we'll call the *switch* dilemma, a runaway trolley is heading down the tracks toward five workmen

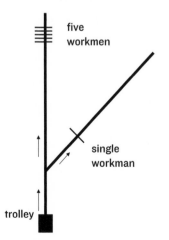

Figure 4.2. The *switch* dilemma.

who will be killed if nothing is done. You can save these five people by hitting a switch that will turn the trolley onto a sidetrack. Unfortunately, there is a single workman on the side-track who will be killed if you hit the switch (see figure 4.2).

Is it morally acceptable to turn the trolley away from the five and onto the one? This seemed morally acceptable to Thomson, and I agreed. As I would learn later, people all over the world agree. Why, then, do we say yes to the *switch* case but no to the *footbridge* case?

This was, for me, the perfect scientific problem. The Trolley Problem brought together, in one beautiful, fruit-fly-like model, all of the things that had been puzzling me since my early teens. First, the Trolley Problem took the big, philosophical problem behind all of those high school debates and boiled it down to its essence: When, and why, do the rights of the individual take precedence over the greater good? Every major moral issue—abortion, affirmative action, higher versus lower taxes, killing civilians in war, sending people to fight in war, rationing resources in healthcare, gun control, the death penalty—was in some way about the (real or alleged) rights of some individuals versus the (real or alleged) greater good. The Trolley Problem hit it right on the nose. In the *footbridge* dilemma, sacrificing one person for the greater good seems wrong, a gross violation of individual rights. In the *switch* dilemma, trading one life for five seems justified, if not ideal. It was Kant versus Mill, all in one neat little puzzle. If I could understand these two simple dilemmas, I could understand a lot.

The Trolley Problem also has a beautiful, Helmholtzian cleanliness to it. How do you figure out the speed of a signal in the nervous system? You don't have to trace the signal's path up the arm, through the inscrutable maze of the brain, and then down the other arm. You just have to switch out an arm for a leg and *subtract*. The Trolley Problem made for a lovely subtraction. These dilemmas have hundreds of potentially relevant attributes, but the differences between the *switch* and *footbridge* cases are fairly minimal. Somewhere on that short list of differences are the differences that matter morally, or at least seem to.

The Trolley Problem is also a decision problem, one that might be illuminated by thinking in terms of "heuristics and biases." Our intuitions tell us that the action in the *footbridge* case is wrong. What springs and levers in the brain make us come to that conclusion? And can we trust that machinery? Our brains, at least sometimes, tell us that cleaning up twenty polluted rivers is no better than cleaning up two polluted rivers. Here we give the same response to two things that ought to get different responses. Perhaps the Trolley Problem involves the opposite bias: treating similar things as if they were very different. Perhaps it's just a quirk of human psychology that trading one life for five seems right in one case and wrong in the other. It was a tantalizing thought, and one that could perhaps vindicate utilitarianism, which otherwise seemed so reasonable.

The summer after "Thinking About Thinking," I got a small grant to do independent research on the Trolley Problem. I read a lot of philosophy books and psychology papers and, to fulfill my obligation as a grant recipient, I wrote a paper entitled "The Two Moralities," in which I distinguished between two different kinds of moral thinking, which I called "abstract" and "sympathetic." This was the beginning of the "dual-process" theory of moral judgment, which I'll describe shortly.

The following spring, I took a behavioral neuroscience class, hoping that brain scientists would have the answers I was looking for. I didn't find my answers in class, but I did come across a recently published book by a neurologist, Antonio Damasio. This book, *Descartes' Error*, was about the role of emotion in decision making. Damasio described the case of Phineas Gage, a famous neurological patient who lived and worked in Vermont during the nineteenth century. Gage was a well-respected railroad foreman whose character changed dramatically after an accidental explosion sent a three-foot tamping iron through his eye socket and out the top of his skull. His injury destroyed much of his medial prefrontal cortex, the part of the brain behind the eyes and forehead, just above the nose. Amazingly, after a few weeks, it seemed that Gage had recovered his cognitive abilities. He could talk, do math, remember the names of people and places, and so on. But Gage was no longer his old self. The diligent, hardworking railroad foreman became an irresponsible wanderer.

Damasio had studied living patients with damage to the ventral (lower) portions of the medial prefrontal cortex (the ventromedial prefrontal cortex, or VMPFC). He saw a consistent pattern. These Gage-like patients perform well on standard cognitive tests, such as IQ tests, but they make disastrous real-life decisions. In a series of studies, Damasio and his colleagues showed that their problems were due to emotional deficits. One such patient, upon viewing pictures of gory car accidents and drowning flood victims, remarked that he felt nothing upon viewing these tragic scenes, but that he knew he used to respond emotionally to such things, before the brain damage. Damasio describes their predicament as "to know, but not to feel."

When I read that passage, I was alone in a hotel room. I got so excited, I stood up on the hotel bed and started jumping. What clicked at that moment was the connection with the Trolley Problem: What's missing in these patients is what drives ordinary people's responses to the *footbridge* case. And there was, of course, a perfect way to test this idea. Give the *switch* and *footbridge* dilemmas to patients with VMPFC damage. If I was right, the patients with VMPFC damage, the patients like Phineas Gage, would give utilitarian answers not just to the *switch* case, but to the *footbridge* case, too. Unfortunately, I didn't know anyone with VMPFC damage.

The following year, I wrote about these ideas in my undergraduate thesis, entitled "Moral Psychology and Moral Progress," an early predecessor of this book. In the fall of 1997, I enrolled in Princeton's philosophy Ph.D. program. I spent my first two years taking seminars and fulfilling requirements, studying a variety of topics, from Plato's *Republic* to the philosophy of quantum mechanics, and generally enjoying the life of the philosopher. In the summer of 1999 I heard about a neuroscientist in the Department of Psychology who was interested in talking to philosophers. Princeton had recruited Jonathan Cohen to direct the new Center for the Study of Brain, Mind & Behavior. I checked his website and saw that his research used brain imaging. Maybe I don't need neurological patients, I thought. Maybe we can see the *footbridge* effect in the brains of healthy people. I made an appointment.

There were still unpacked boxes all over Cohen's lab. In his office, books and papers were piled up in teetering stacks like academic stalagmites. He leaned back in his office chair. "So, what you got?" he said. I started to explain the Trolley Problem—first the *switch* case and then the *footbridge* case. He interrupted me, listing ten different ways in which the two cases differ. "Hold that thought," I said. I then started to describe Damasio's book and Phineas Gage, but before I could finish he blurted out, "I got it! I got it! I got it! Ventral/dorsal! Ventral/dorsal!" I recognized the word *ventral,* but didn't know what he meant by *dorsal.* Like a shark's fin? (*Dorsal* refers to the top half of the brain, the side aligned with the back in a quadruped.) In any case, I was glad that he was excited, too. "Let's do this," he said. "But you're going to have to learn how to do brain imaging." That sounded good to me.

What Cohen got in that first meeting—and what I didn't yet understand—was the other half of the neuroscientific story of Trolley, the part that connected most directly to Cohen's work. I was thinking about the role of emotion in making us say no in the *footbridge* case. But what makes us say yes in the *switch* case? And what is it that remains intact in Phineas Gage and others like him? If these patients can "know" but not "feel," what is it that enables them to "know"? To me, this was just obvious. It's utilitarian, cost-benefit thinking: Saving five lives is better than saving one life. But to a cognitive neuroscientist, nothing about how the mind works is obvious.

Cohen directs the Neuroscience of Cognitive Control Lab. Cognitive control, as Cohen defines it, is "the ability to orchestrate thought and action in accordance with internal goals." A classic test of cognitive control is the color-naming Stroop task, which involves naming the colors in which words appear on a screen. For example, you might see the word "bird" written in blue, and your job would be to say "blue." Things get tricky, however, when the word names a color that is not the color in which that word is printed—for example, the word "red" written in green. Here your job is to say "green," but your first impulse is to say "red," because reading is more automatic than color naming. In these tricky cases, there is an internal conflict, with one population of neurons saying "Read the word!" and a different population of neurons saying "Name the color!" (Of course, these neurons don't speak English. Here and elsewhere I anthropomorphize for illustrative purposes.) What resolves the conflict between these competing commands? And what makes sure it's resolved in the right way ("Name the color") rather than the wrong way ("Read the word")?

This is the job of cognitive control, a hallmark of human cognition, which is enabled by neural circuits in the *dorsal* lateral (dorsolateral) prefrontal cortex, or DLPFC. In the color-naming Stroop task, the DLPFC says, "Hey, team. We're doing color naming now. So, color namers, please step up, and word readers, please step down." The DLPFC can use explicit decision rules ("Name the color") to guide behavior, and it can override competing impulses ("Read the word"). And that's why Cohen exclaimed, "Ventral/dorsal!" As an expert on the neuroscience of cognitive control, Cohen saw immediately that "Save more lives" is like "Name the color." It's an explicit decision rule that can be used to guide one's response to the problem. What's more, Cohen saw that the utilitarian response to the *footbridge* case—approving of pushing the man in order to save more lives—is like naming the color on one of those tricky Stroop trials, such as when the word "red" is written in green. To give the utilitarian answer, one must override a competing impulse.

Putting these ideas together, we have a "dual-process" theory of moral judgment, illustrated by the *footbridge* case and its contrast with the *switch* case. It's a dual-process theory because it posits distinct, and sometimes competing, automatic and controlled responses. (More about dual-process theories in the next chapter.) In response to the switch case, we consciously apply a utilitarian decision rule using our DLPFCs. For reasons we'll discuss later, the harmful action in the *switch* case does not elicit much of an emotional response. As a result, we tend to give utilitarian responses,

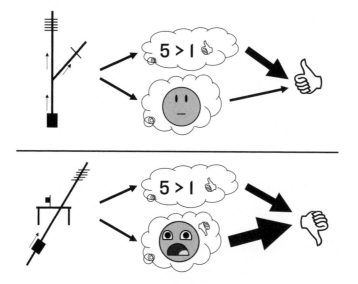

Figure 4.3. Dual-process morality. Turning the trolley away from five and onto one (above) makes utilitarian sense and doesn't trigger much of an opposing emotional response, causing most people to approve. Pushing the man off the footbridge (below) likewise makes utilitarian sense, but it also triggers a significant negative emotional response, causing most people to disapprove.

favoring hitting the switch to maximize the number of lives saved (see the top of figure 4.3). In response to the *footbridge* case, we also apply the utilitarian decision rule using our DLPFC. But here, for whatever reason, the harmful action *does* trigger a (relatively) strong emotional response, enabled by the ventromedial prefrontal cortex (VMPFC). As a result, most people judge that the action is wrong, while understanding that this judgment flies in the face of the utilitarian cost-benefit analysis (see the bottom of figure 4.3). This is the theory we set out to test.

INTO THE SCANNER

For our first experiment, we came up with a set of moral dilemmas like the *switch* case and another set like the *footbridge* case. We called these two sets "impersonal" and "personal." We had people read and respond to these two sets of dilemmas while we scanned their brains, using functional magnetic resonance imaging, or fMRI.** Then we took the brain images from these two different moral judgment tasks and, in Helmholtzian fashion, *subtracted*. As predicted, the "personal" dilemmas, the ones like the *footbridge* case, produced increased activity in the medial prefrontal cortex, including parts of the VMPFC.* In other words, the *footbridge*-like cases elicited increased activity in exactly the part of the brain that had been damaged in Phineas Gage and in Damasio's patients, the ones who "knew" but did not "feel." By contrast, the "impersonal" dilemmas, the ones like the *switch* case, elicited increased activity in the DLPFC, exactly the region that Cohen had seen so many times before in his brain imaging experiments using the color-naming Stroop task. Our second experiment showed that when people give utilitarian responses to dilemmas like the *footbridge* case (endorsing the pushing of one to save five, for example), they exhibit increased activity in a nearby region of the DLPFC. This experiment also showed that a different brain region known for its role in emotion, the amygdala, becomes more active when people contemplate "personal," *footbridge*-like cases, as compared with "impersonal," *switch*-like cases. As you may recall from chapter 2, amygdala activity is associated with heightened vigilance, responding to, among other things, faces of out-group members.

We were delighted with these results, which fit perfectly with our theory, but there was still room for doubt. For one thing, our experiment was not as well controlled as we would have liked. In an ideal world, we would have used just two dilemmas in our experiments, the *switch* dilemma and the *footbridge* dilemma. This would have been a cleaner experiment, because these two cases are so similar. But brain imaging data are noisy, which makes it very hard to compare two individual brain events. Instead you have to generate repeated instances of both kinds of brain events, which you can then average and compare. (This is just like the averaging that Rozin did when he had us squeeze hand-to-hand many times, and then had us squeeze hand-to-ankle many times, to get an estimate of the difference in speed.) This meant that we needed many *footbridge*-like cases and many *switch*-like cases to compare as sets. We couldn't make the cases too similar, or people would stop thinking and just

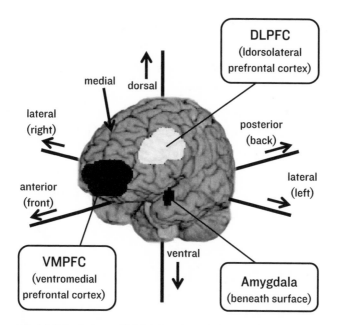

Figure 4.4. A 3-D brain image highlighting three of the brain regions implicated in moral judgment.

give the same answer every time. Even worse, we didn't know what the essential differences between the *footbridge* and *switch* cases were, which made it hard to come up with a set of cases that were *like* the *footbridge* case, and another set of cases *like* the *switch* case. In what way should the cases within a set be alike? (We addressed this question in later work, discussed in chapter 9.) So we took a guess about what the essential differences were, knowing that it would be wrong but hoping that it would be close enough to allow us to test our hypothesis about "ventral" and "dorsal." It worked.

Our initial experiments had another important limitation. Brain imaging data are "correlational," meaning that one can't say for sure whether the brain activity revealed in the images is *causing* people's judgments, or whether it's merely *correlated* with people's judgments. For example, ice cream sales and drownings are correlated, but ice cream doesn't cause drowning. Rather, on hot summer days, people tend to eat more ice cream, and they also tend to go swimming, which leads to more drownings. As scientists often say, "Correlation does not imply causation." (But correlation is still *evidence* for causation, something scientists often forget in their efforts to be tough-minded.) Thus, there was a correlation between contemplating "personal" dilemmas and activity in the VMPFC and the amygdala. But does activity in these regions *cause* people to say no to dilemmas like the *footbridge* case? Likewise, there was a correlation between contemplating "impersonal" dilemmas and DLPFC activity. There was also a correlation between DLPFC activity and giving utilitarian responses to "personal" dilemmas. But was the DLPFC *causing* people to say yes to dilemmas like the *switch* case and, less often, to dilemmas like the *footbridge* case?

EXPERIMENTAL TROLLEYOLOGY
TAKES OFF

One of the joys of science is putting an idea out into the world and then watching other scientists run with it. In the years following our first two brain imaging experiments, many researchers from different fields, using different methods, built on our results, providing further evidence for our theory and taking it in new directions. And we, of course, did some follow-up studies of our own.*

The next key trolley study came from Mario Mendez and colleagues at UCLA, who examined the moral judgments of patients with frontotemporal dementia (FTD). FTD is a degenerative neurological disorder that affects the VMPFC, among other brain regions. Consequently, FTD patients often have behavioral problems similar to those of VMPFC patients like Phineas Gage. In particular, FTD patients are known for their "emotional blunting" and lack of empathy. Mendez and his colleagues gave versions of the *switch* and *footbridge* dilemmas to FTD patients, Alzheimer's patients, and healthy people. The results were striking, and exactly what we had predicted. In response to the *switch* dilemma, all three groups showed the same pattern, with at least 80 percent of respondents approving of turning the trolley to save five lives. About 20 percent of the Alzheimer's patients approved of pushing the man off the *footbridge*, and the healthy control group did about the same. But nearly 60 percent of the FTD patients approved of pushing the man off the *footbridge*, a threefold difference.

This study solved the two problems described above. First, because these researchers were testing patients rather than using brain imaging, they didn't need to average over many dilemmas. This avoided the problem of defining a class of "personal" (*footbridge*-like) and "impersonal" (*switch*-like) dilemmas. They just compared the two original dilemmas, which are nicely matched. Second, this study dealt with the "correlation is not causation" problem. That is, they showed more definitively that emotional responses *cause* people to say no to the *footbridge* dilemma by showing that people with emotional deficits are three times as likely to say yes.

A couple of years later, Damasio himself ran the exact experiment that I had envisioned while jumping on my hotel bed. He and his collaborators, led by Michael Koenigs and Liane Young, gave our full set of dilemmas to their famous Phineas Gage–like patients with VMPFC damage. Sure enough, these patients were about five times as likely as others to give utilitarian answers in response to "personal" moral dilemmas, approving of pushing the man off the footbridge and so forth. The same year, a group of Italian researchers led by Elisa Ciaramelli and Giuseppe di Pellegrino showed similar results. The Italian group also linked the reluctance to give utilitarian answers in healthy people to heightened physiological arousal (indexed by sweaty palms).

A slew of more recent studies point to the same conclusion. Emotional responses make people say no to pushing the man off the footbridge, and likewise for other "personally" harmful utilitarian actions that promote the greater good. Patients with VMPFC damage are more likely than others to approve of turning the trolley onto family members in order to save a larger number of strangers. Low-anxiety psychopaths (known for their social-emotional deficits) make more utilitarian judgments, as do people with alexithymia, a disorder that reduces one's awareness of one's own emotional states. People who exhibit greater physiological arousal (in this case, constriction of peripheral blood vessels) in response to stress make fewer utilitarian judgments, as do people who report relying heavily on their gut feelings. Inducing people to feel mirth (the positive emotion associated with humor, here thought to counteract negative emotional responses) increases utilitarian judgment.

Several studies point to the importance of another emotion-related brain region mentioned above, the amygdala. People who have more psychopathic tendencies exhibit reduced amygdala responses to "personal" dilemmas. Likewise, a study from my lab, led by Amitai Shenhav, shows that amygdala activity correlates positively with ratings of negative emotion in response to *footbridge*-like cases and correlates negatively with utilitarian judgments. This study also indicates that the amygdala functions more like an initial alarm bell, while the VMPFC is responsible for integrating that emotional signal into an "all things considered" decision. In an exciting recent study, Molly Crockett and colleagues gave people citalopram, an SSRI like Prozac, and had them respond to our standard set of dilemmas. A short-term effect of citalopram is the enhancement of emotional reactivity in the amygdala and the VMPFC, among other regions. As predicted, they found that people under the influence of citalopram (as compared with a placebo) made fewer utilitarian judgments in response to "personal" dilemmas like the *footbridge* case. Another study showed that the anti-anxiety drug lorazepam has the opposite effect. A recent study from my lab, led by Elinor Amit, highlights the role of visual imagery in triggering these emotional responses: People who are more visual thinkers, as measured by performance on a visual memory test, make fewer utilitarian judgments. Likewise, interfering with people's visual processing while they make moral judgments makes their judgments more utilitarian.

In short, there's now a lot of evidence—and a lot of different kinds of evidence—telling us that people say no to pushing the man off the footbridge (and other "personally" harmful utilitarian actions) because of emotional responses enabled by the VMPFC and the amygdala. But what about the other side of the dual-process story ("dorsal!")? We had two related hypotheses about utilitarian judgments. First, utilitarian judgments follow from the explicit application of a utilitarian decision rule ("Do whatever will produce the most good").* Second, utilitarian judgments made in the face of competing emotional responses require the application of (additional) cognitive control. Once again, saying yes to the *footbridge* case is rather like naming the color in one of the tricky Stroop cases (e.g., "red" written in green). One has to apply a decision rule in the face of a competing impulse.

We have seen one piece of evidence for this already: When people make utilitarian judgments, they exhibit increased activity in the DLPFC, the brain region most closely associated with applying "top down" rules and the one most closely associated with success in the Stroop task. Other brain imaging studies have since shown similar results, but, as noted above, brain imaging results are "correlational." It would be useful if we could interfere with controlled cognition, much as a brain lesion in the VMPFC interferes with emotional processing.

One way to interfere with controlled cognition is to have people perform an attentionally demanding task along with whatever else they are trying to do. My collaborators and I did this and found, as predicted, that giving people a simultaneous secondary task (i.e., putting people under "cognitive load") slowed down people's utilitarian judgments but had no effect on nonutilitarian judgments. This is consistent with our idea that utilitarian judgments depend more on cognitive control. Another way to modulate cognitive control up or down is to put people under time pressure, or to remove time pressure and encourage deliberation. Renata Suter and Ralph Hertwig did this and found, as predicted, that removing time pressure and encouraging deliberation increases utilitarian judgment. Yet another approach is to put people in a mind-set that favors deliberation over quick, intuitive judgment. One way to put people in a more deliberative mind-set is to give people the experience of being led astray by intuition. Joe Paxton, Leo Ungar, and I did this by having people solve tricky math problems, ones in which the intuitive answer is wrong. As predicted, people who solved these tricky math problems before making moral judgments subsequently made more utilitarian judgments.* Consistent with this, Dan Bartels found that people who generally favor effortful thinking over intuitive thinking are more likely to make utilitarian judgments, and

Adam Moore and colleagues found that utilitarian judgment is associated with better cognitive control abilities.

Finally, we can learn a lot by examining the kinds of moral reasons of which people are conscious when they make their judgments. As I'll explain later, there are many factors that affect people's moral judgments unconsciously. However, in all of my years as a trolleyologist, I've never encountered anyone who was not aware of the utilitarian rationale for pushing the man off the footbridge. No one's ever said, "Try to save more lives? Why, that never occurred to me!" When people approve of pushing, it's always because the benefits outweigh the costs. And when people disapprove of pushing, it's always with an acute awareness that they are making this judgment despite the competing utilitarian rationale. People's reasons for not pushing the man off the footbridge are very different. When people say that it's wrong to push, they're often puzzled by their own judgment ("I know that it's irrational, but . . . "), and they typically have a hard time justifying that judgment in a consistent way. When asked to explain why it's wrong to push, people often say things like "It's murder." But, of course, running someone over with a trolley can be perfectly murderous, and people routinely approve of that action in response to the *switch* case. In short, the utilitarian rationale is always conscious, but people are often in the dark about their own anti-utilitarian motivations. This tells us something important about the way our emotions operate. (More on this in chapter 9.)

THE PATIENT ON THE TROLLEY TRACKS

Philosophers started arguing about trolley dilemmas because they encapsulate a deep philosophical problem: the tension between the rights of the individual and the greater good. In the past decade, we've learned a lot about how our minds/brains respond to these dilemmas, and, as noted above, we're even beginning to understand them on a molecular level. But do the lessons learned from these moral fruit flies really apply to moral thinking in the real world? It's a reasonable question that's difficult to answer. In an ideal scientific world, we would set up controlled experiments in which people make real life-and-death decisions from inside a brain scanner, while under cognitive load, after sustaining a lesion to the VMPFC, and so on. But alas, that's not possible. The next best thing, then, may be to examine the hypothetical judgments of people who make real life-and-death decisions for a living.

With this in mind, Katherine Ransohoff, Daniel Wikler, and I conducted a study examining the moral judgments of medical doctors and public health professionals. We presented both groups with moral dilemmas of the familiar trolleyological sort, as well as more realistic healthcare dilemmas. For example, some of our healthcare dilemmas involve the rationing of drugs or equipment—denying medical resources to some people because they can go further elsewhere. One dilemma involves the quarantining of an infectious patient to protect other patients. Another involves the trade-off between cheap preventive medicine for many people and expensive treatment for a few people who are already sick. These are all problems that healthcare professionals actually face.

First, we found, within both professional groups, a robust correlation between what people say about conventional trolley-type dilemmas and what they say about our more realistic healthcare dilemmas. In other words, someone who approves of pushing the man off the footbridge is more likely to approve of rationing drugs, quarantining infectious patients, et cetera. This suggests that the dual-process psychology at work in trolley dilemmas is also at work in the real world of healthcare decision making.

Next we tested a critical prediction about how the moral judgments of doctors and public health professionals are likely to differ. Doctors aim to promote the health of specific individuals and are duty-bound to minimize the risk of actively harming their patients.* Thus, one might expect doctors to be especially concerned with the rights of the individual. For public health professionals, by contrast, the patient is the society as a whole, and the primary mission is to promote the greater good. (Consistent with this philosophy, the motto of the Johns Hopkins Bloomberg School of Public Health is "Protecting Health, Saving Lives—*Millions at* *a Time.*") Thus, one might expect public health professionals to be espe-cially concerned with the greater good. That is, indeed, what we found. Public health professionals, as compared with doctors, gave more utilitar-ian responses to both the trolley-type dilemmas and to our more realistic healthcare dilemmas. The public health professionals were also more utili-tarian than ordinary people, whose judgments resembled those of the medical doctors. In other words, most people, like doctors, are automati-cally tuned in to the rights of the individual. Giving priority to the greater good seems to require something more unusual.

These findings are important because they indicate that dual-process moral psychology operates in the real world and not just in the lab. The dilemmas in this experiment were hypothetical, but the professional mind-sets of the people we tested are very much real. If people in public health give more utilitarian answers to hypothetical moral dilemmas, it can only be for one of two reasons: Either people with more utilitarian orientations tend to go into public health, or people in public health become more utili-tarian as a result of their professional training. (Or both.) Either way, these are *real-world* phenomena. If the kind of people who go into public health show more concern for the greater good in hypothetical dilemmas, this is almost certainly related to the work they've chosen to do in the real world. Likewise, if training in public health makes people more utilitarian in the lab, it's presumably because such training makes people more utilitarian in the field. After all, the purpose of such training is not to change the way trainees respond to hypothetical dilemmas, but rather to change the way they do their jobs.

We gave our doctors and public health professionals the opportunity to comment on their decisions, and their comments were very revealing. For example, one public health professional wrote, "In these extreme situ-ations . . . I felt that a utilitarian . . . philosophy was most appropriate. Ultimately that is the most moral thing to do. . . . It seems the least murky and the most fair." In contrast, a medical doctor wrote, "To make a life-or-death decision on behalf of someone who is capable of making that decision for themselves (and who has not forfeited that right, for example by knowingly committing a capital crime) is a gross violation of moral and ethical principles." The voices of Mill and Kant, speaking from beyond the grave.

OF TWO MORAL MINDS

We see evidence for dual-process moral psychology in the lab and in the field, in healthy people and in people with severe emotional deficits, in studies using simple questionnaires and in studies using brain imaging, psychophysiology, and psychoactive drugs. It's now clear that we have dual-process moral brains. But why are our brains like this? Why should we have separate automatic and controlled responses to moral questions? This seems especially problematic given that these systems sometimes give conflicting answers. Wouldn't it make more sense to have a unified moral sense?

Back in chapter 2, we saw how a variety of moral emotions and other automatic tendencies—from empathy to hotheadedness to the irrepressible desire to gossip—work together to enable cooperation within groups. If this view of morality is correct, our negative reaction to pushing innocent people off footbridges is just one of our many cooperation-enhancing impulses. (Recall, from chapter 2, Cushman's study in which simulating acts of violence in the lab made people's veins constrict.) It makes sense that we should have these automatic tendencies, honed by thousands of years of biological and cultural evolution. But why aren't they enough? Why bother with conscious, deliberate moral thinking?

In an ideal world, moral intuition is all you need, but in the real world, there are benefits to having a dual-process brain.

5.

Efficiency, Flexibility, and the Dual-Process Brain

When my son was four years old, we read, over and over, a book called *Everything Bug: What Kids Really Want to Know About Insects and Spiders*. It explains:

Young spiders even know how to make perfect webs. They act by instinct, which is behavior they are born with. The good thing about instinct is that it's reliable. It always makes an animal act a certain way. The bad thing about instinct is that it doesn't let the animal act any other way. So, young insects and spiders do fine as long as their environment remains pretty much the same. But if they face a new situation, they can't think their way through it. They must keep doing what their instincts tell them to do.

This account of arachnoid cognition suggests an answer to our question "Why have a dual-process brain?" This answer is one of the central ideas in this book, and one of the most important ideas to emerge from the behavioral sciences in the past few decades.

This idea is summarized by an analogy mentioned in the introduction, which we'll return to again and again: The human brain is like a dual-mode camera with both *automatic settings* and a *manual mode*. A camera's automatic settings are optimized for typical photographic situations ("portrait," "action," "landscape"). The user hits a single button and the camera automatically configures the ISO, aperture, exposure, et cetera—point and shoot. A dual-mode camera also has a manual mode that allows the user to adjust all of the camera's settings by hand. A camera with both automatic settings and a manual mode exemplifies an elegant solution to a ubiquitous design problem, namely the trade-off between *efficiency* and *flexibility*. The automatic settings are highly efficient, but not very flexible, and the reverse is true of the manual mode. Put them together, however, and you get the best of both worlds, provided that you know when to manually adjust your settings and when to point and shoot.

Spiders, unlike humans, have only automatic settings, and this serves them well, so long as they remain in their element. We humans, in contrast, lead much more complicated lives, which is why we need a manual mode. We routinely encounter and master unfamiliar problems, both as individuals and as groups. Our species consists of a single breeding population, and yet we inhabit nearly every terrestrial environment on earth—a testament to our cognitive flexibility. Put a jungle spider in the Arctic and you'll have a cold, dead spider. But an Amazonian baby can, with the right guidance, survive in the frozen North.

Human behavioral flexibility feeds on itself: When we invent something new, such as boats, we create opportunities for new inventions, such as outriggers to stabilize our boats and sails to propel them. The more flexibly we behave, the more our environments change; and the more our environments change, the more opportunities we have to succeed by behaving flexibly. Thus, we reign as the earth's undisputed champions of flexible behavior. Give us a tree and we can climb it, burn it, sculpt it, sell it, hug it, or determine its age by counting its rings. The choices we make depend on the specific opportunities and challenges we face, and our choices need not closely resemble the actions that we, or others, have chosen in the past.

In this chapter, we'll consider how the human brain works in a more general way. We'll see how the dual-process theory of moral judgment described in the last chapter fits into a broader understanding of our dual-process human brains. In nearly every domain of life, our success depends on both the efficiency of our automatic settings and the flexibility of our manual mode. (For a superb, book-length treatment of this idea from its most influential proponent, see *Thinking, Fast and Slow* by Daniel Kahneman.)

EMOTION VERSUS REASON

We sometimes describe our predicaments as pitting "the heart" against "the head." The heart-versus-head metaphor is an oversimplification, but it nevertheless reflects a deep truth about human decision making. Every scientific discipline that studies human behavior has its own version of the distinction between emotion and reason. But what exactly are these things, and why do we have both of them?

Emotions vary widely in their functions, origins, and neural instantiations. For this reason, some have argued that we should get rid of the concept of "emotion" entirely. I think that would be a mistake. Emotions are unified not at the mechanical level, but at the functional level. In other words, the concept of "emotion" is like the concept of "vehicle." At a mechanical level, a motorcycle has more in common with a lawn mower than with a sailboat, but the concept of "vehicle" is still a useful concept. It just operates at a high level of abstraction.

Emotions are automatic processes. You can't choose to experience an emotion in the way that you can choose to count to ten in your head. (At best, you can choose to do something that will likely *trigger* an emotion, such as thinking about someone you love or someone you hate.) Emotions, as automatic processes, are devices for achieving behavioral *efficiency*. Like the automatic settings on a camera, emotions produce behavior that is generally adaptive, and without the need for conscious thought about what to do. And like a camera's automatic settings, the design of emotional responses—the way they map environmental inputs onto behavioral outputs—incorporates the lessons of past experience.

Not all automatic responses are emotional. Low-level visual processing—the grunt work that your visual cortex does to define the boundaries of the objects you see, to integrate information from your two eyes, and so on—is automatic, but it's not emotional.* Your brain does many things automatically, such as coordinating the contractions of your muscles in movement, regulating your breathing, and translating the pressure waves impinging on your eardrum into meaningful messages. Indeed, most of your brain's operations are automatic. What, then, makes some automatic processes *emotional* processes?

There's no universally accepted definition of emotion, but an important feature of some emotions is that they have specific *action tendencies*. Fear, for example, is not just a feeling that one experiences. Fear involves a suite of physiological responses that prepare the body to respond to threats, first by enhancing one's ability to assess the situation and then by preparing the body to flee or fight. The functions of some emotions are revealed in their characteristic facial expressions. Fear expressions widen the eyes and expand the nasal cavity, thus enlarging the field of view and enhancing the sense of smell. Disgust expressions do the opposite, crinkling up the face and thus reducing the likelihood that a pathogen will enter the body through the eyes or nose. Not all emotions have characteristic facial expressions, but, generally speaking, emotions exert pressure on behavior. They are, in short, automatic processes that tell us what to do.

The behavioral advice that we get from our emotions varies in its level of specificity. A fear response triggered by a specific object, such as a snake, tells us very specifically what to do. (Get away from that thing!) Other emotional states, such as the ones we call "moods," exert a more indirect influence on behavior. These emotional states turn up the gain on some automatic settings and turn down the gain on others. For example, in a recent classic study by Jennifer Lerner and colleagues, the experimenters influenced people's economic decisions by influencing their moods.

Some of the people in this experiment were made to feel sad by watching a sad scene from the movie *The Champ*. The saddened people were consequently more willing than others to sell a recently acquired object. Sadness, of course, doesn't directly impel people to sell their possessions in the way that an itch makes people scratch. (Picture a theater full of teary moviegoers texting their stockbrokers.) Rather, sadness sends a more diffuse signal that says something like "Things are not going so well. Let's be open to change." And then, when an opportunity for change presents itself, this signal unconsciously biases behavior in that direction. Thus, some emotions may promote behavioral efficiency, not by telling us what to do directly, but by modulating the automatic settings that tell us what to do.

Reasoning, like emotion, is a real psychological phenomenon with fuzzy boundaries. If one defines "reasoning" broadly enough, it can refer to any psychological process that leads to adaptive behavior. For example, you might say that your automatic system for visual object recognition "reasons" that the thing in front of you is a bird, based on the presence of feathers, a beak, et cetera. At the same time, one can define reasoning so narrowly as to exclude anything other than the conscious application of formal logical rules. For present purposes, we'll adopt a more moderate definition. Reasoning, as applied to decision making, involves the conscious application of decision rules. A simple form of reasoning operates in the Stroop task, in which one consciously applies the decision rule "Name the color." One can view Stroop color-naming in the presence of a red word as applying the following practical syllogism: "The color of the word on the screen is red. My job is to name the color of the word on the screen. Therefore, I should say the word 'red.'" Reasoning can get much more complicated, but this is where it begins. Critically, for our purposes, when one behaves based on reasoning, one knows what one is doing and why one is doing it; one has conscious access to the operative decision rule, the rule that maps the relevant features of the situation onto a suitable behavior.

Although the neural substrates of emotion are quite variable, there is a high degree of unity to the neural substrates of reasoning processes. Reasoning, as you now know, depends critically on the dorsolateral prefrontal cortex (DLPFC). This is not to say that reasoning occurs exclusively in the DLPFC. On the contrary, the DLPFC is more like the conductor of an orchestra than a solo musician. Many brain regions are involved in reasoning, including brain regions that are critical for emotions, such as the ventromedial prefrontal cortex (VMPFC). But there is an asymmetry between reasoning and emotion in terms of how they relate to each other. There are animals that have emotions while lacking the capacity to reason (in our sense), but there are no reasoning animals that do not also have motivating emotions. Although not everyone agrees, I think it's clear that reasoning has no ends of its own, and in this sense reason is, as Hume famously declared, a "slave of the passions." ("Passions" here refers to emotional processes in general, not exclusively to lusty feelings.) And yet, at the same time, the function of reasoning is to free us from our "passions." How can this be?

Reason is the champion of the emotional underdog, enabling what Hume called "calm passions" to win out over "violent passions." Reasoning frees us from the tyranny of our immediate impulses by allowing us to serve values that are not automatically activated by what's in front of us. And yet, at the same time, reason cannot produce good decisions without some kind of emotional input, however indirect.**

THE DUAL-PROCESS BRAIN

The dual-process brain reveals its structure in many ordinary decisions. Take, for example, the ubiquitous problem of *now* versus *later*. Baba Shiv and Alexander Fedorikhin conducted an experiment in which they gave people a choice of snack: fruit salad or chocolate cake. For most of these people, chocolate cake is what they want to eat *now*, while fruit salad is what they will wish they had eaten *later*. Shiv and Fedorikhin put half of these people under cognitive load (see page 127), in this case giving them a seven-digit number to memorize. The other half got a two-digit number to memorize, a lighter load. Each person was instructed to memorize the number, walk down a hallway to another room, and then recall the number for another experimenter in the other room. The two snack options sat on a cart in the hallway, and subjects were instructed to take one. Results: The people who had memorized seven-digit numbers—the ones under higher cognitive load—were about 50 percent more likely to choose the chocolate cake than the ones who had memorized two-digit numbers. And among highly impulsive subjects (as indicated by a questionnaire), the load manipulation more than doubled the number of people who chose cake.

Thus, it seems there are two distinct systems at work in the snacking brain. There is a more basic appetitive system that says *"Gimme! Gimme! Gimme!"* (automatic settings) and then a more controlled, deliberative system that says "Stop. It's not worth the calories" (manual mode). The controlled system, manual mode, considers the big picture, including both present and future rewards, but the automatic system cares only about what it can get right now. And, as we saw in the last chapter, when the manual mode is occupied with other business, the automatic response gets its way more easily.

This hybrid approach to consumption, which makes us simultaneously want to eat and not eat chocolate cake, may look like a shoddy piece of cognitive engineering, but placed in its natural context, it's actually very smart. For nearly all animals in nearly all contexts, it makes sense to consume calorically rich foods as soon as they're available. In a competitive world, a creature that has to stop and think explicitly about whether the benefits of eating outweigh the costs is going to miss lunch. And yet, thanks to modern technology, many of us humans now face an overabundance of lunch. For the sake of our health, not to mention our appeal to other humans, we need the cognitive flexibility to say "No thanks." It's a peculiarly modern problem that we handle with varying degrees of success, but we're all better off having both a decent appetite and the capacity for restraint. This is not to say that only modern humans require restraint: A hungry Paleolithic hunter who can't say no to a patch of ripe berries will miss out on the big game.

In recent years, cognitive neuroscientists have studied the problem of "delaying gratification" in humans, revealing a now familiar cast of neural characters in their characteristic roles. In one study, Sam McClure and colleagues presented people with two different kinds of decisions. Some

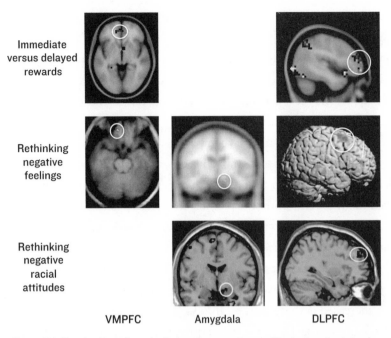

Immediate versus delayed rewards

Rethinking negative feelings

Rethinking negative racial attitudes

VMPFC   Amygdala   DLPFC

Figure 5.I. Results from three brain-imaging experiments illustrating the interplay between automatic emotional responses ("automatic settings") and controlled cognition ("manual mode").

decisions involved options yielding immediate rewards* (two dollars now, or three dollars next week?), while other decisions involved only delayed rewards (three dollars next week, or four dollars the week after?). The prospect of an immediate reward elicited increased activity in a set of brain regions, including the VMPFC. But all of the decisions elicited increased activity in the DLPFC. What's more, when people chose the immediate rewards (*Gimme! Gimme!*), they exhibited relatively more activity in the first set of brain regions, and when they chose the larger, delayed rewards ("Think about the future. . . . "), they exhibited relatively more activity in the second set of brain regions (see the top row of figure 5.1).

Notice, by the way, that choosing a larger delayed reward is, in certain respects, like choosing to push the person into the path of the trolley. In both cases, one uses one's DLPFC to opt for the "greater good," despite a countervailing emotional inclination supported by the VMPFC. There are, of course, important differences. In the now-versus-later dilemma, the emotional signal reflects a self-serving desire, whereas in the *footbridge* dilemma it reflects a moral concern for someone else. Likewise, in the now-versus-later dilemma, the greater good is for oneself (*intra*personal), while in the *footbridge* dilemma it is for several others (*inter*personal). Nevertheless, at the most general functional level, and at the level of functional neuroanatomy, we see the same pattern.

We see a similar pattern when people are attempting to regulate their emotions. Kevin Ochsner and colleagues showed people pictures that elicit strong negative emotions (e.g., women crying outside a church) and asked them to reinterpret the pictures in a more positive way, for example, by imagining that the crying women are overjoyed wedding guests rather than despondent mourners. Simply observing these negative pictures produced increased activity in our old emotional friends, the amygdala and the VMPFC. By contrast, the act of reappraising the pictures was associated with increased activity in the DLPFC (see the middle row of figure 5.1). What's more, the DLPFC's reinterpretation efforts reduced the level of activity in both the amygdala and the VMPFC.

It appears that many of us engage in this kind of reappraisal spontaneously when we encounter members of racial out-groups. Wil Cunningham and colleagues presented white people with pictures of black people's and white people's faces. Sometimes the pictures were presented subliminally— that is, for only thirty milliseconds, too quickly to be consciously perceived. Other times the faces were presented for about half a second, allowing participants to consciously perceive the faces. When the faces were presented subliminally, the black faces, as compared with the white faces, produced more activity in the amygdalas of the white viewers (see the bottom middle of figure 5.1). What's more, this effect was stronger in people who had more negative associations with black people, as measured by an IAT (see pages 51–54).

All of the participants in this study reported being motivated to respond to these faces without prejudice, and their efforts are reflected in their brain scans. When the faces were on the screen long enough to be consciously perceived, activity in the DLPFC went up (see the bottom right of figure 5.1), and amygdala activity went down, just as in Ochsner's emotion regulation experiment. Consistent with these results, a subsequent study showed that, for white people who don't want to be racist, interacting with a black person imposes a kind of cognitive load, leading to poorer performance on the color-naming Stroop task (see page 119).

Thus, we see dual-process brain design not just in moral judgment but in the choices we make about food, money, and the attitudes we'd like to change. For most of the things that we do, our brains have automatic settings that tell us how to proceed. But we can also use our manual mode to override those automatic settings, provided that we are aware of the opportunity to do so and motivated to take it.

GETTING SMART

Based on what I've said so far, you might think that our automatic settings are nothing but trouble—making us fat, sad, and racist. But these unwanted impulses are the exceptions rather than the rule. Our automatic settings can be, and typically are, very smart (see chapter 2). As Paul Whalen and colleagues have shown, the amygdala can respond to a fearful facial expression after being exposed to it for only 1.7 hundredths of a second. To do this, it uses a neat trick. Instead of analyzing the entire face in detail, it simply picks up on a telltale sign of fear: enlarged eye whites (see figure 5.2).

The VMPFC is also very clever. Damasio's group has shown, for example, that the VMPFC helps people make decisions involving risk. In a classic experiment, subjects had to choose cards from four decks. Each card wins or loses money for the player. Two of the decks are good decks, meaning that they, on balance, win the player money. Two of the decks are bad decks, meaning that they deliver large gains, but even larger losses, resulting in net losses. The players don't know at the outset which decks are good or bad. To find out, they have to sample from the decks and see what they get. Healthy people quickly develop negative responses to the bad decks, as revealed in their sweaty palms* as they reach for those decks. Rather amazingly, people start to sweat in response to the bad decks before

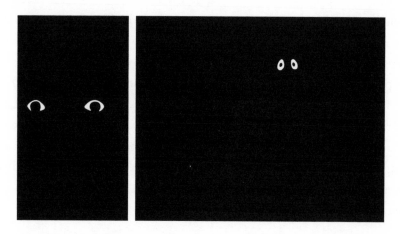

Figure 5.2. *Left*: Fear stimulus used to engage the amygdala. *Right*: The zookeeper's wife discovers that several large African animals have slipped into her bedroom, in the children's book *Good Night, Gorilla*.

they are consciously aware that the bad decks are bad. Patients with VMPFC damage, however, don't get these physiological signals and tend to persist in choosing from the bad decks. In other words, the VMPFC, in healthy people, integrates many pieces of information gained from experience (e.g., many samples from the different decks) and translates that information into an emotional signal that gives the decision maker good advice about what to do. And once again, this advice, this gut feeling, may precede any conscious awareness of what's good or bad and why. This explains why people with VMPFC damage make disastrous real-life decisions, despite their good performance on standard laboratory reasoning tests. They "know," but they don't "feel," and feelings are very helpful.

Thus, we need our emotional automatic settings, as well as our manual mode, and we need them for different things. In photography, the automatic settings work well in situations that have been anticipated by the camera's manufacturer, such as photographing a person from five feet away in moderate indoor light ("portrait") or photographing a mountain from a distance in bright sunlight ("landscape"). Likewise, the brain's automatic settings work best when they have been "manufactured" based on lessons learned from past experience.

Such experience comes in three forms, based on three different kinds of trial and error. First, our automatic settings may be shaped by our genes. Here the design of our brains incorporates lessons learned the hard way by our long-dead relatives, the ones whose genes never made it into our bodies. Second, our automatic settings may be shaped by cultural learning, through the trials and errors of people whose ideas have influenced us. Thanks to them, you need not personally encounter Nazis or the Ku Klux Klan to know in your heart—that is, in your amygdala—that swastikas and men in pointy white hoods are bad news. Finally, there's good old personal experience, as when a child learns the hard way not to touch a hot stove. Our "instincts" need not be innate, like a spider's, but if they are to be useful, they must reflect the lessons learned from someone's experience, whether that someone is you, your biological ancestors, or your cultural "ancestors."

The brain's manual mode (i.e., its capacity for controlled cognition) works fundamentally differently from its automatic settings. Indeed, the function of controlled cognition is to solve precisely those problems that can't be solved using automatic settings. Take, for example, the problem of learning to drive. Obviously, we have no genetically transmitted driving instincts, and, to the chagrin of eager teenagers, cultural familiarity with driving does not enable one to drive (safely, that is). And, of course, a new driver can't rely on personal experience, because personal experience is exactly what a new driver lacks. Learning to drive requires heavy use of the DLPFC. If you try to drive on "autopilot" the first time you get behind the wheel, you'll end up wrapped around a tree.

Thus, getting smart requires three things. First, it requires the acquisition of adaptive instincts—from our biological ancestors, from the people around us, and from our own experiences. Second, getting smart requires a facility with manual mode, the ability to deliberately work through complex, novel problems. Third, it requires a kind of metacognitive skill, analogous to the skills of a photographer. We, unlike cameras, have no masters to tell us when to point and shoot and when to be in manual mode. We have to decide for ourselves, and understanding how our minds work may help us decide more wisely, both as individuals and as herders trying to live together on the new pastures.

PART III

Common Currency

6.

A Splendid Idea

How can we, the herders of the new pastures, resolve our differences? How can we avert the Tragedy of Commonsense Morality? That's the problem we're trying to solve, and we're now ready to start thinking about solutions. Let's retrace our steps.

In chapter 1, we contrasted the Tragedy of Commonsense Morality with the original moral tragedy, the Tragedy of the Commons, in which selfishness threatens cooperation. Morality is nature's solution to the Tragedy of the Commons, enabling us to put Us ahead of Me. But nature has no ready-made solution to the Tragedy of Commonsense Morality, the problem of enabling Us to get along with Them. And therein lies our problem. If we are to avert the Tragedy of Commonsense Morality, we're going to have to find our own, unnatural solution: what I've called a *metamorality*, a higher-level moral system that adjudicates among competing tribal moralities, just as a tribe's morality adjudicates among competing individuals.

In chapter 2, we examined the standard-issue moral machinery installed in our brains. Fortunately for us, we come equipped with automated behavioral programs that motivate and stabilize cooperation within personal relationships and groups. These include capacities for empathy, vengefulness, honor, guilt, embarrassment, tribalism, and righteous indignation. These social impulses serve as counterweights to our selfish impulses. They get us into the magic corner and avert the Tragedy of the Commons.

In chapter 3, we took a second look at the new pastures and the moral machinery we bring to them. Our moral brains, which do a reasonably good job of enabling cooperation *within* groups (Me vs. Us), are not nearly as good at enabling cooperation *between* groups (Us vs. Them). From a biological perspective, this is no surprise, because, biologically speaking, our brains were designed for within-group cooperation* and between-group competition. Cooperation between groups is thwarted by tribalism (group-level selfishness), disagreements over the proper terms of cooperation (individualism or collectivism?), commitments to local "proper nouns" (leaders, gods, holy books), a biased sense of fairness, and a biased perception of the facts.

In these first three chapters (part 1), we described the human brain as a collection of automatic impulses: selfish impulses that make social life challenging and moral impulses that make social life possible. In part 2, we expanded our understanding of the human brain. Like a dual-mode camera, our brains have *automatic settings*, emotional responses that allow us to make decisions *efficiently*, drawing on the precompiled lessons of past genetic, cultural, and individual experience. And our brains have a *manual mode*, a general capacity for conscious, explicit, practical reasoning that makes human decision making *flexible*. The tension between fast and slow thinking is highlighted by moral dilemmas, such as the *footbridge* case, in which gut reactions ("Don't push the man!") compete with conscious, rule-based moral reasoning ("But it will save more lives!"). And as explained in the last chapter, the tension between gut reactions and reasoning isn't specific to morality. It's built into the general architecture of our brains, as revealed in our everyday choices about things like eating chocolate cake versus fruit salad.

Do these ideas suggest a solution to the problems of the new pastures? Yes, they do. In fact, part 1 and part 2 each suggest a solution—one philosophical and one psychological. What's more, these two solutions turn out to be the same solution, a remarkable convergence. Let's start with philosophy.

A SPLENDID IDEA

The individualistic Northerners say that being a good herder is about taking responsibility for one's own actions, keeping one's promises, respecting the property of others, and not much else. The collectivist Southerners say that good herders must do more. A just society, they say, is one in which life's burdens and benefits are shared equally. Tribes have other disagreements—for example, over matters of honor, over who hit first, who hit harder, whose word is infallible, who deserves our allegiance, who deserves a second chance, and which kinds of behaviors are abominations before Almighty God. Given their incompatible visions for moral life, how should the tribes of the new pastures live?

One answer is that there is no right answer. Some tribes do it one way, others do it another way, and that's all there is to say. This is the answer of the proverbial "moral relativist."* The problem with the relativist's answer is that, practically speaking, it's not really an answer. The relativist might be right about something important. Maybe, as the relativist says, there is no ultimate moral truth. However, even if that's correct, people must nevertheless live one way or another. The relativist may not want to choose, but someone must choose. And if the relativist refuses to choose, that, too, is a choice, reflecting a kind of judgment against judging. Even if the relativist is right about the nonexistence of moral truth, there is no escape from moral choice.

If retreating into relativism is no answer, then what is? Here's a natural thought: Maybe the herders should simply do *whatever works best*. If individualism works better than collectivism on the new pastures, then go with individualism. And if collectivism works better, then do the opposite. If a strict code of honor keeps the peace, then let us foster a culture of honor. And if honor culture leads to endless feuding, let us not. And so on.

I think this idea of doing whatever works best is a splendid one, and the rest of this book is devoted to developing and defending it. As you've surely noticed, it's a *utilitarian* idea. (More generally, it's a *consequentialist* idea. More on this shortly.) Presented like so, in the abstract, this idea of doing whatever works best strikes many people as obviously right. After all, who *doesn't* want to do what works best? But as we saw in chapter 4, when we think about specific moral problems, it's not so clear that doing what produces the best consequences is always right: Pushing the man off the footbridge seems wrong to many people, even if we assume that it will produce the best available results. What's more, this kind of cost-benefit thinking clashes with many people's deeply held values about how a society ought to be organized. For illustrative purposes, we'll start with our favorite fictional people.

THE WISDOM OF THE ELDERS

If you ask typical Northerners what they think of our idea—doing whatever works best—nearly all of them will agree that it's splendid. They will tell you that they favor going with the system that works best, which is, of course, individualism. And if you ask typical Southerners what they think of our splendid idea, they will give a similar but incompatible answer. The collectivist Southerners say that herders living on the new pastures should go with whatever system works best, which is, of course, collectivism.

What's going on here? Perhaps the Northerners and the Southerners have, at bottom, the same moral values: They both want to do whatever works best. Their disagreement, then, is just a factual disagreement about what, in fact, works best. To see if that's right, let's try a thought experiment. Suppose that we submit to the individualist Northerners a mountain of evidence indicating that collectivism works best. And we submit to the collectivist Southerners a mountain of evidence indicating that individualism works best. Neither tribe can credibly challenge this evidence, because they have only the faintest knowledge of what life is really like in societies of the opposite sort. How will they respond to this challenge? A few Northerners may be intrigued, but for the most part their tribe will dismiss this so-called evidence as hogwash. (Recall our discussion of tribal biases in chapter 3.) And the Southerners will, for the most part, do the same. This doesn't sound like a simple factual disagreement.

Here's another test. Instead of presenting Northerners and Southerners with evidence favoring the other side's way of life, simply ask them to imagine such evidence. To the individualistic Northerners, we present the following hypothetical: Suppose it turns out that life goes better in collectivist societies, for the following reason: In individualist societies, there are winners and losers. Some have enormous herds, and some have almost nothing. In collectivist societies, there are no winners and losers. Everyone gets the same modest something. There's more total wealth in individualist societies, but things go worse overall, because the losses of the losers in individualist societies weigh heavily against the gains of the winners. In collectivist societies, in contrast, no one's cup overflows, but everyone's cup is full enough, and overall the collectivist society is better off. "If all of this were true," we ask the Northerners, "would you convert to collectivism?"

First, our Northern friends will tell us that this is really a stupid question. It's stupid because everyone knows that collectivism leads to ruin. And then our Northern friends will treat us to a series of lectures detailing the various ways in which collectivism leads inevitably to ruin. They'll tell us that collectivists are lazy people who want others to take care of them, or naive people who don't understand how the world works, or brainwashed people who have spent too much time around collectivists, and so on. We will nod politely and then remind them that we're not asking them about how collectivism fares in the real world. We are, for now, simply asking a hypothetical question: Would you favor collectivism if it did, in fact, work better? After a few more rounds of this, some of our Northern friends will agree to answer the offending hypothetical. Holding their noses, they will offer that if, somehow, in some crazy, upside-down world, collectivism were to produce better results, then it would make sense to favor collectivism.

Then the Great Elder of the North, the wisest of them all, steps forward. He explains that collectivism, in addition to producing ruinous results in practice, is rotten to its philosophical core. He explains that it is simply wrong for the foolish and lazy to demand a share of what they were too foolish and lazy to earn for themselves. A society, he explains, must be measured not by the sum of the benefits it doles out, but by its commitment to justice. And collectivism, he declares, is simply unjust, punishing the best people while rewarding the worst. With that, the Northern crowd erupts with applause, their core values so eloquently affirmed.

To the collectivist Southerners we pose the opposite hypothetical: Would you favor individualism if it turned out to work better? They, like their Northern counterparts, begin by dismissing the question. Everyone knows that a society founded on individualistic greed is bound for ruin. Again, we press our hypothetical question. As in the North, a few hesitantly agree that individualism would be preferable if, somehow, it were to work better. And then this modest concession is countered by the Great Elder of the South, who speaks with the wisdom and authority of the ages. She explains that individualism, in addition to producing widespread misery in practice, is rotten to its philosophical core. A society founded on a principle of greed is an inherently immoral society, she says, and there is no amount of wealth for which noble herders should sell their ideals of love, compassion, sisterhood, and brotherhood. Amen, says the Southern crowd.

These are thought experiments, and I had the privilege of making up the results. But the results I made up are firmly grounded in what we've learned about moral psychology. Northern herders are not committed to individualism because they know that it works best. They did not perform a comparative cost-benefit analysis of social systems. Nor did Southern herders arrive at their collectivist convictions in this way. Rather, the Northerners and Southerners believe what they believe because they've lived their entire lives immersed in their respective tribal cultures. Their moral intuitions are tuned to their respective ways of life, their systems for averting the Tragedy of the Commons. Short of a dramatic cultural shift, collectivism will always seem wrong to the Northerners, and individualism will always seem wrong to the Southerners, no matter how the facts turn out. Both sides truly believe that their respective ways of life produce the best results. However—and this is the crucial point—both sides are more committed to their ways of life than they are to producing good results. The Great Elders understand this. They are the guardians of the local wisdom, and they know better than to take our hypothetical bait. They understand that their values are not, at bottom, about "whatever works." They aspire to live by deeper moral truths.

CONSEQUENTIALISM, UTILITARIANISM, AND PRAGMATISM

People think that doing whatever works best is splendid, until they realize that what they really want is not necessarily what works best. Still, perhaps doing whatever works best *really is* a splendid idea. That's the thought behind utilitarianism, a thoroughly modern philosophy that is easily mistaken for simple common sense.

The idea that we should do whatever works best sounds a lot like "pragmatism" in the colloquial sense. (In Anglo-American philosophy, "pragmatism" often has a different meaning.*) But utilitarianism is more than an injunction to be pragmatic. First, "pragmatism" may imply a preference for short-term expediency over long-term interests. This is not what we have in mind. Utilitarianism says that we should do whatever *really* works best, in the long run, and not just for the moment. Second, "pragmatism" may suggest nothing more than a flexible management style, one that can be deployed in the service of any values. A staunch individualist and a staunch collectivist could both be "pragmatists" in the colloquial sense. Utilitarianism, by contrast, is about core values. It's about taking "pragmatism" all the way down to the level of first principles. It begins with a core commitment to doing whatever works best, whatever that turns out to be, and even if it goes against one's tribal instincts.

For this reason—and for others that we'll discuss later—I prefer to think of utilitarianism as *deep pragmatism*. When your date says, "I'm a utilitarian," it's time to ask for the check. But a "deep pragmatist" you can take home for the night and, later, to meet the parents. The U-word is so ugly, and so misleading, that we're better off ditching it. Still, as a deep pragmatist, I understand that I will not, with a few choice keystrokes, displace a two-hundred-year-old philosophical term. And besides, I need to convince you that deep pragmatism really is what we're looking for, and not just an old pig with new lipstick—or aftershave, as the case may be. Thus, for now, I will bow to tradition and refer to our splendid idea by its ugly, misleading, conventional name. In part 5, we'll return to the idea that utilitarianism, properly understood and wisely applied, is in fact deep pragmatism.

W hat, then, is utilitarianism, and where did this idea come from? To begin with, utilitarianism is a form of *consequentialism*, and everything that I've said so far about utilitarianism actually applies to consequentialism more broadly. Consequentialism says that consequences— "results," as a pragmatist might say—are the only things that ultimately matter. Here the word "ultimately" is very important. It's not that things other than consequences—things like being honest, for example—don't matter, but rather that other things matter, when they do, because of *their* consequences. According to consequentialism, our ultimate goal should be to make things go as well as possible.*

But what do we mean by "well"? What makes some consequences better than others? Utilitarianism gives a specific answer to this question, and that's what differentiates it from consequentialism more generally. Consequentialism sounds a lot like "cost-benefit analysis," and in a sense it is. But usually when people talk about cost-benefit analysis, they're talking about money. Perhaps we herders should measure our success in terms of economic productivity: What works best is whatever maximizes GPP ("gross pasture product"). This would simplify our moral accounting, because material wealth is easy to measure. But is economic productivity what *ultimately* matters? One can imagine a society that is highly economically productive but in which everyone is miserable. Is that a good society?

If the consequences that really matter aren't economic consequences

per se, then what *does* really matter? We might begin by asking ourselves
what we want from economic productivity. Once again, if we're all miserable, our wealth is apparently not doing us any good. On the flip side, if
we're all happy, you might say that it doesn't matter whether we're rich or
poor. Thus, a natural thought is that what really matters is our *happiness*.
Not everyone agrees with this conclusion, but it is, at the very least, a reasonable place to start. If we combine the idea that happiness is what matters with the idea that we should try to maximize good consequences, we
get utilitarianism.

The founding utilitarians, Bentham and Mill, were not just armchair
philosophers. They were daring social reformers, intensely engaged
with the social and political issues of their day. Indeed, many familiar social issues became social issues because Bentham and Mill made them so.
Their views were considered radical at the time, but today we take for
granted most of the social reforms for which they fought. They were
among the earliest opponents of slavery and advocates of free speech, free
markets, widely available education, environmental protection, prison reform, women's rights, animal rights, gay rights, workers' rights, the right
to divorce, and the separation of church and state.

Bentham and Mill were unwilling to accept moral business as usual.
They refused to accept practices and policies as right simply because of
tradition, or because they seemed intuitively right to most people, or because it was "the natural order of things." Nor did the founding utilitarians appeal to God to justify their moral ideas. Instead, they asked the
kinds of questions that we asked above: What really matters, and why? By
what standard can we evaluate our actions and policies, without begging
the questions we're trying to answer? On what grounds, for example, can
one say that slavery is wrong? Bentham and Mill couldn't appeal to God
because their proslavery opponents thought they had God on their side.
Even if the founding utilitarians had wanted to appeal to God, how could
they show that their interpretation of God's will was correct? For similar
reasons, they couldn't appeal to the rights of slaves because the rights of

slaves were precisely what was in question. On what grounds does one determine or discern who has which rights?

Bentham and Mill's answer was utilitarianism. When evaluating laws and social practices, they asked only this: Does it increase or decrease our happiness, and by how much? For example, they argued that slavery is wrong not because God opposes it, but because whatever good it may yield (for example, in terms of economic productivity) is vastly outweighed by the misery it produces. And likewise for the restriction of women's freedom, the brutal treatment of animals, laws against divorce, and so on.

Bentham and Mill introduced a perfectly general standard for measuring moral value and for making hard moral decisions: All actions are to be measured by the sum of their effects on happiness. Utilitarianism is a splendid idea, and it is, I believe, the metamorality that we modern herders so desperately need. But it's also a highly controversial idea, the object of two centuries' worth of philosophical debate: Can all moral value be translated into a single metric? And if so, is aggregate happiness the right metric? In later chapters, especially in part 4, we'll consider the philosophical challenges to utilitarianism. But first, let's get clear about what utilitarianism really is and why some people think that happiness, and happiness alone, is what ultimately matters.

(MIS)UNDERSTANDING UTILITARIANISM

As I said in chapter 4, utilitarianism is widely misunderstood. The trouble begins with its awful name, suggesting a preoccupation with the mundanely functional. (The "utility room" is where one does the laundry.) Replacing "utility" with "happiness" is a step in the right direction, but this, too, is misleading. What utilitarian philosophers mean by "happiness" is far broader than what we think of when we think about "happiness." Once we've understood happiness properly, the idea that we should maximize happiness is also easily misunderstood. One imagines the utilitarian life as one of constant calculation, adding up the costs and benefits for every decision. Not so. Finally, the task of maximizing happiness may appear to be fraught with crippling ambiguity: Whose happiness are we talking about? Doesn't happiness mean different things to different people? How can one possibly measure happiness? Who gets to decide what counts as happiness and how best to increase it? And isn't this idea of maximizing happiness a dangerous kind of utopianism? In what follows, I'll address these questions and attempt to ward off some of the more common misunderstandings about this much-maligned philosophy.

What Do We Mean by "Happiness"?

In middle school, my classmates and I completed a "values project." Each of us identified our ten most important values and produced a book— essentially a scrapbook with drawings, photos, and written explanations— explaining why we hold these values dear. Before starting our individual projects, we did a bit of collective brainstorming. Students called out candidate values, and the teacher listed them on the board: "family," "friends," "religion," "sports," "having fun," "love," "helping others," "learning new things," "my cat." After a bit of conceptual tidying ("Let's put your cat under 'pets' . . . And we'll put 'Disney World' under 'having fun' . . . "), we had a pretty good list. Among the items listed was "happiness."

As a nascent utilitarian philosopher, I was unsure how to fit happiness into my own list. I put "family" first, and "friends" after that. I think happiness ended up being number four. But I couldn't escape the thought that happiness was hidden in all of these other things: In valuing my family and friends, am I not valuing their happiness and the happiness they bring me? Would anyone value "sports" or "love" if these things brought us no happiness? It seemed to me there was further conceptual tidying to be done.

My classmates, apparently, did not share this concern. They saw no problem with treating happiness as just another item on the list. What were they thinking? What they presumably had in mind—what most of us have in mind when we think about happiness—is something like this:

Valuing happiness means valuing things that are directly responsible for putting smiles on our faces. The canonical list of smile inducers appears in the song "My Favorite Things," from *The Sound of Music*:

> *Raindrops on roses and whiskers on kittens*
> *Bright copper kettles and warm woolen mittens*
> *Brown paper packages tied up with strings*
> *These are a few of my favorite things*
> *Cream colored ponies and crisp apple strudels*
> *Doorbells and sleigh bells and schnitzel with noodles*
> *Wild geese that fly with the moon on their wings*
> *These are a few of my favorite things*

Further elaboration appears in "Happiness Is," from *You're a Good Man, Charlie Brown*:

> *Happiness is two kinds of ice cream*
> *knowing a secret*
> *climbing a tree*

The contemporary adult version features the lifestylish leisure activities depicted in glossy catalogs: reading the newspaper on the iPad while reclining in a gently swaying hammock, chatting with the neighbors before embarking on an afternoon of mountain biking, clinking glasses on the deck at sunset . . . We'll call this the "favorite things" conception of happiness. If what we mean by happiness is enjoying our favorite things, then happiness is indeed just another item on the list of values. Moreover, if happiness is enjoying our favorite things, then calling happiness the ultimate value, the standard by which to measure all actions, seems incredibly shallow.

The "favorite things" conception of happiness, however, doesn't hold up to scrutiny. The problem, once again, is that our happiness is greatly affected by things that we don't think of when we think of "happiness." Replacing the brake pads on my car is not one of my favorite things, but if I don't replace the brake pads, many people (myself, my family, other motorists, their families) might have their happiness significantly diminished. Consider someone who labors for years on a worthy but arduous task. Within the framework of the values project, this sounds more like "hard work," "perseverance," and "discipline" than happiness. But of course, this hard work is presumably being done to enhance someone's well-being—if not the hard worker's, then someone else's. Likewise, consider one who volunteers at a homeless shelter, not because she particularly enjoys it, but because she thinks it's important to help others who are less fortunate. This sounds more like "helping others," "charity," or "social responsibility" than happiness. But again, our volunteer presumably hopes to increase these poor people's prospects for living a good life, and their living a good life presumably includes being happy and contributing to the happiness of others. This last example highlights a further problem with the "favorite things" conception of happiness, which is that it ignores the negative side of the happiness scale. If our goal is to maximize happiness, lifting people out of misery is generally more important than placing cherries atop people's sundaes. But when we think of "happiness," we're more likely to think of cherries than homeless shelters.

Thus, happiness is embedded in many other values that, upon first blush, strike us as deeper and more meaningful than happiness. Above I mentioned family, friends, and love. Add to this the intellectual values (knowledge, truth, education, the arts), civic values (freedom, justice), and the values of character (bravery, honesty, creativity). All of these things make the world happier, and this is not unrelated to our valuing them. More generally, it's hard to think of values worth valuing that are not intimately connected to happiness. What this means, at the very least, is that happiness, as a moral value, is easily misunderstood and underestimated. To think about happiness properly, one has to think *abstractly* and, more specifically, *counterfactually*. Valuing happiness, in the utilitarian sense, is not merely about valuing the things that come to mind when we think of "happiness." It's about valuing all of the things whose *absence* would diminish our happiness, and that includes almost everything that we value.

But utilitarianism is saying more than that happiness is valuable. It's

saying that happiness—properly understood—is the *only thing* that ultimately matters. Why would one think that? To see why, start with the
things that you care about most immediately and then work backward,
repeatedly asking yourself, "Why do I care about that?" until you run out
of answers: You went to work today. Why did you bother? Perhaps because
you enjoy your work, but also to earn money. Why do you want money?
To buy things like food. Why do you want food? Because you and your
family enjoy eating food, and don't enjoy being hungry. Food also keeps
you and your family alive. But why do you and your loved ones want to
remain alive? Because you all enjoy being alive, and enjoy living with one
another in particular. Why do you care about your and your family's enjoyment of life? *Uhhhh . . .*

Following this logic over and over, you might conclude that everything you do is ultimately done to enhance the quality of someone's
experience. Even things like punishment, which are immediately aimed at
delivering unpleasant experiences, are related to the enhancement of experience: We punish people to make them feel worse, which deters them and
others from doing bad things, which improves the experience of their
would-be victims. Likewise, we generally regard things that have no experience, things like rocks, as beyond the scope of moral concern.

Thus, it's plausible that the goodness and badness of everything ultimately cashes out in terms of the quality of people's experience. On this
view, there are many worthy values: family, education, freedom, bravery,
and all the rest of the values listed on the chalkboard. But, says utilitarianism, these things are valuable because, and only because, of their effects on
our experience. Subtract from these things their positive effects on experience and their value is lost. In short, if it doesn't affect someone's experience, then it doesn't really matter.

This is the central idea behind the utilitarian conception of happiness.
Happiness is not (just) ice cream and warm summer evenings at the lake
house. One's happiness is the overall quality of one's experience, and to
value happiness is to value everything that improves the quality of experience, for oneself and for others—and especially for others whose lives leave
much room for improvement. From a utilitarian perspective, it's not that

happiness *beats out* the other values on the list. Happiness, properly understood, *encompasses* the other values. Happiness is the *ur*-value, the Higgs
boson of normativity, the value that gives other values their value.

You may or may not agree with this. As I'll explain later, I think this is,
in fact, an overstatement. (If happiness is the one and only ultimate value,
then how can values ever conflict?) For now, what I want to convey is two-
fold: First, the utilitarian conception of happiness is very broad, encom-
passing all positive aspects of experience as well as the removal of negative
aspects. This is what we mean by "happiness." Second, in light of this, it's
not unreasonable to think of happiness as occupying a special place among
our values, as more than just another item on the list. Unlike the most ar-
dent utilitarians, I don't think that happiness is the one true value. Instead,
what makes happiness special—and this is Bentham and Mill's real insight, in my opinion—is that happiness is the *common currency* of human
values. We'll develop this idea further in the next two chapters.

Are Not Some Kinds of Happiness More Valuable than Others?

Above I said that the utilitarian conception of happiness is very broad, but
there is in fact a debate among utilitarians about how broad it should be.
Bentham conceived of utility rather narrowly, in terms of pleasure and
pain. Mill took a broader view, regarding some pleasures as qualitatively
different from others, and more valuable. Mill famously wrote:

> It is better to be a human being dissatisfied than a pig satisfied; bet-
> ter to be Socrates dissatisfied than a fool satisfied. And if the fool, or
> the pig, are of a different opinion, it is because they only know their
> own side of the question. The other party to the comparison knows
> both sides.

On the one hand, it seems foolish to limit our conception of well-
being to mere pleasure and pain, and to count all pleasures as equally

worthy. On the other hand, Mill's privileging of the "higher" pleasures seems unprincipled and, perhaps, elitist: "My dear piggy fool, if only you could appreciate the joys of the mind, you, too, would prefer them to beer." Fortunately, I think we can reconcile these two views using an argument that Mill dashes off in passing, one that is better than his primary argument, in my opinion, and bolstered by more recent psychology.

As Barbara Frederikson has argued in her "broaden and build" theory of positive emotion, the things that we find pleasurable are often things that build resources. Tasty food provides nutritional resources. Spending time with friends builds social resources. Learning builds cognitive resources. It seems that Mill's "higher pleasures" are pleasures derived from activities that build *durable* and *shareable* resources. This opens up a more principled utilitarian argument in favor of Mill's "higher pleasures."

Mill wants to say that philosophy is better than beer, despite the latter's greater popularity. His solution is to insist that those who know both favor philosophy, indicating that it offers a better pleasure, a *higher* pleasure. According to Mill, the drunken fool is really missing out. I suspect that this is not the best way to defend the life of the mind, and the higher pleasures more generally. Instead I suggest that being the satisfied fool is—or may be—better *for the fool*, while being Socrates is better *for the rest of us*. (Sound familiar?) Likewise, being a satisfied fool may be better for the fool *now* but not so good for the fool *later*. On this view, it's not that reading Plato produces a better *pleasure* than beer. Instead, it's better—if, in fact, it is—because it's a pleasure that leads to *more* pleasure, not just for oneself but for others as well. In defending the noble life, Mill appealed to immediate self-interest ("Really, it's a much better high!"), when he should have instead appealed to the greater good: The higher pleasures are higher because of their characteristic long-term consequences, not because of how they feel.

There is, however, a downside to this reconciliation between Mill and Bentham: One must conclude that a life of sex, drugs, and rock and roll *would* be better than a life of placid intellectual contemplation, provided that the long-term consequences were held constant. I have mixed feelings about this conclusion. When it comes to other people's debauchery, I'm happy to say "Good for them!" so long as they're not making themselves or others worse off. But when I contemplate this as a personal choice for myself, it's harder. If I knew there were no broader consequences for me or for others, would I give up my pleasant professorial existence for a nonstop party? Probably not. But maybe I would—hypothetically—just need to get over my hang-ups and dive in.

In any case, the point for now is that utilitarianism need not be construed as "swine morality." We have perfectly good utilitarian reasons for valuing at least some "higher" pleasures over some "lower" pleasures. The higher pleasures are better (at least sometimes) not because they are better *pleasures*, but because they are pleasures that serve us better in the long run.

Whose Happiness Are We Talking About?

Everyone's. The second defining feature of utilitarianism, in addition to its focus on experience, is that it's *impartial*. Everyone's happiness counts the same. This doesn't mean that, in a utilitarian world, everyone gets to be equally happy. As the Northern herders will tell you, a world in which everyone gets the same outcome no matter what they do is an idle world in which people have little incentive to do anything. Thus, the way to maximize happiness is not to decree that everyone gets to be equally happy, but to encourage people to behave in ways that maximize happiness. When we measure our moral success, we count everyone's happiness equally, but achieving success almost certainly involves inequality of both material wealth and happiness. Such inequality is not ideal, but it's justified on the grounds that, without it, things would be worse overall.

There is another way to understand this question, "Whose happiness?" We may be asking: Whose *conception* of happiness shall prevail? Happiness for me is two kinds of ice cream. Happiness for you is reading Plato. Happiness for someone else is being tied up and whipped by a three-hundred-pound woman in a Bo Peep costume. *Whose happiness?*

This is mostly a verbal problem. We can say that happiness is different things for different people, but that's needlessly confusing. It's clearer to say that happiness is the *same thing* for everyone, and that different people are made happy and unhappy by different things. Two kinds of ice cream does it for me, but not for you, and so on.

I say this is *mostly* a verbal problem because one could doubt that happiness really is the same thing for all people. However, to doubt that humans the world over have something in common when they experience "happiness" is in fact a very radical doubt. Consider these sentences, about a boy living in Japan during the eighth century: "Kammu went down to the well and found, to his surprise, that the water had returned. This made him very happy." Do you find this confusing? Of course you don't. That's because you've understood these sentences exactly as intended. Kammu is feeling more or less what you feel when you describe yourself as happy. How about this? "Kammu found two dead ladybugs pressed against a rock. This made him very happy." This is weird, but it's weird precisely because you are applying *the same* understanding of happiness to yourself and to Kammu: What you get from two kinds of ice cream is, apparently, what he gets from dead ladybugs. Your experiences and Kammu's may differ in myriad interesting ways, due to the cultural gulf that separates you. Nevertheless, and despite these differences, your experiences have something in common, which is that they're both to some extent *positive*. (And to some extent negative.) Happiness is common currency.

How Can One Possibly Measure Happiness?

Having agreed that people everywhere are capable of having positive (and negative) experiences, we turn to the problem of measurement. Measuring happiness is a complicated business that has, over the past few decades, occupied some of the best minds in social science. However, the point I wish to make here requires no fancy scientific analysis: Measuring happiness is easy. What's hard is measuring happiness as *accurately* as we'd like. Our inability to measure happiness with pinpoint accuracy poses formidable practical challenges, but it poses no deep philosophical problems.

Take Ricardo. He's in the hospital with a broken kneecap—very painful. Take Beatriz. She's reclining in a gently swaying hammock ($315), reading the news on her iPad ($499). One might suppose that, at this moment, Beatriz is feeling better than Ricardo. But how can we know? We can ask them: *On a scale from 1 to 10, how good do you feel right now?* Ricardo says "2." Beatriz says "8." We've just measured their respective levels of happiness.

Did we get it exactly right? We don't know. Maybe Ricardo actually feels great but doesn't want to alarm us. Maybe this is the worst moment of Ricardo's life, but for fear of sounding whiny he said "2" instead of "1." Maybe Beatriz is experiencing deep inner torment but won't admit this to us, or even to herself. Or maybe she's reluctant to use the high end of the scale, which makes her rating artificially low. We've asked Ricardo and Beatriz about how they are feeling right now, but we could also ask them about how their lives are going overall. Here the measurement problems are even worse. Maybe Ricardo's life is actually going great, better than Beatriz's, but at the moment it doesn't seem that way to him.

These are serious problems. But the existence of these problems doesn't imply that we can't measure happiness. It just means that our measurements of happiness are inevitably *estimates*. Whether our estimates are good enough depends on what we're trying to do with them. If we want to know exactly how happy a given person is, or how a given person's happiness compares with that of another person in similar circumstances, our estimates may not be good enough. But fortunately, when it comes to the big decisions that we must make as a society, it's not necessary to measure the happiness of any given individual with great accuracy. Instead we need to understand general patterns: What kinds of policies tend to increase happiness? What kinds of policies tend to decrease happiness?

It's here that the new science of happiness excels. For example, we've learned that unemployment is often emotionally devastating, with psychological costs that far exceed its economic costs. By contrast, making a bit less money, if you're already wealthy, is unlikely to have much of an effect on your happiness. Of course, some people lose their jobs and it turns out to be a godsend. And some people go through hell (maybe) when their annual income drops from $220,000 to $200,000. But overall, the general effects of these economic variables on happiness are clear, and that enables us to make more informed policy decisions—for example, about the trade-off between raising taxes and creating jobs. And this is true despite our inability to measure happiness in any given person with great accuracy.

When people worry about our ability to measure happiness, they may have something else in mind. It's not that it didn't occur to them that you could ask people how they're feeling. Their worry is that merely asking people isn't good enough. They want a "real" measure of happiness, a direct measure that bypasses our subjective impressions, much as a thermometer bypasses our feelings of hot and cold. With the advent of functional brain imaging, we may soon have such measures.* But when the neuro-happiness-o-meters arrive, they won't be game changers. We may use them to calibrate people's verbal reports up or down, or to catch people who, for whatever reason, are lying about their happiness. But for most purposes, simple answers to simple questions are all we need. As Dan Gilbert has observed, the optometrist doesn't scan your brain to figure out which lenses produce the clearest visual percepts; she just asks you, "How does it look now?"

Measuring happiness is not an insurmountable problem, and to the extent that it's a problem, it's a problem for *everyone*, not just utilitarians. No one thinks that the effects of our choices on our happiness are *irrelevant*. Thus, even if you reject the utilitarian idea that happiness is all that ultimately matters, as long as you think that happiness matters to some extent, you need to measure it, too!

Are Utilitarians Always "Calculating"?

If there's one word that sums up people's misunderstanding of utilitarianism, it's "calculating." The stereotype of the "calculating" utilitarian has two related features.

First, a "calculating" person is a *bad* person, a *selfish* person, one who is forever working out how best to serve *himself*. This stereotype of the "calculating" utilitarian is undeserved. The utilitarian ideal is *impartial*. An ideal utilitarian values the well-being of others no less than she values her own well-being—a perfect embodiment of the Golden Rule. Far from being a selfish philosophy, utilitarianism faces the objection that it requires too much *selflessness*. (More on this shortly.)

There is, however, a grain of truth to this charge. Moral calculation, even with the best of intentions, may lead one astray. One might attempt to calculate on behalf of the greater good but, through various forms of self-deception, end up calculating on behalf of oneself. ("All for the love of Rome!") To calculate at all is to distrust, at least temporarily, the moral machinery described in chapter 2, the social instincts that put Us ahead of Me. The worry, then, is that once you're doing any kind of moral calculating, going off moral autopilot, you're likely to get into trouble. This view is supported by the experiments with Public Goods Games described at the end of chapter 2: More thinking leads to more free riding and less cooperation.

This worry about the pitfalls of moral calculation brings us to the second part of the utilitarian stereotype, according to which the utilitarian is constantly making moral calculations. One imagines a utilitarian standing in the aisle at the store, adding up the costs and benefits of shoplifting. Fortunately, most of us don't engage in this kind of moral calculation, but it might seem that this is what utilitarianism recommends. If you think about it, however, this is a decidedly un-utilitarian way to be. Why? Because constant moral calculation about what will serve the greater good is *clearly not going to serve the greater good*. Were we to allow ourselves to do whatever we wanted, so long as we could convince ourselves that what we were doing was for the "greater good," it would be a disaster. We humans are notoriously biased in our own favor (see chapters 2 and 3) and not especially good at calculating the long-term, global effects of our actions. Thus, in everyday life, we're much better off listening to our moral instincts, rather than trying to work out whether petty theft, for example, might serve the greater good. Our moral instincts evolved, both biologically and culturally, to help us put Us ahead of Me. In everyday life, we attempt to outcalculate these instincts at our peril.

At this point, you might wonder whether utilitarianism has been defended into obsolescence: If our moral instincts reliably guide us toward the greater good, then why bother with moral philosophy, utilitarian or otherwise? Here it's important not to confuse our two tragedies. Once again, our moral instincts do well with the Tragedy of the Commons (Me vs. Us), but not so well with the Tragedy of Commonsense Morality (Us vs. Them). The utilitarian thing to do, then, is to let our instincts carry us past the moral temptations of everyday life (Me vs. Us) but to engage in explicit utilitarian thinking when we're figuring out how to live on the new pastures (Us vs. Them). I'll say more about how this works in part 5.

Is Utilitarianism a Dangerous Kind of Utopianism?

History offers no shortage of grand utopian visions gone bad, including the rise and (nearly complete) fall of communism during the twentieth century. Communists such as Stalin and Mao justified thousands of murders, millions more deaths from starvation, and repressive totalitarian governments in the name of the "greater good." Shouldn't we be very wary of people with big plans who say that it's all for the greater good?

Yes, we should. Especially when those big plans call for big sacrifices. And *especially*, especially when the people making the sacrifices (or being sacrificed!) are not the ones making the big plans. But this wariness is perfectly pragmatic, utilitarian wariness. What we're talking about here is *avoiding bad consequences*. Aiming for the greater good does not mean blindly following any charismatic leader who says that it's all for the greater good. That's a recipe for disaster.

We began this chapter by comparing the ideals and rhetoric of the individualist Northern herders and the collectivist Southern herders. The utilitarian is not, as you might expect, necessarily on the side of the collectivists. Nor is she necessarily an individualist. Our splendid idea is for herders of all tribes to put aside their respective ideologies and instead figure out what actually works best—what actually maximizes happiness—on the new pastures. And what works best may turn out to be more individualistic or more collectivist. Figuring out what works best requires putting our prejudices aside and instead gathering and evaluating *evidence* about how various policies and practices fare in the real world. Utilitarianism is, as I've said, pragmatism taken all the way down to first principles.**

This worry about dangerous utilitarian utopianism exemplifies a whole class of confused, though nonetheless tempting, objections to utilitarianism: If the utilitarian world that you're imagining sounds like a generally miserable place, then, by definition, you're imagining the wrong thing. Your objection is a utilitarian one, and what you're objecting to is not really utilitarianism.

Who Gets to Decide How to Maximize Happiness?

By now, I hope that you've gotten the hang of utilitarian thinking and can answer these kinds of questions yourself. But we'll do one more, for the sake of completeness.

From a utilitarian perspective, deciding who gets to decide is a decision like any other. There is no official Utilitarian Decider with a fancy hat. From a utilitarian perspective, a good decision-making system is one in which the decision makers are more likely than otherwise to make decisions that produce good results. In principle, this could be one in which all decision-making power is vested in a single philosopher king. But everything we know of history and human nature suggests that this is a bad idea. Instead, it seems we're better off with representative democracy, coupled with a free press and widely accessible education, and so on, and so forth.

To Summarize . . .

Utilitarianism combines two reasonable and universally accessible ideas. We can think of these ideas as answers to two questions: *What* really matters? *Who* really matters?

According to utilitarianism, what ultimately matters is the quality of our experience. Utilitarianism is not about maximizing "utility" in the laundry room sense, favoring the mundanely functional over things that glimmer. Nor is it about favoring things that glimmer—our "favorite things"—over things that are more deeply meaningful or important. Utilitarianism embraces nearly all of the values we hold dear, including values associated with personal relationships (family, friends, love), personal virtues (honesty, perseverance), noble pursuits (truth, art, sports), and good governance (freedom, justice). But according to utilitarianism, all of these values ultimately derive their value from their effects on our experience. Had they no impact on our experience, they would not be valuable.** This idea may or may not be correct. We've not yet considered the arguments against it. But it's a plausible idea, and—just as important—it's an idea that any thoughtful person, independent of his or her tribe, can understand and appreciate.

The second utilitarian ingredient is impartiality, the universal essence of morality that's distilled in the Golden Rule. Having added this second ingredient, we can summarize utilitarianism thus: Happiness is what matters, and everyone's happiness counts the same. This doesn't mean that everyone gets to be equally happy, but it does mean that no one's happiness is inherently more valuable than anyone else's.

Happiness can be measured, though measuring happiness accurately poses challenges. Often, however, we can learn what we need to know about happiness not by studying the happiness of specific individuals with great accuracy but by studying the happiness of populations, to draw general conclusions about what tends to increase or decrease happiness.

Knowing what will maximize happiness in the long run is obviously impossible. Some people see this as a fatal flaw for utilitarianism, but if you think about it, this makes no sense. *Everyone* needs to take some sort of guess—educated or otherwise—about what will produce the best long-term consequences. (Everyone, that is, except people who don't care about long-term consequences.) Utilitarianism isn't distinguished by its concern for long-term consequences. It's distinguished by its giving ultimate priority to long-term consequences.

Utilitarianism is not, at the most fundamental level, a *decision procedure*. It is, instead, a theory about what matters at the most fundamental level, about what's worth valuing and why. Utilitarianism does not require us to constantly calculate the expected costs and benefits of our actions. On the contrary, it requires us to trust our moral intuitions most of the time, because that's more likely to serve us well than constant moral calculation.

Utilitarianism does not require us to march behind anyone who claims to be serving the greater good. Instead, it asks us to make decisions in ways that are likely to lead to good outcomes, taking into account the limitations and biases inherent in our natures. And given the history of utopian politics, utilitarianism requires us to be skeptical of leaders who claim to have the greater good all worked out.

In sum, utilitarianism combines the Golden Rule's impartiality with the common currency of human experience. This yields a moral system that can acknowledge moral trade-offs and adjudicate among them, and it can do so in a way that makes sense to members of all tribes.

A REMARKABLE CONVERGENCE

Looking down on the new pastures from ten thousand feet, watching embattled herders from different tribes with different moral systems and different moral instincts go at one another, the utilitarian's pragmatic solution seems so obvious: They should put their tribal ideologies aside, figure out which way of life works best on the new pastures, and then live that way. This is the conclusion suggested by our analysis of moral problems in part 1. But there is, as I've said, another line of reasoning, suggested by the psychological analysis presented in part 2.

Once again, we have dual-process brains, with automatic settings that make our thinking *efficient* and a manual mode that makes our thinking *flexible*. This analogy between our moral brains and a dual-mode camera is useful not only because it provides an apt description of moral psychology, but because it suggests an answer to our big, practical question: How can modern herders resolve their disagreements? Above, we posed this as a philosophical question: What philosophy could serve as our metamorality? But we can also pose our question in psychological terms: What *kind of thinking* is right for the new pastures? Here the camera analogy offers good guidance.

What's better for photography: automatic settings or manual mode? The answer, of course, is that neither is better in an absolute sense. These two ways of taking photos are relatively good and relatively bad at different things. If you're facing a typical photographic situation, the kind that the camera's manufacturer has anticipated ("portrait," "landscape"), then automatic settings are probably all you need—point and shoot. But if you're facing a situation that the camera's manufacturer never imagined, or if your aesthetic preferences differ from the manufacturer's, then you probably need manual mode.

Our question now is this: Morally speaking, which situation are we in? Do the problems of the new pastures call for *automatic settings* or *manual mode*?

The Tragedy of the Commons is averted by a suite of *automatic settings*—moral emotions that motivate and stabilize cooperation within limited groups. But the Tragedy of Commonsense Morality arises *because* of automatic settings, because different tribes have different automatic settings, causing them to see the world through different moral lenses. The Tragedy of the Commons is a tragedy of selfishness, but the Tragedy of Commonsense Morality is a tragedy of *moral inflexibility*. There is strife on the new pastures not because herders are hopelessly selfish, immoral, or amoral, but because they cannot step outside their respective moral perspectives. How should they think? The answer is now obvious: They should *shift into manual mode*.

But what does that mean? We got a hint in chapter 4. There seems to be a connection between manual-mode thinking and *utilitarian* thinking.** In response to the *footbridge* dilemma and others like it, manual mode advises us to maximize the number of lives saved, while our gut reactions tell us to do the opposite. The parts of the brain that support the utilitarian answer, most notably the DLFPC, are the same parts of the brain that enable us to behave flexibly in other domains—to stick to our diets and to be less racist, for example. And the parts of the brain that work against the utilitarian answer in moral dilemmas, most notably the amygdala and the VMPFC,** are the parts of the brain that inflexibly respond with heightened vigilance to things like the faces of out-group members. This doesn't prove that utilitarian thinking is right, or that un-utilitarian thinking is wrong. As we'll see later, the human manual mode can implement nonutilitarian principles, too. And we wouldn't want to blindly condemn our moral intuitions with "guilt by neural association." Nevertheless, it's a remarkable convergence.

If I'm right, this convergence between what seems like the right moral philosophy (from a certain perspective) and what seems like the right moral psychology (from a certain perspective) is no accident. If I'm right, Bentham and Mill did something fundamentally different from all of their predecessors, both philosophically and psychologically. They transcended the limitations of commonsense morality by turning the problem of morality (almost) entirely over to manual mode. They put aside their inflexible automatic settings and instead asked two very abstract questions. First: What really matters? Second: What is the essence of morality? They concluded that *experience* is what ultimately matters, and that *impartiality* is the essence of morality. Combining these two ideas in manual mode we get utilitarianism: We should maximize the quality of our experience, giving equal weight to the experience of each person. Thus, the original utilitarians took the famously ambiguous Golden Rule**—which captures the idea of impartiality—and gave it teeth by coupling it with a universal moral currency, the currency of experience.

But is this the right currency? And is this really the best philosophy for us? As I've said, utilitarianism is highly controversial. In fact, most experts believe that utilitarianism is deeply flawed. As noted above, it seems to give the wrong answers in some cases: Pushing the man off the footbridge seems wrong, even if doing so will produce better consequences and increase the total amount of happiness. And this is just one of many intuitively compelling objections to utilitarianism. We'll discuss these objections at length in part 4. But first, we'll delve deeper into this idea of common currency. Are there other philosophies that can bridge the gap between Us and Them? Might one of these philosophies be better than utilitarianism? Is there a moral philosophy that's actually correct—the *moral truth*? If so, is it utilitarianism, or something else? In the next two chapters, we'll consider our options (chapter 7) and see why utilitarianism is uniquely suited to serve as the modern world's metamorality (chapter 8).

7.

In Search of
Common Currency

*Democracy demands that the religiously motivated translate their concerns
into universal, rather than religion-specific, values. It requires that their
proposals be subject to argument, and amenable to reason. I may be
opposed to abortion for religious reasons, but if I seek to pass a law banning
the practice, I cannot simply point to the teachings of my church or
[invoke] God's will. I have to explain why abortion violates some principle
that is accessible to people of all faiths, including those with no faith at all.*

—*Barack Obama*

As Obama's remarks suggest, modern herders need a common currency, a universal metric for weighing the values of different tribes. Without a common currency there can be no metamorality, no system for making compromises, trade-offs. Finding a common currency is challenging. Some say impossible.

The most fundamental challenge comes from tribal loyalists. Obama urges religious moral thinkers to translate their concerns into "universal" rather than "religion-specific" values. But what if you firmly believe that your *specific* religion delivers the *universal* moral truth? In that case, the distinction between universal and religion-specific values makes no sense. (Obama is aware of this problem.*) Rick Santorum, a socially conservative senator who sought the Republican presidential nomination in 2012, declared that Obama's position makes him sick to his stomach. "What kind of country do we live [in] that says only people of non-faith can come into the public square and make their case?"* Santorum is overstating. No one's said that religious people can't make their case. Instead, says Obama, they must make their moral cases *in secular terms*. But to many religious moralists, this is like telling a ballerina to dance in one of those padded sumo wrestling suits. Try translating "The gay lifestyle is an abomination against God" into secular terms. No wonder Santorum feels queasy.

Another challenge, mentioned earlier, comes from the proverbial "relativist," the "communitarian," and others who doubt the existence of universal values. According to them, there simply is no universal moral currency, and people who say otherwise are, like religious fundamentalists, simply projecting their tribal values onto the rest of the world. Here the bumper stickers read "All morality is local."

Yet another challenge comes from modern moralists who are bullish about universal secular morality but bearish about the kind I'm advocating. Once again, many contemporary moral thinkers believe that morality is, at its core, about *rights*. They say that the moral truth—the secular and universal moral truth—is fundamentally a truth about who has which rights and which rights take precedence over others. It's not just philosophers who think this way. This is how most of us talk when asked to justify our moral convictions, to make our proposals "subject to argument, and amenable to reason." When we argue about abortion, for example, we talk about a woman's "right to choose" or a fetus's "right to life." We insist that one of these rights outweighs the other, or deny that the other right exists.

Utilitarians, too, can talk about rights and weigh one right against another: The "right to choose" outweighs the "right to life" if preserving the right to choose at the expense of the right to life maximizes happiness. But this isn't how most of us think of rights. Recall the Trolley Problem: Pushing the man off the footbridge may maximize happiness, but it still seems like a gross violation of his rights. Rights, as we ordinarily conceive of them, do not "reduce" to consequences. They *trump* consequences.

If facts about which rights we have (or ought to have) aren't facts about what produces good consequences, then what kind of facts are they? One traditional model for moral facts is *mathematics*: What's the one hundredth prime number? You don't know, but, gosh darn it, you could work it out if you wanted to. Likewise, if we think hard enough about morality, perhaps we can work out the moral facts from first principles. This would give us a different kind of common currency: facts about which rights exist and their relative priorities and weights. On this view, we can work out whether the right to choose outweighs the right to life just as we can work out which number is the one hundredth prime. Of course, no one thinks that moral facts *are* mathematical facts, to be worked out through calculation; rather, the idea is that the moral facts are *like* mathematical facts, abstract truths that we can work out if we think sufficiently hard, objectively, and carefully. For many modern moral thinkers, this is the dream.

Another model for moral facts comes from natural science: Some tribes say that earthquakes are caused by the thrashing of a giant catfish. Other tribes say that earthquakes are caused by the earth's shivering when it's ill. But science tells us that earthquakes are caused by large, platelike pieces of the earth's crust floating on molten rock and rubbing together. The modern scientific understanding of earthquakes isn't just another tribal myth. It's based on evidence—evidence that, with enough time and patience, can be appreciated by members of any tribe. More generally, when it comes to understanding the natural world, science provides a kind of common currency. The modern theory of plate tectonics, for example, is accepted by scientists working on every continent, with diverse cultural backgrounds. With this in mind, one might hope that science will reveal the hidden *essence* of morality and, thus, not only *describe* morality (as we did in chapters 2, 3, and 4), but also *prescribe*. Perhaps science can tell us which rights really exist and how their respective weights compare—a periodic table of the moral elements. That, too, would give us the common currency that we need.

In this chapter, we'll explore our options for finding a moral common currency and corresponding metamorality. There are two ways to think about our search. If we're feeling metaphysically ambitious, we may seek the *moral truth*: universal principles that tell us how we, herders of the new pastures, truly ought to live, what rights and duties we truly have.

With this in mind, we'll start with the three approaches to moral truth outlined above: the religious model, the mathematical model, and the scientific model. I'll explain why none of them is likely to give us what we need. Without a moral truth imposed on us from the outside—by God, by Reason, or by Nature—we'll have to settle for a more modest metamorality, an intertribal system that works for us, whether or not it's the moral truth.** In the next chapter, I'll draw on the dual-process theory of moral psychology presented in part 2 to explain why utilitarianism is uniquely qualified for this job.

DOES OUR COMMON CURRENCY COME FROM GOD?

For many people, there is only one source of universal moral rules, and that is God. There are, however, at least two major problems with appealing to God's moral authority, one concerning the *scope* of God's authority, the other concerning the *accessibility* of God's will.

The first problem goes back to Plato, who questioned the relationship between moral authority and divine will. Translating Plato's question into modern theological idiom, it reads thus: Are bad things bad because God disapproves of them, or does God disapprove of them because they're bad? Take, for example, the case of rape. Rape is bad. God thinks so, and so do we. But could God have said otherwise? Is it within God's power to make rape morally acceptable? If you, like Plato, think the answer is no, then what you're really saying is that God doesn't get to make the moral rules— at least not all of them. Some moral rules, including very important ones, such as rules against rape, are independent of God's will. And if that's right, then we really do need some sort of secular account of why some things are right and other things are wrong. Alternatively, you might think that God really could make the moral rules any way he wants: lifting the ban on rape, for example, or maybe even requiring it. But if God's will is so unconstrained, if he can make anything right or wrong, then in what sense is God's will moral? If the most we can say for God's will is "Because he said so," then it's just an arbitrary set of rules, which happen to be endorsed by a supreme power.

Plato's argument has been around for a while, but it's not put religious morality out of business. For one thing, Plato's argument may not be as compelling today as it was in its original, polytheistic context. The gods of ancient Greece were a rather rambunctious bunch and, for the most part, not exactly models of virtue. With gods like that, it's easy to think that morality is independent of divine will. But a more modern and sophisticated conception of divinity may stand up to Plato's argument. Modern theologians might say that it's impossible to separate morality from God's will. And while it's true that God could not approve of rape, this is not because of some inconvenient limit on God's power. God is an all-encompassing force, outside of space and time, whose actions are not discreet events that may succeed or fail, but rather features of reality that we humans can comprehend only imperfectly with our finite minds. The fact that God could not approve of rape simply reflects the eternal and essential perfection of his will. (Not bad for an atheist, huh?) In short, sophisticated theologians can dismiss Plato's challenge as depending on a simple-minded conception of God. Or so they say. To me, this sounds like a fancy version of "It's a mystery." But we need not resolve this issue here. For us, the more serious problem with divine moral truth is our inability to know what the moral truth is without begging our questions.

Let's grant that God exists and that his will authoritatively defines the moral truth. How can we know God's will? Many Christians say, for example, that gay sex is immoral. How do we know whether they're right? The first stop is the Old Testament, where it states in Leviticus 18:22, "Do not lie with a man as one lies with a woman; it is an abomination." Next stop is Leviticus 20:13: "If a man lies with a man as one lies with a woman, both of them have done what is detestable. They must be put to death; their blood will be on their own heads." These passages certainly require

some interpretation, but let's suppose, as many do, that these are clear denunciations of gay sex. The challenge for inquiring moralists is to figure
out whether to take these denunciations seriously, given that the Old Testament denounces many things that seem perfectly fine to us today and
condones many things that today seem, well, abominable.

The message below highlights this interpretive challenge. It was written as an open letter to Dr. Laura Schlessinger, a conservative commentator and radio host who once cited the Old Testament as her basis for
denouncing homosexuality.

Dear Dr. Laura,

*Thank you for doing so much to educate people regarding God's
Law. I have learned a great deal from your show, and I try to share
that knowledge with as many people as I can. When someone tries to
defend the homosexual lifestyle, for example, I simply remind him that
Leviticus 18:22 clearly states it to be an abomination. End of debate.*

*I do need some advice from you, however, regarding some of the
specific laws and how to best follow them.*

*When I burn a bull on the altar as a sacrifice, I know it creates a
pleasing odour for the Lord (Lev. 1:9). The problem is my neighbors.
They claim the odour is not pleasing to them. Should I smite them?*

*I would like to sell my daughter into slavery, as sanctioned in Exo-
dus 21:7. In this day and age, what do you think would be a fair price
for her?*

*I know that I am allowed no contact with a woman while she is in
her period of menstrual uncleanliness (Lev. 15:19–24). The problem is,
how do I tell? I have tried asking, but most women take offense.*

*Lev. 25:44 states that I may indeed possess slaves, both male and
female, provided they are purchased from neighboring nations. A
friend of mine claims that this applies to Mexicans, but not Canadians.
Can you clarify? Why can't I own Canadians?*

*I have a neighbor who insists on working on the Sabbath. Exodus
35:2 clearly states he should be put to death. Am I morally obligated to
kill him myself?*

A friend of mine feels that even though eating shellfish is an abomination (Lev. 11:10), it is a lesser abomination than homosexuality. I don't agree. Can you settle this?

Lev. 21:20 states that I may not approach the altar of God if I have a defect in my sight. I have to admit that I wear reading glasses. Does my vision have to be 20/20, or is there some wiggle room here?

Most of my male friends get their hair trimmed, including the hair around their temples, even though this is expressly forbidden by Lev. 19:27. How should they die?

I know from Lev. 11:6–8 that touching the skin of a dead pig makes me unclean, but may I still play football if I wear gloves?

My uncle has a farm. He violates Lev. 19:19 by planting two different crops in the same field, as does his wife by wearing garments made of two different kinds of thread (cotton/polyester blend). He also tends to curse and blaspheme a lot. Is it really necessary that we go to all the trouble of getting the whole town together to stone them? (Lev. 24:10–16) Couldn't we just burn them to death at a private family affair like we do with people who sleep with their in-laws? (Lev. 20:14)

I know you have studied these things extensively, so I am confident you can help. Thank you again for reminding us that God's word is eternal and unchanging.

> *Your devoted disciple and adoring fan,*
> *J. Kent Ashcraft*

As you might expect, the argument doesn't end here. Sophisticated biblical interpreters have things to say about which biblical passages offer straightforward moral guidance and which ones do not. But Mr. Ashcraft is certainly correct to point out that a simple appeal to scripture is insufficient for establishing moral truth, even within a single religious tradition. The problems multiply when religious traditions compete for scriptural authority. If we want to find moral truth in scripture, we have to decide which interpretations of which passages in which texts from which religious traditions are truly authoritative. Because people on opposite sides of a religiously fraught moral disagreement are unlikely to agree on which texts, passages, and interpretations are authoritative, appeals to scripture are unlikely to settle all but the most narrow and scholarly moral disagreements.

The same problems apply to alleged moral truths revealed in dreams, waking visions, cosmic signs, and other forms of divine communication. As Obama explains in the same speech quoted above:

> Abraham is ordered by God to offer up his only son, and without argument, he takes Isaac to the mountaintop, binds him to an altar, and raises his knife, prepared to act as God has commanded. . . . But it's fair to say that if any of us leaving this church saw Abraham on a roof of a building raising his knife, we would, at the very least, call the police and expect the Department of Children and Family Services to take Isaac away from Abraham. We would do so because we do not hear what Abraham hears, do not see what Abraham sees, true as those experiences may be. So the best we can do is act in accordance with those things that we all see, and that we all hear, be it common laws or basic reason.

In the end, there may be no argument that can stop tribal loyalists from heeding their tribal calls. No argument will convince Senator Santorum and Dr. Laura that their religious convictions, untranslated into secular terms, are unfit bases for public policy. At most, we can urge moderation, reminding tribal loyalists that they are not acting on "common sense," but rather imposing their tribe's account of moral truth onto others who do not hear what they hear or see what they see.

My purpose here, however, is not to formulate an argument against the existence of divine moral truth. For all I've said, the moral truth may lie with one tribe's interpretation of God's will. Our task here is to find a common currency, and for that, God is not much help. (Because of this, it's perhaps not surprising that reflection discourages belief in God.)

The religions of the world have much in common. They tell us to be kind to our neighbors, not to lie, not to steal, not to make moral exceptions of ourselves. In short, the world's religions enable their adherents to avert the Tragedy of the Commons, to put Us ahead of Me. What religions don't do—most of them, at least*—is help us avert the Tragedy of Common-sense Morality. They exacerbate, rather than ease, conflicts between the values of Us and the values of Them. For our common currency, we must look elsewhere.

IS MORALITY LIKE MATH?

So much for faith. Next stop: *reason*. I'm a big fan of reason. This whole book—indeed, my whole career—is devoted to producing a more reasoned understanding of morality. But there is a rationalist vision of morality that, in my opinion, goes too far. According to the hard-line rationalists, morality is like math: Moral truths are abstract truths that we can work out simply through clear thinking, the way mathematicians work out mathematical truths. Kant, for example, famously claimed that substantive moral truths, such as the wrongness of lying and stealing, could be deduced from principles of "pure practical reasoning." Today, few people explicitly endorse this view. Nevertheless, many of us appear to have something like hard-line Kantian rationalism in mind when we insist that our own moral views, unlike those of our opponents, are backed up by *reason*. Many people state, or imply, that their moral opponents hold views that can't be rationally defended,** the moral equivalents of 2 + 2 = 5.

What would it take for morality to be like math? For it to be thoroughly reasoned? Mathematicians are in the business of proving theorems. All proofs begin with assumptions, and the assumptions of mathematical proofs come from two sources: previously proven theorems and *axioms*. Axioms are mathematical statements that are taken as *self-evidently* true. For example, one of Euclid's axioms for plane geometry is that it's possible to connect any two points with a straight line. Euclid doesn't argue for this claim. He just assumes that it's true and that you, too, can see that this must be correct. Because all theorems derive from previous theorems and axioms, and because theorems do not go back indefinitely, all mathematical truths ultimately follow from axioms, from foundational mathematical truths taken as self-evident.

If morality is like math, then the moral truths to which we appeal in our arguments must ultimately follow from moral axioms, from a manageable set of self-evident moral truths.* The fundamental problem with modeling morality on math is that, after centuries of trying, no one has found a serviceable set of moral axioms, ones that (a) are self-evidently true and (b) can be used to derive substantive moral conclusions, conclusions that settle real-world moral disagreements.*** Now, you may think it's obvious that morality cannot be axiomatized, and that morality is therefore not like math. But it's worth pausing to consider the implications of this obvious fact.

Take the case of abortion, which we'll discuss further in chapter 11. Does a fetus's right to life outweigh a woman's right to choose? We can't settle the matter by appeal to religious doctrine. (Which doctrine? Which interpretation?) And—for now—we're reluctant to think of rights in purely utilitarian terms. According to the pro-lifers, a fetus, like the man on the footbridge, has an absolute right to life. (One that doesn't depend on the net costs and benefits of abortion.) Likewise, the pro-choicers say that a woman has an absolute right to choose. What's going to settle this debate?

Here's what's not going to settle this debate: *Reason*. By *Reason* with a capital *R* I mean *reason alone* or, as Kant would say, "pure practical reason." As noted above, I'm a big fan of reasoning, and of reasoning about moral problems. But if morality isn't like math, then reason alone can't do it. Reason can't tell us which rights we have and which rights take precedence over others. Once again, this is because all reasoning requires *premises*. If Joe's premises are not self-evidently true, and Jane dislikes the conclusions that follow from Joe's premises, then Jane is free to simply reject one or more of Joe's premises, and with it Joe's conclusions.

My point is this: Without self-evident premises, pure reasoning doesn't answer our questions. What reason can do is force us to achieve greater *consistency* among our factual and moral beliefs, and that's important. (More on this in chapter 11.) But moral reasoning can't tell you what to think about abortion in the way that mathematics can tell you what to think about 439,569 > 3 x 17 x 13. That's because mathematics begins with a small number of shared, self-evident assumptions, while morality begins with a huge number of interconnected assumptions, largely unquestioned, all of which sound reasonable to the assumption maker and precious few of which are truly self-evident. (In other words, moral epistemology is *coherentist* rather than *foundationalist*.)

As a brainy person, you might hope that Reason can cut clean through the morass of competing human values. Alas, it can't. People often talk as if this were possible: "My views, unlike yours, are grounded in *Reason*." But this is half-true at best. Reasoning, once again, can make our moral opinions more consistent, both within ourselves and, I believe, across tribes. And, as I will explain in the next chapter, our shared capacity for reasoning plays a critical role in my argument for a utilitarian metamorality. But reason *by itself* doesn't tell us how to make trade-offs among the competing values of different moral tribes, and it doesn't tell us which rights we have or how people's competing rights weigh against one another. For our common currency, we'll have to look elsewhere—again.

DOES SCIENCE DELIVER THE MORAL TRUTH?

If neither religion nor pure reasoning can settle our moral disagreements, then perhaps we should consult our favorite source of objective, unbiased facts: *Science.* Perhaps science can supply us with moral premises that, while not self-evident, are still *evident*, backed up by things we've discovered about ourselves and the world around us.

In chapter 1, I summarized a general scientific theory of morality, reflecting a consensus that's been building since Darwin:

> Morality is a set of psychological adaptations that allow otherwise selfish individuals to reap the benefits of cooperation.

Suppose that this theory is correct. (And, as I've said, it's really the only game in town.) Does our understanding of morality's natural function give us any purchase on the problem of moral truth? If the function of morality is to promote cooperation, then why can't we say that the moral truth is the metamorality that best promotes cooperation? It's an intriguing idea, but it has some serious problems.

If we're right, morality evolved to promote cooperation, but that's not the whole story. Once again, morality evolved (biologically) to promote cooperation *within groups* for the sake of *competition between groups*.* The only reason that natural selection would favor genes that promote cooperation is that cooperative individuals are better able to outcompete others. This highlights a more general point about the function of morality, which is that its ultimate function, like that of all biological adaptations, is to spread genetic material. Evolution is not aimed at promoting cooperation per se. It promotes cooperation only insofar as cooperation helps propagate the genes of the cooperators. Evolution might favor people who are nice to their neighbors, but it might also favor people with genocidal tendencies, and for the same underlying reason. Thus, if you're looking to evolution for moral truth, you're barking up the wrong tree. (Note that this argument, which is controversial, also applies to cultural evolution.**)

The problem with looking to evolution for moral truth exemplifies a more general problem known as the "is-ought" problem, sometimes referred to (somewhat incorrectly) as the "naturalistic fallacy."* The fallacy is to identify that which is natural with that which is right or good. This fallacy was most famously committed by so-called social Darwinists, who saw the ruthless competitiveness of nature—weeding out the weak, promoting the strong—as a model for human society. Today we know that natural selection can promote nice behavior as well as nasty behavior (see chapter 2), and thus, looking toward evolutionary theory for insights into moral truth doesn't sound quite so fascist. But to say that an action is right because it's consistent with the evolved function of morality, or wrong because it's inconsistent with the same, is still fallacious. It simply doesn't follow that something is good because it's doing what it evolved to do.

Looking for moral truth in evolutionary function is, in fact, just a version of the idea that morality is like math. Behind the evolutionary approach to moral truth is a hidden moral axiom: What's right is what best fulfills the purpose for which morality evolved. This axiom is not self-evidently true. Nor is it supported by scientific evidence. It's just an assumption. To see how dubious this assumption is, imagine trying to settle an argument with it. Suppose it turns out that a pro-life abortion policy will help us spread our genes. If you're pro-choice, is this going to change your mind? It won't, and it shouldn't. You believe in a woman's right to choose, not in the spreading of human genetic material. Nor should pro-lifers change their minds if the evolutionary facts were to go the other way. (Here's something pro-choicers and pro-lifers can chant together outside the clinic: "Evolution, schmevolution! What's right is not necessarily that which spreads our genes!")

Still, you might think that there's something to this idea of finding moral truth in the natural function of morality. Let's see if we can make this work. We'll dispense with the not especially moral goal of spreading genes and focus instead on the more proximate goal of cooperation. Is cooperation the ultimate moral good? And can we *assume* without evidence or further argument that it's the ultimate moral good? Consider the interstellar menace known as the Borg, from the later *Star Trek* serieses. For the uninitiated, the Borg is not a Swedish tennis champion but rather a collection of former humans and humanoid aliens who have been "assimilated" into a vast collective of cybernetic (part machine, part biological) drones. The Borg function like a superintelligent, high-tech ant colony, swallowing up other life forms and incorporating them into their roving hive. Two points about the Borg: First, they are as cooperative as can be. Second, being in the Borg doesn't seem like much fun. And yet, if cooperation were the ultimate moral good, then the triumph of the Borg would be the best possible ending for life in the universe. And this would be true even if Borg membership were completely miserable, so long as it is highly cooperative.

Thus, as the Borg remind us, it's implausible to think that cooperation is the ultimate moral good. But if cooperation is not what ultimately matters, then what does ultimately matter? One might think that cooperation is valuable not as an end it itself, but because of the benefits it yields and, ultimately, because of the happiness and relief from suffering that cooperation brings. Now *that* sounds like a splendid idea.

But notice how far from evolutionary theory we've come. Through a series of tweaks, we've turned our evolutionary theory of moral truth into a moral theory that predates Darwin. In other words, by the time we've sufficiently cleaned up our evolutionary theory of moral truth, it's ceased to be evolutionary. Rather than starting with evolution's "values" and then modifying them to suit us, we might as well just start with our own values and seek our common currency there.

PLAN B: IN SEARCH OF SHARED VALUES

If there were a God whose will we could discern without begging our questions, or if we could deduce substantive moral truths from self-evident first principles, or if we could discover moral truths the way we've discovered the causes of earthquakes, we'd be in good shape. But instead we're thrown back on the morass of competing moral values. (Henceforth I'll call it "the morass," for short.)

Does this mean that there's no moral truth? I remain agnostic. Once upon a time, I thought this was *the* question, but I've since changed my mind. What really matters is whether we have direct, reliable, non-question-begging access to the moral truth—a clear path through the morass—not whether moral truth exists. For the reasons given above, I'm confident that we don't have this kind of access. (If there are authoritative ways to resolve moral disagreements that don't rely on divine revelation, pure reasoning, or empirical investigation, I've not heard of them.) Once we've resigned ourselves to working with the morass, the question of moral truth loses its practical importance.

(In a nutshell, the problem of moral truth becomes a question about how to describe what we're left with once we've subjected our moral beliefs to as much objective improvement as possible. Do we call what's left "the moral truth"? Or do we just call it "what's left"? I no longer think that this question has a clear answer, but I also no longer think that we have to answer it.***)

Resigned to the morass, we've no choice but to capitalize on the values we share and seek our common currency there. Identifying our shared values is harder than it seems, however, because words—pretty ones especially—can be misleading. For example, two families may both value "family," but this need not be a source of moral agreement. If the issue is promoting family-friendly policies in the workplace, then the shared value of "family" may be a source of agreement between our two families. But if the issue is "what your kid did to my kid," the fact that both families value "family" might make things worse. Thus, moral abstractions such as "family" can provide an illusion of shared values. And, as I'll explain later, the same goes for values such as "freedom," "equality," "life," "justice," "fairness," "human rights," and so on. Identifying the values that we truly share is harder than it looks, because deep moral differences can be cloaked in shared moral rhetoric. What, then, is our true moral common ground?

As you now well know, I believe that the values behind utilitarianism are our true common ground. Once again, we herders are united by our capacity for positive and negative *experience*, for happiness and suffering, and by our recognition that morality must, at the highest level,** be *impartial*. Put them together and our task, insofar as we're moral, is to make the world as happy as possible, giving equal weight to everyone's happiness.

I do not claim, however, that utilitarianism is the moral truth. Nor do I claim, more specifically, and as some readers might expect me to, that science proves that utilitarianism is the moral truth. Instead, I claim that utilitarianism becomes uniquely attractive once our moral thinking has been *objectively improved* by a scientific understanding of morality. (Whether this makes it the "moral truth" I leave as an open question.*) Although we may not be able to establish utilitarianism as the moral truth, I believe that we can nevertheless use twenty-first-century science to vindicate nineteenth-century moral philosophy against its twentieth-century critics.

In the next chapter, I'll make the case that utilitarianism is indeed based on shared values. We'll consider utilitarianism from a psychological, neural, and evolutionary perspective: What is it? And why do utilitarianism's values make for such excellent common currency?

8.

Common Currency Found

In a provocative episode of *The Twilight Zone*, a couple receives a tempting offer. A mysterious stranger delivers a small box with a button on top. He explains that if one of them pushes the button, two things will happen: They will receive $200,000, which they desperately need, and someone they don't know will die. After much moral agonizing and rationalizing, one of them pushes the button. The mysterious man returns to deliver the money. He explains—spoiler alert!—that he will now give the box, on the same terms, to another a person, someone "you don't know."

Most of us, we hope, would not push the button, but some people certainly would. As a way of delineating our shared values, let us think about the buttons that we would and would not push. We'll start with some nonmoral buttons and work our way up to moral ones.

Question 1: The Happiness Button. Next week, you will accidentally trip on an uneven sidewalk and break your kneecap. This will be extremely painful and will significantly reduce your happiness for several months. However, if you press this button, a little bit of magic will make you more attentive as you're walking along, and you won't break your kneecap. Will you push? Of course you will. This tells us something rather obvious: *If all else is equal,** people prefer being more happy to being less happy. Next question.

Question 2: The Net Happiness Button. Here, too, you're headed for a broken kneecap, which you can avoid by pressing the button. But in this case, pressing the button will also cause a mosquito to bite your arm, giving you a mildly irritating itch for a couple of days. Will you press the button? Of course you will. The irritation of the mosquito bite is well worth avoiding a broken kneecap. Lesson: We're all willing to make *trade-offs* in which we accept a loss in happiness in return for a greater gain in happiness. More generally, *if all else is equal*, we prefer more *net* happiness to less net happiness.

Question 3: The Other's Happiness Button. This is like Question 1, except that now we're in moral territory. Instead of sparing yourself a broken kneecap, pushing the button will spare someone else from this hardship. Will you push? Certainly you'll push if it's someone you know and like, or a member of a tribe with which you identify. And maybe you won't push if it's someone you dislike. But let's suppose that it's just someone "you don't know." My assumption is that you would push.

Would all people do this? Unfortunately, probably not. Some people are psychopaths who care for others not at all. And within the normal population, people exhibit varying degrees of altruism, indifference, and antipathy toward strangers (see chapter 2). But let us bear in mind the context of our question. We're looking for a metamorality based on shared values. For our purposes, shared values need not be perfectly universal. They just need to be shared widely, shared by members of different tribes whose disagreements we might hope to resolve by appeal to a common moral standard. If you're so selfish that you're not willing to lift a finger to spare another human from serious suffering, then you're simply not part of this conversation.

You're not part of the "we" who are interested in answering our question. With this in mind, *we* can say this: *If all else is equal*, we prefer that others be more happy rather than less happy. What's more, I think we can assume that we care about other people's *net* happiness: You would still push the button even if sparing the stranger a broken kneecap also means giving her a mosquito bite.

Question 4: The More People's Happiness Button. Now there are two buttons. Button A will spare one person from a broken kneecap, while button B will spare ten people from the same fate. Would you push A or B? Or would you flip a coin? My assumption is that you would push button B. Lesson: *If all else is equal*, we prefer increasing more people's happiness to increasing fewer people's happiness.

Question 5: The Utilitarian Button. Now for our final question. Button A will spare two people from mosquito bites. Button B will spare one person from a broken kneecap. A, B, or flip a coin? My assumption is that you would push button B and thus prevent the greater hardship, even though button A is better for more people. Lesson: *If all else is equal*, we prefer more total happiness across people to less total happiness across people.

Perhaps you've found these questions boring. If so, that's good. We're laying the foundation for a utilitarian metamorality, and the more obvious the answers to these questions are, the more solid our foundation. What we've established, first, is that *if all else is equal,* we prefer more happiness to less happiness, not only for ourselves but for others. Second, we've established that, when it comes to others, we care not only about the amount of happiness within individuals but also about the number of individuals affected. And finally, we've established that we care about the sum of happiness across individuals, taking into account both the amount of happiness for each person and the number of people affected. *If all else is equal,* we prefer to increase the total amount of happiness across people.

Once again, the "we" here is not every human who has every lived. What matters is that this "we" is very large and very diverse, including members of all tribes. I hereby conjecture that there is not a tribe in the world whose members cannot feel the pull of the utilitarian thinking described above, and in what follows I'll explain why. (Experimental anthropologists are encouraged to go out and check.**) If that's right—if we all have (or can be made to have) an "if all else is equal" commitment to increasing happiness—this has profound moral implications. It means that we have some very substantial moral common ground.**

Of course, what we have is only a *default* commitment to maximizing happiness, reflected in the qualifier *"if all else is equal."* It's this qualifier that makes the questions above so boring. Suppose that the couple from *The Twilight Zone* had been given this option: Push the button and one stranger dies, while a different stranger gets $200,000. This would have been a very boring episode. That's because, in this version, *all else is equal.* It's not a question of a big gain for *Us* versus an even bigger loss for one of *Them.* In the boring version, there's no salient difference between the winner and the loser, and, thus, impartial happiness maximization prevails in a predictable, boring way.

Now consider this choice: If you choose button A, five people live and one person dies. If you choose button B, five people die and one person lives. Which do you choose? *If all else is equal,* you'd choose button A, of course. But suppose that we're actually talking about the *footbridge* dilemma. (Or the *transplant* dilemma.) Now, all else is *no longer equal.* With these details filled in, button A (pushing the man off the footbridge) seems wrong.

As this example suggests, a lot hangs on the qualifier *"if all else is equal."* If we keep this qualifier, our commitment to maximizing happiness is perfectly agreeable, but also toothless. We're all happy to maximize happiness, as long as maximizing happiness doesn't conflict with something else that we care about. If we drop the *"if all else is equal"* qualifier, we get utilitarianism. We get a complete moral system, a metamorality that can (given enough factual information) resolve any moral disagreement. But when we drop the qualifier, what we gain in "toothfulness" we lose in agreeability. Maximizing happiness means (or could in principle mean) doing things that seem morally wrong, such as pushing people in front of trolleys.

Our question now is this: Should we drop the *"if all else is equal"* clause and simply aim for maximum happiness? Or is this just too simple? We'll tackle this question in part 4. But first, let's take a closer look at the part of the moral brain that thinks maximizing happiness is splendid. Let's try to understand the evolutionary and cognitive mechanics behind our utilitarian common ground.

WHAT IS UTILITARIANISM?

We already know the philosophy textbook's answer, but that's not the whole story. In fact, it's just the philosophical tip of a much deeper psychological and biological iceberg.

We're not all utilitarians—far from it. But we all "get" utilitarianism. We all understand why maximizing happiness is, at least on the face of it, a reasonable thing to do. Why do we all get it? Why is there a systematic moral philosophy that makes sense to everyone? And why does this universally gettable philosophy also manage, at one point or another, to offend everyone's moral sensibilities? It's as if there's one part of our brain that thinks utilitarianism makes perfect sense and other parts of our brain that are horribly offended by it. This should sound familiar.

Understanding utilitarianism requires the dual-process framework laid out in part 2. If I'm right, utilitarianism is the native philosophy of the human *manual mode*, and all of the objections to utilitarianism are ultimately driven by *automatic settings*. Utilitarianism makes sense to everybody because all humans have more or less the same manual-mode machinery. This is why utilitarianism is uniquely suited to serve as our metamorality, and why it gives us an invaluable common currency.

Our automatic settings, in contrast, are far less uniform. As explained in chapter 2, we herders all have the same *kinds* of automatic settings, the same moral emotions, which include feelings of empathy, anger, disgust, guilt, shame, and discomfort with certain forms of personal violence. But the precise triggers of our moral emotions vary from tribe to tribe, and from person to person. Despite this variety, the automatic settings of all human tribes have something in common: No one's gut reactions are consistently utilitarian. Thus, our dual-process moral brains make utilitarianism seem partially, but not completely, right to everyone. We all *get* utilitarianism because we all have the same manual mode, and we're all *offended by* utilitarianism because we all have nonutilitarian automatic settings. Why is this?

Based on what we learned in chapter 2 ("Moral Machinery"), it's no surprise that our automatic settings are nonutilitarian. Our moral brains evolved to help us spread our genes, not to maximize our collective happiness. More specifically, our moral machinery evolved to strike a biologically advantageous balance between selfishness (Me) and within-group cooperation (Us), without concern for people who are more likely to be competitors than allies (Them). Thus, we should expect our moral intuitions to be, on the whole, more selfish and more tribalistic than utilitarianism prescribes. But even if our brains had evolved to maximize collective happiness, automatic settings are, by their very nature, blunt instruments—efficient but inflexible. Thus, it would be very surprising if our automatic settings were to consistently guide us toward the maximization of happiness, even if that were their biological purpose, which it isn't. Our automatic settings are not, and could not be expected to be, utilitarian.

Our automatic settings are not utilitarian, but, if I'm right, our manual modes are. Why is that? First, let's recall from chapter 5 ("Efficiency, Flexibility, and the Dual-Process Brain") what the human manual mode is and why we have it in the first place. Consider once again the problem of deciding when and what to eat. Eating is generally good for animals, and that's why we have automatic settings that incline us to eat. But of course, sometimes it's better, all things considered, not to eat, depending on the situation. Once again, you might be a hunter tracking a large animal, with no time to stop for berries, yummy though they are. Manual mode is what gives us the flexibility to ignore our automatic settings (*Berries! Yum!*) and instead do what best serves our long-term goals (*Mastodon! A week's food for the whole village!*). Manual mode also gives us the ability to make big plans, to envision possibilities that are not automatically suggested by whatever happens to be in front of us. I want to emphasize that what I'm calling the "manual mode" is not an abstract thing. It is, once again, a set of neural networks, based primarily in the prefrontal cortex (PFC), that enables humans to engage in conscious and controlled reasoning and planning. It's what enables us, unlike spiders, to solve complex, novel problems.

But what exactly is problem solving? In the lingo of artificial intelligence, solving a behavioral problem is about realizing a *goal state*. A problem solver begins with an idea (a representation) of how the world could be and then operates (behaves) on the world so as to make the world that way. One very simple kind of problem solver is a thermostat. It has a goal state (the desired temperature), and it has mechanisms that allow it to operate on the world so as to realize its goal state. A thermostat is fairly flexible. It can heat things up or cool things down. It can heat or cool for just the right amounts of time, adjusting its behavior according to fluctuating temperature conditions. But as problem-solving systems go, a thermostat is still very simple and inflexible. You can easily fool a thermostat by putting something hot or cold on its sensor, for example.

Problem-solving systems vary widely, but at the most abstract level, they all share certain properties. First, they all deal with *consequences*. A goal state is a consequence, one that may be actual or merely desired. All problem solvers perform *actions*, where actions are selected based on their *causal relationships* to desired and undesired consequences. Thus, a thermostat may turn up the heat (an action) because turning up the heat will *cause* a desired consequence (the room's being seventy-two degrees).

A thermostat is a simple problem-solving system for several reasons. At any given time, it has only one goal: to make the actual temperature equal the desired temperature. Likewise, it has only one "belief" about the current state of the world: its representation of the current temperature. A thermostat can execute only four different actions (engage or disengage the furnace, engage or disengage the air conditioner), and it "knows" of only two causal relationships: (1) Engaging the furnace raises the temperature. (2) Engaging the air conditioner lowers the temperature. Finally, a thermostat can perform some internal computations, most notably to determine whether the current temperature is higher than, lower than, or equal to the desired temperature.

As simple as a thermostat is, there are useful devices that are even simpler. A motion sensor, for example, has only one "belief" (concerning whether or not something is moving) and one action (*Hey! Something's moving!*). A motion sensor can contribute to solving a problem (securing the art in the museum), but a motion sensor is not itself a problem solver. It doesn't represent a goal state, and it does not operate on the world until it senses that a goal state has been realized. A motion sensor just flips on, or doesn't, depending on what it senses. It's a reflex system, an automatic setting.

The human manual mode is, needless to say, much more complicated than a thermostat. A single human has many different goals, many different beliefs about the current state of the world, many different available actions, and many different general beliefs about causal relationships between actions and events in the world. Nevertheless, human problem solving, at the most abstract level, shares the basic "ontology" of a thermostat: consequences, actions, beliefs about the current state of the world, and general beliefs about how the world works—that is, about the causal relationships between possible actions and consequences. To see how this works, let's consider a problem that is, at present, solvable only by a human brain.

Suppose that I offer you $10,000 in return for meeting me at the post office in Oxford, Maine, next Friday at noon. Prior to receiving my offer, your brain encodes three different things: a set of standing goals, a set of possible actions that you can perform, and an elaborate model of how the world works. In comes my offer, and your brain goes to work. Using your model of the world, you infer that, by coming to possess an additional $10,000, you will more easily achieve some of your existing goals. Your model of how the world works also tells you that my offer is credible, that doing as I ask will indeed cause you to acquire an additional $10,000. Thus, your PFC transfers the value of acquiring $10,000 (minus anticipated costs) to the consequence of arriving in Maine at the appointed time and place.

Having established your new goal, you next rely on your understanding of causal relationships to construct a plan of action. To cause the relocation of your body to Oxford, Maine, you'll need a car, and perhaps some plane tickets. This may involve turning on your computer and pressing a sequence of buttons, visiting an auto mechanic, or imploring your sister to lend you her car. The particular sequence of body movements that you perform will be very elaborate and highly situation-specific. No set of instincts, no collection of automatic settings, could pull this off, because no creature in the history of our planet—not even you—has had trial-and-error experience getting you from your house to the post office in Oxford, Maine, with your unique set of logistical constraints. And that's why you need a fundamentally different kind of cognitive system to pull this off, a general-purpose action planner that can adopt an arbitrary goal and use a complex model of the world to map out a sequence of more specific actions that will realize that goal. That, in a nutshell, is what the human PFC does.

So why is the human manual mode utilitarian? I don't think that it's *inherently* utilitarian. Rather, I think that utilitarianism is the philosophy that the human manual mode is predisposed to adopt, once it's shopping for a moral philosophy. So let's rephrase: Why is the manual mode predisposed toward utilitarianism? The manual mode's job is, once again, to realize goal states, to produce desired *consequences*. For a simple problem solver, such as a thermostat, there's not much to the evaluation of consequences. All that matters to a thermostat is whether the current temperature is too hot, too cold, or just right. Critically, a thermostat doesn't face any *trade-offs*. For a thermostat, there's no such thing as a temperature that's good in some ways but bad in other ways. Human decision making, in contrast, is all about trade-offs.

Like a thermostat, your PFC selects actions that bring about desired consequences. But your PFC doesn't simply propel you into action as soon as it traces out a behavioral pathway from where you are to where you want to be. If it were to do that, you might spend $600 on a glass of iced tea. The PFC takes into account both the value of the goal and the costs of attaining it. But that is not enough for adaptive behavior, either. If you're very thirsty, you might be willing to pay $8 for a glass of iced tea. But it would be foolish to do so if you could buy the same drink for $2 next door. Thus, your PFC not only compares the benefits of doing X to the costs of doing X. It also compares the net costs/benefits of doing X to the net costs/benefits of doing Y. But even *that's* not enough for adaptive behavior. Suppose that the $8 iced tea comes with a free sandwich. Or suppose that the local crime boss strongly prefers that you buy iced tea from his cousin's store instead of the cheaper shop next door. Not only do you have to consider the direct costs and benefits associated with buying iced tea at both locations; you also have to consider the costs and benefits associated with any *side effects* of choosing one option over the other. And things get even more complicated when we acknowledge that outcomes in the real world are uncertain.

Thus, your general-purpose action planner is, by necessity, a very complex device that thinks not only in terms of consequences but also in terms of the trade-offs involved in choosing one action over another, based on their expected consequences, including side effects. In other words, the human manual mode is designed to produce *optimal consequences*, where "optimal" is defined by the decision maker's ultimate goals, taking into account the expected effects of one's actions, which include both intended effects and foreseeable side effects. (Of course, people's decisions are not always optimal, and can often deviate from optimality in systematic ways. However it's usually our automatic settings that lead to these systematic errors, and it's our manual-mode thinking that enables us to recognize these errors as errors.) From here, getting to utilitarianism is a two-step process, corresponding to utilitarianism's two essential ingredients.

FROM GENERAL RATIONALITY TO UTILITARIAN MORALITY

The human manual mode, housed in the PFC, is a general-purpose problem solver, an optimizer of consequences. But what counts as optimal? This question can be broken down into two questions. First: Optimal for whom? Second: What counts as optimal for a given person? Let's start with the first question.

Suppose that you're completely selfish, and suppose that you and nine other selfish people have stumbled upon something valuable and fungible—say, a chest containing one thousand identical gold coins. And suppose that you're all equally skilled fighters. No one has an advantage. You would like all of the gold for yourself, of course. What should you do? You could start fighting, attempting to disable as many of your competitors as possible. But if you do that, your competitors will fight back, and others may start fighting, too. By starting a fight, you could end up with a large share of the coins, but you could also end up with no coins, severely injured, or even dead.

There is, of course, an obvious solution: Everyone divides up the coins equally and goes their separate ways. Why equally? Because if the division is unequal, that gives the people who got smaller shares a reason to fight: If it's possible to get a larger share—and it clearly is—why not fight for it, or threaten to fight for it? If there are no power asymmetries, an equal division is the only stable solution. In other words, what we would call a "fair" distribution of resources naturally emerges among people—even people who don't care about "fairness"—when there is no power imbalance. This is one way to get utilitarianism's first essential ingredient, *impartiality.*

Here's another way to get impartiality, traced out by Peter Singer in his book *The Expanding Circle.* People are not naturally impartial. People care most of all about themselves, their family members, their friends, and other in-group members. People, for the most part, don't care very much about complete strangers. But at the same time, people may come to appreciate the following fact: Other people are, more or less, just like them. They, too, care most of all about themselves, their family members, their friends, and so on. Eventually, people may make a cognitive leap, or a set of cognitive leaps, culminating in a thought like this: "To me, I'm special. But other people see themselves as special just as I do. Therefore, I'm not really special, because even if I'm special, I'm not especially special. There is nothing that makes my interests objectively more important than the interests of others."

Of course, this recognition does not, by itself, entail a commitment to impartiality. The ten rogues with the gold coins may understand that their positions are symmetrical and yet still remain rogues. In other words, recognizing that there is no *objective* reason to favor oneself over others does not entail abandoning one's *subjective* reasons for favoring oneself.* But it seems that, somehow, we do manage to translate this intellectual insight into a preference, however weak, for genuine impartiality. I suspect that this translation has something to do with *empathy*, the ability to feel what others feel. Human empathy is fickle and limited, but our capacity for empathy may provide an emotional seed that, when watered by reasoning, flowers into the ideal of impartial morality.

To be perfectly honest, I don't know how the ideal of impartiality took hold in human brains, but I'm fairly confident of two things. First, the ideal of impartiality has taken hold in us (we who are in on this conversation) not as an overriding ideal but as one that we can appreciate. None of us lives perfectly by the Golden Rule, but we all at least "get" it. Second, I'm confident that the moral ideal of impartiality is a manual-mode phenomenon. This ideal almost certainly has origins in automatic settings, in feelings of concern for others, but our moral emotions are themselves nowhere near impartial. Only a creature with a manual mode can grasp the ideal of impartiality. As Adam Smith observed in the eighteenth century, the thought of losing your little finger tomorrow would keep you up all night, but you might sleep soundly knowing that tomorrow thousands of faraway people will die in an earthquake. And yet—and this is Smith's point—we recognize that thousands dying in an earthquake is far worse than losing one's finger, and that it would be monstrous to choose one's finger over the lives of thousands of innocent people. This kind of moral thinking requires manual mode.*

Now, you might wonder how feelings, such as feelings of empathy, can get translated into a motivating abstract ideal. I wonder, too. But in any case, this kind of process may be more familiar than it sounds. Consider, for example, the familiar contrast between shopping for food while hungry and shopping while full. Human food acquisition decisions *can* be motivated directly by automatic settings, as they are in other animals. But it's also possible to acquire food while completely stuffed, at a time when the thought of eating anything, even Nutella, is thoroughly unappealing. Under such conditions, you'll make somewhat different choices, maybe better choices, but in any case the job can get done. How is this possible? Certainly, the shopping decisions that you make while stuffed are related to your automatic settings, to your brute appetites. Even while stuffed, you tend to buy things that you like and avoid things that you don't like. But at the same time, shopping while stuffed means that you can't rely directly on your brute appetites. Instead, the "hot" preferences generated by your automatic settings are translated into "cool" cognitions that can be represented more matter-of-factly in manual mode. When you're stuffed, you know that you need to buy next week's Nutella in the same way that you know that Tallahassee is the capital of Florida.

Your shopping decisions can be distanced from your brute appetites in more complex ways. You may shop for other people, substituting their preferences for your own. In shopping for other people, your decisions will be affected not just by what you and others like, but by mathematical calculations concerning the number of people you need to feed. And likewise, when you shop for yourself, you must consider the time frame for which you're shopping. (Do I need a week's worth of Nutella or a year's supply?) Somehow, the human brain can take values that originate with automatic settings and translate them into motivational states that are susceptible to the influence of explicit reasoning and quantitative manipulation. We don't know exactly how it works, but it clearly happens.

L et's recap: First, the human manual mode is, by nature, a cost-benefit reasoning system that aims for optimal consequences. Second, the human manual mode is susceptible to the ideal of impartiality. And, I submit, this susceptibility is not tribe-specific. Members of any tribe can get the idea behind the Golden Rule. Put these two things together and we get manual modes that aspire, however imperfectly, to produce consequences that are optimal from an impartial perspective, giving equal weight to all people.

Now we face our second question: What counts as optimal for a given person? What makes a consequence good or bad for you, or me, or anyone else? We attempted to answer this question back in chapter 6 ("A Splendid Idea") by playing an iterated game of "Why do you care about that?" For example, most people care about money. But what good is money? It allows you to buy things, things like Nutella and nifty gadgets. But why do you want those things? Once again, a natural thought is that, if you trace these chains of value all the way to the end, you'll find a concern for the quality of experience—for happiness, broadly construed, whether it's your happiness or someone else's. As I've said, this conclusion is not inevitable, but it's certainly a natural conclusion. Moreover, it's a conclusion that anyone can "get." Whether or not *all* of our value chains end with happiness, it's certainly true that all of us have *a lot* of value chains that end with happiness. We all do things simply because we enjoy them, and we all avoid things simply because we don't. In other words, we all place *intrinsic value* on our own happiness and that of at least some other people, as demonstrated by the buttons we're willing to push. And this, too, is a manual-mode phenomenon, even if it begins with automatic settings. We all consciously endorse the idea that happiness is intrinsically valuable. None of us says, "Increase someone's happiness? Why would you want to do *that*?"

Utilitarianism can be summarized in three words: *Maximize happiness impartially*. The "maximize" part comes from the human manual mode, which is, by nature, a device for maximizing. This, I claim, is universal—standard issue in every healthy human brain. The "happiness" part comes from reflecting on what really matters to us as individuals. Happiness—yours and that of others—might not be the only thing that you value intrinsically, as an end in itself, but it's certainly one of the primary things that you value intrinsically. This, too, I claim, is universal, or near enough. Everyone "gets" that happiness matters, and everyone can, with a little reflection, see that happiness lies behind many of the other things we value, if not all of them. Finally, the ideal of "impartially" comes from an intellectual recognition of some kind. It may come from recognizing that impartial solutions are often stable. Or it may come from making a kind of moral-cognitive leap, when empathy collides with the recognition that no one is objectively special. None of us is truly impartial, but everyone feels the pull of impartiality as a moral ideal. This, too, is universal, or near enough.

Thus, if I'm right, utilitarianism is special, and Bentham and Mill did something unprecedented in intellectual history. They wrested moral philosophy away from the automatic settings, away from the limitations of our biological and cultural histories, and turned it over, almost entirely, to the brain's general-purpose problem-solving system. The manual mode doesn't come with a moral philosophy, but it can create one if it's seeded with two universally accessible moral values: happiness and impartiality. This combination yields a complete moral system that is accessible to members of all tribes. This gives us a pathway out of the morass, a system for transcending our incompatible visions of the moral truth. Utilitarianism may not be the moral truth, but it is, I think, the metamorality that we're looking for.

The vast majority of experts, however, strongly disagree. Most moral philosophers think of utilitarianism as a quaint relic of the nineteenth century. Utilitarianism, they say, is far too simple. It captures something important about morality, for sure, but in its imperialism, in its reduction of right and wrong to a one-line formula, they say, utilitarianism goes terribly, horribly wrong.

WHAT'S WRONG WITH UTILITARIANISM?

We've encountered one potent objection already: Sometimes the action that produces the best consequences (measured in terms of happiness or otherwise) seems just plain wrong. Here the classic example is the *footbridge* dilemma, in which using someone as a human trolley-stopper can promote the greater good.

As noted earlier, you can attempt to wiggle out of this problem by rejecting the assumptions of the dilemma: Maybe pushing won't work. Maybe it will set a very bad precedent. And so on. If this is what you're thinking, you're certainly on to something. There really is something fishy and unrealistic about the *footbridge* case. But please resist this temptation for now, because the *footbridge* case is meant to illustrate a broader point that we ought to take seriously: Sometimes doing what produces the best consequences seems very clearly wrong. Assuming that this is at least sometimes true, what does it tell us?

Many moral philosophers think that the *footbridge* dilemma highlights a fundamental flaw in utilitarian thinking. Once again, the most common complaint about utilitarianism is that it undervalues people's *rights*. Many critics say that it's simply wrong to use someone as a human trolley-stopper, and this is true even if doing so will produce better consequences. This is a pretty compelling argument, and it doesn't end there.

According to John Rawls, utilitarianism's most influential critic, utilitarianism is a poor principle for organizing a society, for much the same reason that it's a poor principle for deciding whether or not to push. Bentham and Mill were among the first to oppose slavery, and that's to their credit. But according to Rawls, they're not opposed *enough*. Utilitarians oppose slavery because they believe that, as a matter of fact, slavery greatly reduces the overall sum of happiness. But, Rawls asks, what if slavery were to maximize happiness? Then would slavery be right? Suppose that 90 percent of us could increase our happiness by enslaving the remaining 10 percent. And suppose that the gains in happiness to the slaveholding 90 percent are large enough to offset the losses to those enslaved. Utilitarianism, it seems, would endorse this gross injustice, just as it endorses pushing the man off the footbridge for the sake of the other five. Not so splendid, it seems.

Along similar lines, utilitarianism endorses what some regard as gross miscarriages of criminal justice. Recall the Magistrates and the Mob case, from chapter 3: What if the only way to prevent a violent riot is to frame and convict an innocent person? Many people—Americans more than Chinese people, apparently—think that doing this would be horribly wrong. But a utilitarian would say that it might well be the best thing to do, depending on the details.

And it gets worse. In the cases described above, utilitarianism seems too morally lax, allowing us to trample on other people's rights. In other cases, however, utilitarianism seems too morally *demanding*, trampling on our own rights. And these utilitarian demands are not hypothetical. In fact, you're almost certainly living in one of these cases right now.

As you read this, there are millions of people who desperately need food, water, and medicine. Many more lack access to education, protection from persecution, political representation, and other important things that affluent people take for granted. For example, as I write, Oxfam America, a highly regarded international aid organization, is providing clean water, food, sanitation, and other forms of economic support to more than 300,000 civilians caught in the conflict in the Darfur region of Sudan. A small donation to Oxfam America—less than $100—can make a big difference to one of these people. You often hear that you can save someone's life for "just a few dollars." According to GiveWell, an organization founded by financial analysts that evaluates the cost-effectiveness of charities, such estimates dramatically underestimate the true (average) cost of saving a life. But they say that one can expect to save a life for about $2,500, taking into account all of the costs and all of the uncertainties. That's not "a few dollars," but it's well within reach of middle-class people and, over time, even some poor people.

Let's say that you could save someone's life by giving $500 once per year over the next five years, or that you and four friends could do the same right now. And let's say that this year you have $500 that you were planning to spend on yourself, not for something you really need, but something just for fun—for example, a skiing trip instead of a less expensive camping trip. Why not give it instead to Oxfam or the Against Malaria Foundation (AMF), GiveWell's top-rated charity?

I emphasize that the personal details presented here (vacation upgrade vs. $500 to charity) are not essential. If you don't have $500 of truly disposable income, then call it $50, or $10. (You can do a lot of good, even if you don't manage to save someone's life all by yourself.) If you don't care for skiing, substitute your own nice-but-not-necessary luxury: sushi instead of a simple sandwich, replacing your perfectly functional old dresser with something more stylish, et cetera. Likewise, if Oxfam and AMF are not high on your list of charities, substitute any charity that serves people whose needs are more desperate than your own. The critical point is this: If you're reading this book, odds are that somewhere in your budget there is truly disposable income, money that you spend on yourself that you could instead spend on people who, through no fault of their own, have much greater needs than you. So why not spend that money on them?

This question was originally posed by Peter Singer, a utilitarian philosopher and heir to the legacy of Bentham and Mill. The utilitarian argument for giving is straightforward: Going skiing instead of camping (or whatever) may increase your happiness, but it's nothing compared with the increase in happiness that a poor African child gains from clean water, food, and shelter. Not to mention the happiness the child's mother gains from not watching her child starve or die from a treatable disease. Thus, says utilitarianism, you should spend that money helping desperately needy people rather than on luxuries for yourself.

This may sound like a good argument. In fact, I think it *is* a good argument, and I'll defend it in chapter 10. But taking this argument seriously suggests a radical conclusion that's very hard to swallow. Let's agree that you should donate your $500 to Oxfam or AMF instead of spending that money on yourself. What about the next $500? The same argument still applies. The world is still full of desperate people, and you could still write another check. The utilitarian bleeding will continue until you've given away all of your disposable income—where "disposable" refers to all of the income that you don't need to maximize your ability to donate to people who are worse off than you. Utilitarianism apparently requires you to turn yourself into a *happiness pump*. To most people, this does not sound so splendid.

(And these, I'm sorry to report, are not the only un-splendid implications attributed to utilitarianism.**)

As an abstract idea, utilitarianism sounds perfectly reasonable, if not obviously correct. Why would we ever want to make the world less happy than it could otherwise be? Likewise, when we consider the new pastures from a detached and impartial perspective, it seems obvious that the warring tribes should put aside their respective ideologies, figure out which way of life will work best on the new pastures, and then live that way. However, when we apply utilitarian thinking to certain specific problems, it seems downright absurd.* Utilitarianism would ask us, at least in principle, to use people as human trolley-stoppers, enslave people, perpetrate miscarriages of criminal justice, and turn ourselves into happiness pumps.

What, then, should we make of utilitarianism? Is it the metamorality that we're looking for, a rational standard grounded in shared values with which we can resolve our moral disagreements? Or is it a misguided oversimplification of morality that, if taken seriously, leads to moral absurdities? To answer these questions, we need a better understanding of moral psychology. Our gut reactions tell us that utilitarianism sometimes gets things terribly wrong. Do these gut reactions reflect deeper moral truths? Or do they simply reflect the inflexibility of our automatic settings? In other words, is the problem with utilitarianism or with *us*? The new science of moral cognition can help us answer these questions.

Readers, be warned: The next two chapters are a heavy lift, but they are necessary for completing the argument of this book. The classic objections to utilitarianism are very intuitive. Unfortunately, the best replies to these objections are not at all intuitive. They require a scientific understanding of the moral machinery behind these objections and a fair amount of philosophical argument. These chapters also take us deep into the world of hypothetical philosophical dilemmas, a world that some readers prefer not to visit. (Alas, the value of hypothetical questions as tools for illuminating the real world is widely underappreciated,* but that's a metaphilosophical topic for another time.)

If you're satisfied that utilitarianism is a good metamorality for modern herders, you can skip the next two chapters and go right to part 5. There, in the book's two final chapters, we'll return to the real-world problems that divide us and apply the lessons we've learned. But first, if you want to see utilitarianism defended against the convictions of its critics, please read on, and be prepared to exercise your manual mode.**

PART IV

Moral Convictions

9.

Alarming Acts

In that fateful high school debate in Jacksonville, Florida, I was confronted with an ugly truth about utilitarianism: Promoting the greater good can, at least in principle, mean doing things that seem terribly, horribly wrong: harvesting people's organs against their will, pushing innocent people in front of speeding trolleys, and so on. No surprise, then, that many people, especially thoughtful philosophers, have concluded that there's more to right and wrong than maximizing happiness.

In this chapter and the next, we'll face this challenge head-on, employing two general strategies, which we'll call *accommodation* and *reform*. By *accommodation*, I mean showing that maximizing happiness does not, in fact, have the apparently absurd implications that it seems to have. In other words, the answers that utilitarianism actually gives in the real world are generally consistent with common sense. For example, there are many good reasons to think that pushing people off footbridges and stealing people's organs, even with the best of intentions, is unlikely to promote the greater good in the real world, in the long run.*

Utilitarianism, however, is more than just a philosophical affirmation of common sense. It couldn't be. The world's moral tribes have different versions of common sense—hence the Tragedy of Commonsense Morality. The world's tribal moralities can't all be equally good from a utilitarian perspective, and that means that utilitarianism must be at odds with at least some versions of common sense. Recall that utilitarianism came of age in nineteenth-century Britain as a justification for social reform. To call for reform is to challenge the conventional wisdom, and to challenge the conventional wisdom effectively, a reformer must explain why the conventional wisdom is wrong, despite its appeal. For example, Mill argued, against the conventional wisdom of his day, that women are the intellectual equals of men. And he argued, more specifically, that the apparent intellectual inferiority of women can be explained by their limited access to education.

Mill's argument is an example of a *debunking* argument, one that explains how something could appear to be true even if it isn't. In this chapter and the next, our reforming strategy will mirror Mill's, explaining away apparent moral truths. More specifically, we'll use science to get underneath our anti-utilitarian moral intuitions, to understand why they are useful and why they are too inflexible to serve as the ultimate arbiters of right and wrong.*

Our automatic settings, our moral intuitions, can fail us in two ways. First, they can be *oversensitive*, responding to things that, upon reflection, don't seem to be morally relevant. For example, research has shown, sadly, that the judgments of juries in death penalty cases are often sensitive to the race of the defendant, a factor that we (participants in this conversation) today regard as morally irrelevant. Automatic settings can also be *undersensitive*, failing to respond to things that, upon reflection, do seem to be morally relevant. For example, a jury might render its judgment without adequately accounting for the defendant's age at the time of the crime, a factor that we today regard as morally relevant.

In what follows, we'll see evidence for both kinds of unreliability in our anti-utilitarian moral intuitions. We'll start with our favorite moral fruit flies, trolley dilemmas. Later, I'll explain how our gut reactions to hypothetical cases are related to real-world problems. (We've done some of this already. Recall the study of moral judgment in medical doctors versus public health professionals from chapter 4, pages 128–31.)

PUSHING MORAL BUTTONS

Let's review the basic trolley facts: In response to the *switch* case, most people approve of hitting a switch that will turn the trolley away from five people and into one. In response to the *footbridge* case, most people disapprove of pushing the man off the footbridge and into the trolley's path, thus trading one life for five. The psychological question: Why do we find it acceptable to trade one life for five in the *switch* case but not in the *footbridge* case?

We got a partial answer back in chapter 4. We have an automatic emotional response that makes us disapprove of pushing the man off the *footbridge*, but we have no comparable emotional response to the thought of hitting the switch. In both cases, we engage in utilitarian, cost-benefit thinking ("Better to save five lives at the cost of one"). But only in the footbridge case is the emotional response (typically) strong enough to trump our utilitarian thinking (see figure 4.3, page 121).

Once again, this dual-process explanation is supported by studies employing functional brain imaging, neurological patients with emotional deficits, physiological measures of emotional arousal, emotional inductions, manipulations that disrupt manual-mode thinking (time pressure, distracting secondary tasks), manipulations that disrupt visual imagery, personality questionnaires, cognitive tests, and pharmacological interventions (see pages 121–28). This explanation, however, is only partial: The *footbridge* dilemma is more emotionally engaging than the *switch* dilemma, but *why*? What is it about the *footbridge* dilemma that pushes our emotional buttons?

Before we get to the right answer, it's worth reviewing the evidence against a very appealing wrong answer about what's going on in the

Trolley Problem, one that we've mentioned before: In the *switch* case, the action that saves the five is likely to *work*. But in the *footbridge* case, there are a million things that can go wrong. Can a person's body really stop a trolley? What if the guy doesn't land on the track? Third, what if he fights back? And so on. In other words, and as noted above, there are several good *utilitarian* reasons for hitting the switch—but not for pushing the man—in the real world. While this is undoubtedly true, the evidence suggests that this is not why people say no to pushing. If saying no to pushing were based on hard-nosed, realistic cost-benefit thinking, then why are people who score higher on a test of "cognitive reflection" less likely to say no to pushing? Why does putting people under time pressure make them more likely to say no? Why are people with emotional deficits and people with compromised visual imagery less likely to say no? And so on. These results tell us that our negative reaction to pushing comes from a gut reaction and not from extra-realistic cost-benefit calculations. (Further evidence comes from the experiments that I'm about to describe, in which we controlled for people's real-world expectations.**) So if it's a gut reaction, rather than realistic utilitarian thinking, that makes us object to pushing, then what triggers this gut reaction?

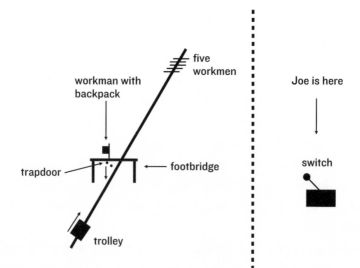

Figure 9.I. The *remote footbridge* dilemma.

Back in chapter 4, I called the cases like the *footbridge* case "personal" and the ones like the *switch* case "impersonal." This suggests a theory that we can test by comparing versions of the footbridge case that differ in the "personalness" of the harmful action.* We can start by comparing the original *footbridge* case with one in which the harm is caused by hitting a switch from a distance, as in the *switch* case. We'll call this the *remote footbridge* case.

In the *remote footbridge* case, our protagonist, Joe, can hit a switch that will open a trapdoor on the footbridge, dropping the workman into the trolley's path, thus blocking the trolley and saving the five (see figure 9.1). In one study using the original *footbridge* case, 31 percent of people approved of pushing in order to save the five. We gave the *remote footbridge* case to a separate group of otherwise identical people, and 63 percent approved, roughly doubling the number of utilitarian judgments. This suggests that, indeed, something like "personalness" is involved.

The *remote footbridge* case differs from the original in that the agent is farther from the victim. It also differs in that the agent doesn't *touch* the victim. So is it about distance, touching, or both? To find out, we can use the *footbridge switch* case (see figure 9.2). This is like the *remote footbridge* case, except that here the switch is on the footbridge, next to the victim.

Here 59 percent of people approved of the utilitarian action. This

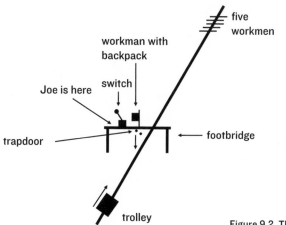

Figure 9.2. The *footbridge switch* dilemma.

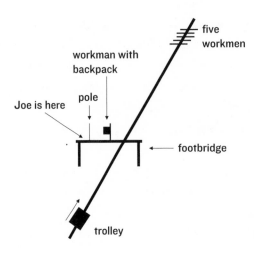

Figure 9.3. The *footbridge pole* dilemma.

closely matches what we saw in the *remote footbridge* case and is not statistically different. Thus, it seems that spatial distance has little or no effect. Instead, what seems to matter is touching.*

But even here there are multiple interpretations. In the *footbridge* case, but not in the *footbridge switch* case, the agent touches the victim. But the agent also does something more subtle. He directly impacts the victim with the force of his muscles. He *pushes* the victim. Call this the application of *personal force*. To distinguish between touching and the use of personal force, we'll use the *footbridge pole* case (see figure 9.3). This one is just like the original *footbridge* case, except that here the agent pushes the victim with a pole. Thus, we have the application of personal force, but without touching.*

Here 33 percent of people approve of pushing—a big drop. That's about half the number of people who approve in the *remote footbridge* and *footbridge switch* cases. What's more, this 33 percent approval is not statistically different from the 31 percent approval we got for the original *footbridge* case.

Thus, it seems that one important psychological difference between the *footbridge* and *switch* cases has to do with the "personalness" of the harm and, more specifically, with personal force—pushing versus hitting a switch.

From a normative perspective, the interesting thing about personal force is that it's not something that we, upon reflection, would regard as morally relevant. Someone's willingness to cause harm by personal force might be relevant in assessing that person's character—that is, you would rightly think less of someone who's willing to kill someone with his bare hands as opposed to more indirectly (*accommodation**)—but that doesn't mean that the application of personal force actually makes the action more wrong (*reform*). Think of it this way: Suppose that a friend calls you from a footbridge, seeking moral advice: "Should I kill one to save the five?" You wouldn't say, "Well, that depends . . . Will you be pushing this person, or can you drop 'em with a switch?" Clearly, the physical mechanism is not, in and of itself, morally relevant. But it does seem to be *psychologically* relevant.

This is exactly what the dual-process theory predicts. We know from all of the science described in chapter 4 that it's an automatic setting, an intuitive emotional response, that makes us disapprove of pushing in the *footbridge* case. And we know from chapter 5 that automatic settings are heuristic devices that are rather inflexible, and therefore likely to be unreliable, at least in some contexts. But in what way is this automatic setting unreliable?

Based only on what we've seen so far, we can't say for sure. Here we may have an automatic setting that's undersensitive: Perhaps dropping the man through the trapdoor really is wrong, but because this action doesn't involve pushing, we're insufficiently alarmed. Or perhaps this automatic setting is oversensitive: Pushing is really right, but our automatic setting is overly concerned with the harm that befalls the pushee, relative to the fivefold greater harm that can be averted by pushing. We'll return to this question later. The point, for now, is that our automatic settings are, in one way or another, leading us astray.

MEANS AND SIDE EFFECT

There is another important difference between the *switch* and *footbridge* cases, one I alluded to back in chapter 4. This is the distinction between harm caused as a *means* to an end and harm caused as a *side effect*. In the *footbridge* case, we're talking about literally *using* someone as a trolley stopper, but in the *switch* case the victim is killed as a side effect—"collateral damage." One way to think about this difference is by imagining what would happen if the victim were to magically disappear. In the *footbridge* case, a disappearing victim would foil the plan: no trolley stopper. But in the *switch* case, the disappearance of the lone person on the sidetrack would be a godsend.

The means/side-effect distinction has a long history in philosophy, going back at least as far as St. Thomas Aquinas (1225–1274 CE), who framed the "Doctrine of Double Effect," which is essentially the "Doctrine of Side Effect." According to the Doctrine of Double Effect, it's wrong to harm someone as a means to an end, but it may be permissible to harm someone as a side effect in pursuit of a good end. Likewise, as noted in chapter 4, Kant says that the moral law requires us to treat people "always as an end and never as a means only."

The means/side-effect distinction plays an important role in the real world, from criminal law to bioethics to the international rules of war. For example, the means/side-effect distinction is the basis for the distinction between "strategic bombing" and "terror bombing." If one bombs civilians as a way of reducing enemy morale, that's terror bombing, which is forbidden by international law. However, if one bombs a munitions factory, knowing that nearby civilians will be killed as "collateral damage." that's strategic bombing, which is not strictly forbidden. Likewise, the American Medical Association distinguishes between intentionally ending a chronically ill patient's life by administering high doses of painkillers (forbidden) and doing the same with the intention of reducing pain, while knowing that the drugs will end the patient's life (not necessarily forbidden).

Are our automatic settings sensitive to the means/side-effect distinction? And could that explain our responses to different trolley dilemmas? To find out, we can compare the original *footbridge* case with a similar case in which the harm is caused as a side effect. Consider the *obstacle collide* case:

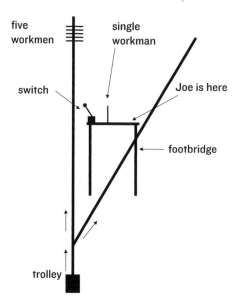

Figure 9.4. The *obstacle collide* dilemma.

Here the trolley is headed toward five people, and there is a side-track with one person. As in the *switch* case, one can save the five by hitting a switch that will turn the trolley onto the sidetrack. Our protagonist, Joe, is on a high and narrow footbridge over the sidetrack. The switch that turns the trolley is at the opposite end of the footbridge, and, unfortunately, a single workman is in between Joe and the switch. To save the five, Joe needs to get to the switch very quickly. To do this, he must run toward the switch as fast as he can. He knows that if he does this, he will collide with the workman and knock him off the footbridge, causing his death. In this case, as in the original *footbridge* case, the harmful action is fully *personal*. Joe knocks the workman off the footbridge with his personal force. But in this case, unlike the *footbridge* case, the victim is harmed as a *side effect*, as collateral damage. If the single workman were to magically disappear, that would be great news for everyone.

Here 81 percent of our subjects approved of Joe's saving the five while knowing that this would cause the single workman's death as a side effect. That's a very high approval rating, much more than the 31 percent

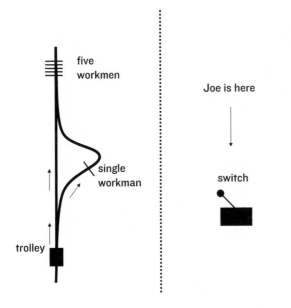

Figure 9.5. The *loop* case.

approval for *footbridge*. What's more, this 81 percent approval rating is not so far from (and not statistically different from) the 87 percent approval rating that we got for the original *switch* case. Thus, it seems that our automatic settings are highly sensitive to the means/side-effect distinction, and this factor can explain why people say yes to the *switch* case and no to the *footbridge* case.

This is starting to sound like a vindication of our automatic settings. The means/side-effect distinction is widely regarded as morally relevant. And our moral intuitions seem to be tracking this distinction pretty well, saying no when the victim is harmed as a means (*footbridge*, *footbridge pole*) and yes when the victim is harmed as a side effect, as collateral damage (*switch*, *obstacle collide*). But there's a glitch. People mostly approve of the actions in the *remote footbridge* and *footbridge switch* cases, voting yes about 60 percent of the time, even though these cases involve using a workman as a trolley stopper. And it gets glitchier.

In the early days of philosophical trolleyology, Judith Jarvis Thomson presented a version of the following case, which we'll call the *loop* case. It's like the *switch* case, except that here the sidetrack loops back to the main track as shown in figure 9.5.

In this case, if the person were not on the sidetrack, the trolley would return to the main track and run over the five people. In other words, when one hits the switch in this case, one uses the victim as a *means* to save the five people, as a trolley stopper. (If the single workman weren't there, there'd be no point in hitting the switch.*) Nevertheless, 81 percent of the people we tested approved of hitting the switch. Thus, at least sometimes, using someone as a trolley stopper seems morally acceptable.

Here's another case that makes trouble for means/side-effect distinction. (Hard-core trolleyologists, take note. This case also makes trouble for the "Doctrine of Triple Effect.") In this case, which we'll call the *collision alarm* case, the mechanism of harm is identical to that of the original *switch* case, but the victim is harmed as a means (see figure 9.6).

Here's how it works: The first trolley is headed for five people, and if nothing is done, these people will die. The second trolley is on a different track with nothing in front of it. Joe can hit a switch that will turn the second trolley onto a sidetrack. On this sidetrack there is a person, and next to this person is a sensor that is connected to an alarm system. If Joe

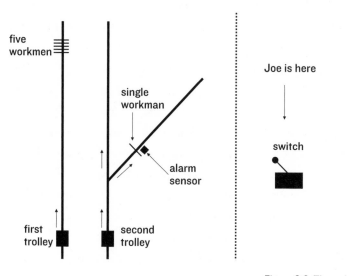

Figure 9.6. The *collision alarm* case.

hits the switch, the trolley will turn onto the sidetrack and collide with the person there. The sensor will detect this collision and set off an alarm. This will cut the power to the whole trolley system. Cutting the power to the whole trolley system will cut the power to the first trolley and thus prevent it from killing the five. The critical point, once again, is that we're now running the victim over as a *means* to save the five.

Eighty-six percent of the people in our sample approved of the utilitarian action in this case, which is almost identical to (and not statistically different from) the 87 percent who approved of the action in the original *switch* case. Thus, here, too, people approve of killing one to save five, even though the victim is killed as a *means* to save the others.

What's going on? We've identified two factors that affect people's intuitive judgments in the Trolley Problem: whether the victim is harmed through the direct application of *personal force* (pushing vs. hitting a switch) and whether the victim is harmed as a *means versus side effect* (used as a trolley stopper vs. collateral damage). But the influence of these factors is inconsistent. Sometimes personal force matters a lot (as in *footbridge switch* vs. *footbridge pole*), but sometimes it doesn't (as in *switch* vs. *obstacle collide*). Likewise, sometimes the means/side-effect factor matters a lot (as in *obstacle collide* vs. *footbridge*), but sometimes it doesn't (as in *switch* vs. *loop* and *collision alarm*). Why do these factors matter sometimes but not always?

If you look closely, you'll see that it's the *combination* of these two factors that matters. If you harm someone using personal force, but as a side effect, that doesn't seem so bad (*obstacle collide*, 81 percent approval). And if you harm someone as a means, but without the use of personal force, that doesn't see so bad (*loop*, 81 percent; *collision alarm*, 86 percent). But if you harm someone as a means *and* you use personal force, then the action seems wrong to most people (*footbridge,* 31 percent; *footbridge pole,* 33 percent). Thus, it seems that harm as a means of using personal force is a magic combination.** (In technical terms, this is an *interaction*, like the interaction between two medications: The effect of taking both medications at once is more than the sum of the effects of taking the two medications separately.)

That's a lot of trolleys. Let's take a moment to reflect on the signifi-cance of these findings. We're trying to figure out whether our moral intu-itions are reliable in these cases. Are they? Not so much, it seems. For one thing, our judgments are (sometimes) sensitive to personal force (pushing vs. hitting a switch), and this factor seems morally irrelevant. Our moral intuitions do seem to track the means/side-effect distinction, but in a very imperfect way. It seems that our sensitivity to the much beloved means/ side-effect distinction is bound up with our not beloved sensitivity to per-sonal force.*

There is an important further question here. Clearly, the mere differ-ence between pushing and hitting a switch is morally irrelevant. However, if you ask a well-credentialed moral philosopher, she will tell you that the difference between harm as a means and harm as a side effect is highly morally relevant. But why? Why should killing someone as collateral dam-age, knowing full well what's going to happen, be any better than killing someone as a means to an end? After all, when you get killed as collateral damage, you're just as dead. And the person who killed you knew just as well that you were going to die by his hand. (Note that we're talking about *foreseen* side effects, as in the *switch* case, not accidents.) The idea that kill-ing someone as a means is worse than killing someone as a side effect has been around for a long time and is widely respected. But as far as I know— as far as anyone knows—the venerable Doctrine of Double Effect has no justification beyond the fact that it's supported (imperfectly) by some of our intuitions. Indeed, people all around the world make judgments that are (imperfectly) consistent with the Doctrine of Double Effect while having no knowledge of the doctrine. This tells us that intuitive judgments come first, and that the doctrine is just an (imperfect) organizing summary of those intuitive judgments. In other words, the "principled" distinction be-tween harm as a means and harm as a side effect doesn't justify our intui-tive judgments. Rather, it's our moral intuitions that justify the principle.

So where do these intuitive judgments come from? Why does harming someone as a means (often) feel worse than harming someone as a foreseen side effect? In what follows, I'll present a theory that explains why our moral brains are sensitive to the means/side-effect distinction. And if this theory is right, it casts serious doubt on the moral legitimacy of the hallowed Doctrine of Double Effect, which, as noted above, is the basis for many real-world policies, ones by which people live and die every day.

MODULAR MYOPIA

I call this theory the *modular myopia hypothesis*. It synthesizes the dual-process theory of moral judgment with a theory about how our minds represent actions. The modular myopia hypothesis is the most complicated single idea that I will present in this book. To help it go down easier, I'll first summarize the general idea and then, in the sections that follow, lay out the theory in more detail.

Here's the general idea: First, our brains have a cognitive subsystem, a "module," that monitors our behavioral plans and sounds an emotional alarm bell when we contemplate harming other people.** Second, this alarm system is *"myopic,"* because it is *blind to harmful side effects.* This module inspects plans of action, looking for harm, but the inspecting module, for reasons I will explain shortly, can't "see" harms that will occur as side effects of planned actions. Rather, the module sees only harmful events that are planned as a means to achieve a goal. Thus, the modular myopia hypothesis explains our intuitive tendency to draw the means/side-effect distinction in terms of the limitations of a cognitive subsystem, a module, that is responsible for warning us against committing basic acts of violence. These limitations make us emotionally blind—but not cognitively blind— to certain kinds of harm. This idea, that we can be *emotionally* blind but not *cognitively* blind, should sound familiar. As I will explain, this duality is the duality of the "dual-process" theory of moral judgment.

This summary description of the modular myopia hypothesis raises two big questions. First: Why should the human brain have a system that inspects action plans, looking for harm? In the next section, I'll explain why it would make sense for us to have such a device in our heads. Second: Why should this module be myopic in precisely this way? As I will explain shortly, the beauty of the myopic module hypothesis is that it follows naturally from the dual-process theory of moral judgment, combined with a simple theory about how our brains represent plans for action.

WHY AREN'T WE PSYCHOPATHS?

Why would human brains need an action-plan inspector? My hypothesis is as follows.

At some point in our history, our ancestors became sophisticated action planners who could think about distant goals and dream up creative ways of achieving them. In other words, we acquired manual-mode reasoning and planning. This was a fantastic advance. It enabled our ancestors to kill large animals through coordinated attacks and by setting traps, to build better housing structures, to plant seeds with the intention of harvesting crops months later, and so on. But this general ability to dream up ways of achieving distant goals came with a terrible cost. It opened the door to *premeditated violence*. Violence no longer had to be motivated by a *present impulse*. Violence could be deployed as a general-purpose tool for getting the things one wants: Tired of taking orders from that jerk? Wait until the right moment and then get rid of him! Fancy the female next door? Wait until she's all alone and then have your way with her! A creature that can plan for the future, a creature that can dream up new ways of achieving its goals, is a very dangerous kind of creature, especially if that creature can use *tools*.

It's pretty hard for a single chimp to kill another chimp, especially if the other chimp is bigger and stronger. But an interesting—and scary— fact about primates like humans is that, as Hobbes observed, any healthy adult is capable of killing any other member of the species, and without anyone else's help. A five-foot-three woman can sneak up on a six-foot-five man while he is sleeping and smash his head in with a rock, for example.* Thus, as we humans became more adept at planning actions and using tools to our advantage, we acquired an enormous capacity for violence.

What's wrong with having an enormous capacity for violence? Nothing, perhaps, if you're a solitary, territorial animal such as a tiger. But humans survive by living together in cooperative groups. Humans who are attacked tend to seek revenge ("tit for tat"), and this makes violence dangerous for both the attacker and the attacked. This is especially true if one's intended victim is a tool-using action planner. Even if the violent attacker is twice the victim's size, as long as the victim survives, he can wait for the right moment to retaliate, with a rock to the head or a knife to the back. And if the victim doesn't survive, he may have relatives or friends who are motivated to exact vengeance on his behalf. In a world in which people are vengeful, and in which anyone is capable of killing anyone else, you have to be very careful about how you treat others. What's more, even if a reluctance to be violent were to confer no individual advantage, it might confer an advantage at the group level, such that groups that are (internally) more docile are more cooperative and thus have a survival advantage. In short, individuals who are indiscriminately violent are likely to suffer from the payback they receive from their group mates and may disrupt their group's ability to cooperate, putting their group at a disadvantage in group-level competition.

To keep one's violent behavior in check, it would help to have some kind of internal monitor, an alarm system that says "Don't do that!" when one is contemplating an act of violence. Such an action-plan inspector would not necessarily object to all forms of violence. It might shut down, for example, when it's time to defend oneself or attack one's enemies. But it would, in general, make individuals very reluctant to physically harm one another, thus protecting individuals from retaliation and, perhaps, supporting cooperation at the group level. My hypothesis is that the myopic module is precisely this action-plan inspecting system, a device for keeping us from being casually violent.

Why would such a module be myopic? Because all modules are, in one way or another, myopic. What we're hypothesizing is a little alarm system, an automatic setting that provides a check on the potentially dangerous plans drawn up by the outcome-maximizing manual mode. (Which, you'll recall, tends to be far from impartial when there are personal gains and losses at stake.) All automatic settings are heuristic, and therefore myopic in one way or another. Take, for example, the cognitive system housed in the amygdala that automatically recognizes fear based on enlarged eye whites (see page 141). This system is blind to the fact that the "eye whites" to which it's responding may be pixels on a computer screen rather than the eyes of a real person in a real dangerous situation. All automatic settings rely on specific cues that are only imperfectly related to the things they're designed to detect. In the same way, our hypothesized antiviolence alarm will, however it works, work by responding to some limited set of cues. So the question is not *whether* an automatic action-plan inspector would be myopic. The question is: In *what ways* would this device be myopic?

So far, this is all just theory. Is there any evidence that we have such a module in our brains? Indeed, there is. We know that we have automatic emotional responses to certain kinds of violent actions, such as pushing people off footbridges. And we know that this system is at least somewhat "modular." That is, this system's internal operations are closed off from the rest of the brain, or at least closed off from the parts of the brain that enable conscious thought. (Which is why we can't figure out how our trolley intuitions work through introspection and instead must do experiments like those described above to understand them.) Experimental trolleyology indicates that there is something like an automated antiviolence system in our brains.

The modular myopia hypothesis makes three further predictions. First, this module didn't evolve to respond to trolley dilemmas. What should really get this alarm system going is real violence. Second, if this module responds to *cues* related to violence, then it shouldn't require real violence to set it off, just the right cues. In other words, merely *simulating* violence of the right sort should be enough to set it off, even if the person simulating the violence knows (in manual mode) that there's no real violence taking place. Third, if the function of this alarm system is to stop *oneself* from committing violence (without provocation), then this system should respond most strongly to simulating violent actions *oneself*, as opposed to watching others simulate violence or simulating physically similar, but nonviolent, actions oneself.

With all of this in mind, Fiery Cushman, Wendy Mendes, and colleagues conducted an experiment, which I mentioned back in chapter 2.

Cushman and colleagues had people simulate violent actions, such as smashing someone's leg with a hammer and smashing a baby's head on a table (see figure 2.2, on page 36). Once again, people's peripheral veins constricted (giving them "cold feet") when they performed these pseudo-violent actions *themselves*, but not when they watched others do the same, and not when they performed physically similar actions that were not pseudo-violent. And this happened despite their knowing full well (in manual mode) that these actions were harmless. Thus, Cushman and colleagues observed exactly what the modular myopia theory predicts: an automatic aversion to personally performing actions that are superficially (but not too superficially) similar to acts of violence.

However, the myopic module hypothesis goes further than this. Not only do we have an alarm system that responds to cues related to being violent. According to this theory, this system is myopic in a specific way. It's blind to harm caused as a foreseen *side effect*. Why would it be like that?

BLINDNESS TO SIDE EFFECTS

This is where things get a bit complicated. This part of the theory begins with a theory of action representation proposed by John Mikhail, building on earlier proposals from Alvin Goldman and Michael Bratman. Mikhail's idea is that the human brain represents actions in terms of branching *actions plans* like the ones in figure 9.7, which describe the action plans for the agents in the *switch* and *footbridge* dilemmas.

Every action plan has a primary chain, or "trunk," that begins with the agent's body movement and ends with the agent's goal (the intended outcome). The primary chain consists of the sequence of events that are *causally necessary* for the achievement of the goal. For example, in the *switch* case, the agent's moving his hands (the body movement) causes the switch in his hands to move, which causes the switch on the track to align for the turn, which causes the trolley to go onto the sidetrack instead of staying on the main track, which saves the five people on the main track (the goal).

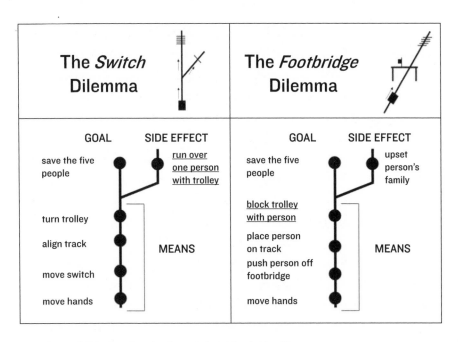

Figure 9.7. Action plans for the *switch* and *footbridge* dilemmas.

Likewise, in the *footbridge* case, the agent's moving his hands (the body movement) causes the person to fall from the footbridge, which causes him to land on the track, which causes the trolley to collide with him, which stops the trolley, which saves the five people (the goal). You can see these sequences of events in figure 9.7 by starting at the bottom of each of the two primary chains and working your way up. Any event that is represented on the primary chain of the action plan is thus represented as a *means* to the agent's goal, as a necessary step in causing the goal to be achieved.

The action plans in figure 9.7 also have secondary chains that branch off the primary chains. In the *switch* case, the turning of the trolley has two effects (that is, a "double effect"). It causes the five on the main track to be saved (the goal), but it also has a notable side effect: killing the person on the sidetrack. This event is represented on a secondary chain because it is a foreseen side effect. It's an event that the agent expects to occur but that is not causally necessary for the achievement of her goal. (Once again, if the person on the sidetrack were to disappear, the goal would still be achieved.) Likewise, there are foreseen side effects in the *footbridge* dilemma. Using the man as a trolley stopper will save five lives, but one can expect it to have other effects, such as upsetting the family of the man who was used as a trolley stopper. Thus, figure 9.7 depicts that effect as an event on a secondary chain, as an event that is foreseen but not causally necessary for achieving the goal. Here the plan still works even if the victim's family is perfectly happy with the outcome.

Mikhail's theory, then, is relatively straightforward: We think that it's wrong to push the man off the *footbridge* because this involves using him as a means, and we think that it's acceptable to turn the trolley in the *switch* case because here we are "merely" killing the man as a foreseen side effect. Mikhail's idea is that a certain kind of mental representation, an asymmetrically branching action plan, can serve as a natural format for representing the means/side-effect distinction. It's a truly elegant idea.

When I first encountered Mikhail's theory, I thought it was interesting, but I didn't think it could be right. First, there was already plenty of evidence for the dual-process theory, according to which emotional responses from one part of the brain compete with utilitarian judgments from a different part of the brain. In Mikhail's theory, there is no emotion, and no competition between competing systems. Instead there is a single system, a "universal moral grammar," that does all the work by emotionlessly representing and analyzing branching action plans. Thus, I thought that Mikhail's theory, however intriguing, was a step in the wrong direction. Second, and more immediately compelling, I knew that the means/side-effect distinction could not adequately explain the data. The original showstopper for the means/side-effect theory is Judith Thomson's *loop* case, which we met earlier in this chapter (page 220). In that case, the single workman gets used as a trolley stopper, but people seem to think it's fine.

Then, on a beastly hot summer day, while standing in front of the window-unit air conditioner in my old Philadelphia apartment, I had a *Eureka!* moment. (At least *I* think it's *Eureky*. You can judge for yourself in a moment.) I had just seen some data from other researchers showing that the means/side-effect distinction *does* work for a wide range of cases, with people approving more of the cases involving harmful side effects and less of the cases in which the harm is a means. (We saw this, for example, when we compared the *footbridge* case with the *obstacle collide* case.) These studies suggested that the means/side-effect distinction really does matter.

But how to reconcile these findings with the dual-process theory? Once again, the dual-process theory says that the actions we don't like, actions like pushing the man off the *footbridge*, are the ones that trigger a negative emotional response. So, I thought, if the means/side-effect distinction matters, it must matter by influencing the system that triggers these emotional responses. In other words, we must respond more emotionally to cases in which the harm is caused as a means. But then why don't we mind the harm in the *loop* case, which clearly involves using the victim as a trolley stopper? Could there be something funny about the *loop* case? Is there something about that case that would prevent the harm from bothering us? That's when I realized that Mikhail's theory of action representation and the dual-process theory could be integrated.

There *is* something funny about the *loop* case. The *loop* case is a means case, but it's an unusually complicated one. More specifically, it's complicated because you have to *keep track of multiple causal chains* to see that the victim gets used as a means. In a simpler case, like the *footbridge* case, you only have to keep track of a single causal chain to see that someone is going to get hurt. All you need to know is what you see in figure 9.8.

That is, all you need to know is that the harmful event—blocking the trolley with the person—is a necessary step along the way from the body movement (moving hands) to the goal (saving the five). But in the *loop* case, you have to keep track of *two* causal chains in order to see that the harmful event is necessary for achieving the goal. This is because there are two ways that the trolley can harm the five people: (1) by going down the main track and (2) by going around the loop. Both of these causal chains must be disrupted in order to save the five. The disruption of the first causal chain is diagrammed in figure 9.9.

In the *loop* case, the trolley is headed for the five people, but this can be avoided by turning the trolley. In other words, the turning of the trolley disrupts the first causal chain, the one whereby the trolley proceeds down the main track and kills the five. Now, if this were the *switch* case, there would be nothing more to say about events that are necessary for achieving

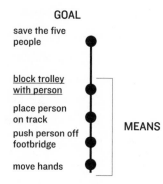

Figure 9.8. Primary causal chain of the action plan for the *footbridge* case.

the goal. In the *switch* case, you turn the trolley, and nothing else needs to happen to save the five. In other words, figure 9.9 is a more or less complete diagram of the events that are necessary for saving the five in the *switch* case. But in the *loop* case, what you see in figure 9.9 is only part of the story. In the *loop* case, there is a second causal chain that must be disrupted in order to save the five. Turning the trolley puts the trolley onto the side loop, where it threatens the five once more, but from a different direction— that is, by way of a different causal chain. To stop the trolley from getting a second shot at the five, there has to be something on the side loop to stop

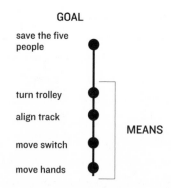

Figure 9.9. Primary causal chain for the *loop* and *switch* cases. Hitting the switch prevents a different causal chain from being realized, one in which the trolley runs over the five.

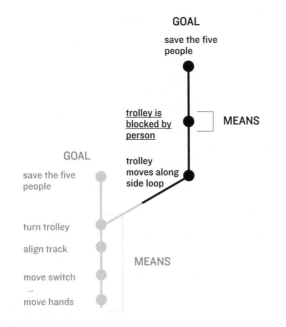

Figure 9.10. Diagram highlighting the secondary causal chain in the *loop* case.

it. And, of course, there is: the hapless victim. The disruption of the second causal chain is depicted in figure 9.10.

If you look only at the primary causal chain, as depicted in figure 9.9, you won't see any harm at all. The trolley is turned away from the five, and that's it. To see that there is a harm in the *loop* case, and that there is a harm that is necessary to achieving the goal, you have to see the secondary causal chain, the one that is highlighted in figure 9.10.

Now, back to the dual-process theory for a moment. According to the dual-process theory, there is an automatic setting that sounds an emotional alarm in response to certain kinds of harmful actions, such as the action in the *footbridge* case. Then there's manual mode, which by its nature tends to think in terms of costs and benefits. Manual mode looks at all of these cases—*switch*, *footbridge*, and *loop*—and says "Five for one? Sounds like a good deal." Because manual mode always reaches the same conclusion in these five-for-one cases ("Good deal!"), the trend in judgment for each of these cases is ultimately determined by the automatic setting—that is, by whether the myopic module sounds the alarm.*

But what determines whether this alarm system sounds the alarm? We know that our judgments are, at least sometimes, sensitive to the means/side-effect distinction. And yet, our judgments are not *always* sensitive to the means/side-effect distinction, as is demonstrated by the *loop* case. What's going on? Above, we got what would appear to be an adequate answer: The missing ingredient is personal force. We say no to pushing the man off the footbridge and yes to turning the trolley onto the loop because pushing the man off the footbridge involves *pushing*. However, if that were a complete explanation, then adding some pushing to the loop case should make the utilitarian action in that case seem as wrong as pushing the man off the footbridge, and that does not appear to happen.** Given that, is there any reason why our brains would treat the *loop* case more like a *side-effect* case and less like a typical *means* case, even though it *is* a means case?

Here's a clue: According to the dual-process theory, the system that sounds the emotional alarm is supposed to be a relatively simple system, an action monitor that sounds an alarm when we contemplate committing violent acts.

Here's the next clue: The *loop* case is, as explained above, an exceptionally complicated means case. To see that it's a means case, it's not enough to look at the primary causal chain, the one in figure 9.9. To see that achieving the goal requires causing harm, you have to look at the secondary causal chain, the one highlighted in figure 9.10.

Have you got it yet? If I'm right, the solution to the mystery of the *loop* case—and much more—goes like this: We have an automatic system that "inspects" action plans and sounds the alarm whenever it detects a harmful event in an action plan (e.g., running someone over with a trolley). But (drumroll, please . . .) this action-plan inspector is a relatively simple, "single-channel" system that *doesn't keep track of multiple causal chains*. That is, *it can't keep track of branching action plans*. Instead, when it's presented with an action plan for inspection, it *only sees what's on the primary causal chain*.

Why would it be like that? Think about how you remember song lyrics: What's the third sentence of "I've Been Working on the Railroad"? Even if you know the answer, you probably can't pop right out with it. You have to start from the beginning and work your way forward: *I've been working on the railroad all the live-long day. I've been working on the railroad just to pass the time away. Can't you hear the whistle blowing?* . . . Instead of processing the entire song at once, you work your way through the song, counting on the fact that each segment pulls along the one behind it. The idea, then, is that the little action-plan inspector in your head unconsciously processes action plans the way that you consciously process song lyrics: by working its way down a chain. When inspecting an action plan, this processor starts with the body movement (e.g., pushing) the way that you start with *I've been working on* . . . , and it continues straight ahead to the goal (e.g., saving the five) in the same way that you continue all the way through to the end of the song. The action monitor can't see the secondary branches on the action plan because all it knows how to do is work its way up the main branch.

Thus, when this system looks at the action plan for the *footbridge* dilemma, it doesn't see what's on the right in figure 9.7. Instead, it only sees what's in figure 9.8. But that's enough for it to sound the alarm, because the harmful event, the squashing of the man with the trolley, is right there in the primary chain.

However, when this system looks at the action plan for the *switch* dilemma, it doesn't see what's on the left in figure 9.7. Instead, all it sees is what's in figure 9.9. And here, there is *no harm to be found.* As far as this system is concerned, here's what's happening in the *switch* case: Move hand → move switch → align track → turn trolley → save five. In other words, it sees only what's in figure 9.11.

Figure 9.11. Spatial diagram of the primary causal chain of the *switch* and *loop* cases.

This system, because it sees only the primary causal chain, is blissfully unaware of the fact that this action will kill someone. In other words, because the harm is a side effect, a harm on a secondary causal chain, the alarm never gets sounded.

What about the *loop* case? Our anomalous reactions to the *loop* case are what provide the critical (albeit preliminary) evidence for the modular myopia hypothesis. Once again, our responses to this case are "anomalous," because most people approve of turning the trolley in the *loop* case, even though this involves using someone as a trolley stopper, as a means. The funny thing about the *loop* case is that it is *structured like a side-effect case*, even though it's a means case. More specifically, in the *loop* case, the harmful event occurs on a secondary causal chain, just as in a side-effect case, even though the harm is causally necessary, a means to achieve the goal of saving the five. As explained above, the primary causal chain in the *loop* case is identical to the primary causal chain in the *switch* case. So in the *loop* case as well, all the action-plan inspector sees is this: Move hand → move switch → align track → turn trolley → save five. And, as in the *switch* case, the alarm bell never goes off, because there is no harm on the primary causal chain. (And why, you ask, is the causal chain with the harm on it the *secondary* one?**) And thus, because the harmful event is not on the primary causal chain, the myopic module can't see it either. Once again, the harmful event in the *loop* case is a means that is *structured like a side effect*, and thus invisible to the myopic module.*

According to this theory, the module is myopic because it is blind to side effects, but that doesn't mean that *we* are blind to side effects. On the contrary, we are perfectly capable of recognizing that the *switch* case is a side-effect case and that the *footbridge* and *loop* cases are means cases. If we can see these side effects, but the myopic module can't, that must mean that there is some other part of our brains that can see (i.e., represent) side effects. So where are the side effects represented?

Enter, once more, the dual-process theory of moral judgment. The myopic module is just an automatic setting, a gizmo that determines whether the emotional alarm is sounded. But there is also the other side of the dual-process story, the brain's manual mode. As explained in the last chapter, this

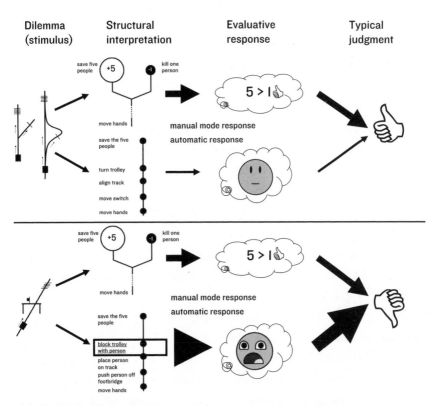

Figure 9.12. Diagram of the dual-process responses to the *switch*, *loop*, and *footbridge* dilemmas. All three dilemmas elicit utilitarian thinking in manual mode (rows I and 3), but only the killing in the *footbridge* dilemma gets "seen" by the system that triggers automatic negative emotional responses to (certain kinds of) harmful actions (row 2 vs. row 4).

system is designed for general-purpose maximizing, for tallying costs and benefits. As such, it is perfectly capable of seeing side effects. Manual mode knows what's a side effect and what's not, but it doesn't "care" about whether an event, such as running someone over with a trolley, is a means or a side effect. (That is, unless it's gone to graduate school in philosophy.) The maximizing manual mode says the same thing in response to all of these cases: "Five for one? Good deal." And, once again, manual mode tends to prevail as long as there are no alarm bells going off, sending an opposite message. This is why we tend to give utilitarian answers in response to the *switch* and *loop* cases but not in response to the *footbridge* case.

For your convenience, this whole story is summarized in figure 9.12.

The myopic module inspects action plans and responds to any harm that it can see by sounding an emotional alarm. But it's blind to harmful side effects because it represents only the events that are causally necessary for the achievement of the goal, namely, the events that constitute the action plan's primary causal chain. This system has no problem with the action in the *switch* and *loop* cases because it can't see the harm (second row). But it sounds the alarm mightily in response to the *footbridge* case because, in that case, it can see the harmful event, which is right there on the primary causal chain (fourth row). Manual mode can represent harmful events that are side effects, as well as events that are causally necessary for achieving its goals. But manual mode doesn't "care" about whether a harm is a means to its end or whether it's merely a foreseen side effect, meaning that it doesn't attach more emotional weight to harm caused as a means. All manual mode cares about is the bottom line: Which action has the best net consequences? Thus, manual mode is generally happy to trade one life to save five (rows 1 and 3). These two systems interact as follows. When the emotional alarm is silent, the manual mode gets its way (rows 1 and 2). But when the emotional alarm goes off, the manual mode's reasoning tends to lose (rows 3 and 4). (Note that it doesn't always lose. Manual mode will likely override the emotional response, favoring its own cost-benefit reasoning, if the cost-benefit rationale is sufficiently compelling.) Thus, the myopic module hypothesis integrates Mikhail's theory of action plans with the dual-process theory of moral judgment, explaining why we care less about harms caused as side effects.

This theory depends on the supposition that the myopic module can see only what's on the primary causal chain of an action plan, the chain of events that are causally necessary for achieving the goal. But why should it be myopic in this way? Wouldn't it function better if it could read secondary causal chains as well? Perhaps it would, but this would be much more cognitively demanding, for two reasons. First, in our discussion so far, we've assumed that there are only two causal chains: a primary one and a secondary one. This is an enormous oversimplification. With the exception of weird, loopy cases, there is only one causal chain of events connecting the agent's body movement to the goal—that is, only one candidate for being the primary causal chain. But for any given action, there may be *many* secondary causal chains radiating out from the body movement. For example, when one hits the switch in the *switch* case, one kills the person on the sidetrack, but one can foresee causing many other things as well: One disturbs the air surrounding the sidetrack. One disturbs the friends and family of the person who gets run over. One causes oneself to have a memory of these events. One will perhaps get oneself in trouble with the law. And so on. For any action there are many, many foreseeable side effects. And thus, for a module to keep track of side effects as well as specifically intended effects, it would have to do many, many times the amount of work. That would prevent it from operating efficiently, which is a must, given its function.

The second thing that makes it hard for an action-plan inspector to inspect all of the branches is that this would require a more sophisticated kind of memory system, something like what computer scientists call a "queue," a system for storing items to be processed in a particular order. Recall the idea that the myopic module reads action plans in a linear fashion, much as we recall song lyrics, with each link in the chain pulling along the next. This kind of processing is possible only if the chain is truly linear. If, however, the chain *branches*, then there has to be a memory store that keeps one task on hold (going down the second branch of the chain) and remembers to return to it once the first task (going down the first branch of the chain) is completed. The ability to engage in this kind of nested multitasking, with subordinate and superordinate goals, is easy for computers, but it's a real challenge for animal brains. It is, however, something that the human manual mode does very well. Still, it would be very hard, perhaps impossible, for a simple cognitive module, an automatic setting, to pull this off.

Thus, the supposition that the myopic module is blind to side effects, far from being undermotivated, actually makes perfect cognitive sense. It would be very hard for an efficient, automated action inspector to "review" all of an action's foreseeable side effects, because there are too many of them, and because it would require a rather sophisticated kind of memory system.

We've covered a lot of rather technical ground here, but the bottom line is this: If the modular myopia hypothesis is correct, then the intuitive moral distinction we draw between harm caused as a means and harm caused as a side effect may be nothing more than a cognitive accident, a by-product. Harms caused as a means push our moral-emotional buttons not because they are objectively worse but because the alarm system that keeps us from being casually violent lacks the cognitive capacity to keep track of side effects. More on this shortly. But first, let's consider another classic moral distinction that stands in the way of maximizing happiness.

DOING AND ALLOWING

Once, in college, on my way out of the dining hall, I tried unsuccessfully to throw my napkin away. I crumpled it up and lobbed it at the overflowing trash bin, but it bounced off and landed on the floor. Being a decent person, I didn't want my actions to make the mess worse, so I went over to pick up my napkin and put it in the bin. But I couldn't tell which of the many crumpled napkins on the floor was mine. I spent an embarrassingly long time staring at this scattering of napkins, trying to figure out which one's location was most consistent with the terminus of my own napkin's trajectory. Eventually I decided that this was stupid. *A napkin is a napkin is a napkin*, I thought, and there's no point in trying to distinguish the one that arrived through my actions from the others. I selected a napkin, more or less at random, and put it in the bin. But then I had a new problem: Why stop at one napkin? I had already crossed the line from clean-up-your-own-mess to clean-up-someone-else's-mess. I'd already touched an icky, *other-person* napkin. So why not grab a few more? But how many more? I picked up a handful of other napkins, put them in the bin, and went on my way.

The conviction that one bears a greater moral responsibility for the napkins one throws oneself, as compared with napkins that arrive by other means, has a distinguished philosophical history. This conviction has been canonized as the "Doctrine of Doing and Allowing," which says that harms caused by actions, by things that we actively do, are worse than harms of omission. This idea is intuitively compelling and plays an important role in real-world moral decision making. For example, according to the American Medical Association's ethical guidelines, it is never accept-able for a doctor to actively (and intentionally) bring about a patient's death, but it is acceptable for a doctor to (intentionally) *allow* a patient to die, under certain circumstances. Our sensitivity to this distinction also affects our responses to preventable suffering. You wouldn't cause a deadly earthquake, but you are willing to allow earthquake victims to die by fail-ing to contribute to the rescue effort. You wouldn't kill people in Rwanda or Darfur, but you would allow others to kill them by failing to actively support their defense. And so on.

According to utilitarians, the distinction between doing and allowing is morally irrelevant, or at least has no independent moral force. *A harm is a harm is a harm*, we say, and there is no fundamental moral distinction between harms that we actively cause and harms that we merely allow to happen. (There are, however, nonfundamental, practical differences. More on this shortly.) Given our values and our circumstances, does it make sense to draw a moral distinction between what we do and what we allow to happen? As with the means/side-effect distinction, I believe that we can explain our tendency to draw a fundamental moral distinction between actions and omissions in terms of more basic cognitive mecha-nisms, ones that have nothing to do with morality per se. In other words, I think that the (independent) moral authority of the doing/allowing dis-tinction can be *debunked*.

Forget, for a moment, about morality. Why would an animal's brain distinguish between things that it actively causes to happen and things that it merely allows to happen? Right now, as you read this book, you are ac-tively causing your eyes to move across the page, actively causing the pages to turn, and so on. That's what you're *doing*. But think of all the things that you are *not doing*. You are not teaching a poodle to dance, not writing a fan letter to Rod Stewart, not juggling flaming torches, and not installing a hot tub in your basement. And that's just the beginning. At any given moment, there are infinitely many things that you are not doing, and it would be impossible for your brain to represent all of them, or even a significant frac-tion of them. (Sound familiar?) What this means is that an agent's brain *must*, in some sense, privilege actions over omissions. We have to represent actions in order to perform them, in order to make sure they go as planned, and to understand the actions of others. But we simply can't keep track of all the things that we and others *don't* do. This doesn't mean that we can't think about omissions, but it does mean that our brains have to represent actions and omissions in fundamentally different ways, such that representations of actions are more basic and accessible.

The fact that action representations are more basic can be seen in infants. In a pioneering attempt to understand the cognitive roots of the action/omission distinction, Fiery Cushman, Roman Feiman, and Susan Carey conducted an experiment in which six-month-old infants watched an experimenter choose between pairs of objects. The infants were trained to recognize the experimenter's preferences. In front of the infant, there would be, say, a blue mug on the right and a red mug on the left. The experimenter would then take the blue mug. Next time, the choice might be between a blue mug on the right (as before) and a green mug on the left, and once again the experimenter would choose the blue mug on the right. Trial after trial, the experimenter would choose the blue mug on the right over some other mug on the left, where the other mug is a different color every time. Then, in the critical test trial, the blue mug would appear on the *left* and a new, other-colored mug would appear on the right. For some infants, the experimenter would choose the blue mug on the left. For other infants, the experimenter would choose the other mug on the right. The key question: Which choice was more surprising to the infants? On the one hand, the experimenter had so far always chosen the blue mug. On the other hand, the experimenter had so far always chosen the mug on the right. So what would the infant expect: choosing blue or choosing right? It turned out that the infants looked longer when the experimenter chose the nonblue mug on the right, an indication that the infants were *surprised* by this. In other words, the infants expected the experimenter to choose the blue mug. What this means is that these six-month-old infants' brains represented the fact that the experimenter wanted the blue mug and acted so as to acquire it.

That was just half of the experiment. In the other half, infants once again watched the experimenter choose between a blue mug and a different-colored mug, over and over. But this time the experimenter always chose the *other mug*, that is, not the blue one on the right. And then, in the critical test trial, the experimenter once again had to choose between a blue mug on the left and a new, different-colored mug on the right. The question, once again, concerned the infants' expectations. The first time, when they saw the experimenter choose the blue mug over and over, they had expected the experimenter to choose the blue mug. But this time they saw the experimenter *not* choose the blue mug over and over. (A repeated omission.) Would the infants now expect the experimenter to *not* choose the blue mug?

No. That is, the infants showed no sign of being surprised when the experimenter chose the blue mug. What this means is that the infants could grasp the idea of "choosing the blue mug," but they couldn't grasp the idea of "not choosing the blue mug." Note that these infants could *distinguish* between choosing the blue mug and not choosing the blue mug. If they couldn't do that, then they wouldn't have been surprised originally when the experimenter had chosen the nonblue mug, after choosing blue many times. But all that means is that the infants could expect something and know that their expectation had not been met. What these infants apparently could not do was represent the idea of "not choosing the blue mug" as a distinct behavior. They couldn't watch the experimenter and think to themselves, *There he goes again, not choosing the blue mug.*

What this means, then, is that representing a specific goal-directed action, such as choosing a blue mug, is a fairly basic cognitive ability, an ability that six-month-old infants have. But representing an omission, a failure to do some specific thing, is, for humans, a less basic and more sophisticated ability. Note that this is not because representing an omission necessarily requires substantially more complex information processing. If there are only two possibilities—choosing A and not choosing A—then representing what is not done is not much harder than representing what is done. If you were programming a computer to monitor and predict someone's two-alternative mug selections, you could program the computer to represent "didn't choose the blue mug" almost as easily as "chose the blue mug." (All you'd need is a little "not" operator to turn the latter representation into the former.) Nevertheless, it appears that humans find it much easier to represent what one does rather than what one doesn't do. And that makes sense, given that in real life, it's more important to keep track of the relatively few things that people do, compared with the millions of things that people could do but don't.

The fact that babies represent doings more easily than nondoings makes a prediction about adults: When human adults distinguish between harmful actions and omissions (nondoings) in their moral judgments, it's the result of automatic settings, not the manual-mode application of the Doctrine of Doing and Allowing. Cushman and I tested this prediction in a brain-imaging study in which people evaluated both active and passive harmful actions. As predicted, we found that *ignoring* the action/omission distinction—treating passive harm as morally equivalent to active harm—requires more manual-mode DLPFC activity than abiding by the action/omission distinction.** This makes sense, given that representations of omissions are inherently abstract. An action, unlike an omission, can be represented in a basic sensory way. It's easy, for example, to draw a picture of someone running. But how do you draw a picture of someone *not* running? You can draw a picture of someone standing still, but this will convey something like "person" or "woman" or "standing" rather than "not running." The conventional way to represent what something is not is to use an abstract symbol, such as a circle with a slash through it, conjoined with a conventional image. But a conventional image can't do the job by itself. You need an abstract symbol.

Actions, in addition to having natural sensory representations, also have natural motor representations. Reading words like "lick," "pick," or "kick" automatically increases activation in the subregions of the motor cortex that control, respectively, the tongue, fingers, and feet. But there is no part of the brain that ramps up when people think about actions that do not involve the tongue (etc.) because there is no part of the brain specifically devoted to performing actions that do not involve the tongue.

As we saw earlier, our emotions, and ultimately our moral judgments, seem to be sensitive to the sensory and motor properties of actions, to things like pushing. (And to visual imagery of pushing; see pages 46–48.)

Omissions, unlike actions, have no distinctive sensory and motor properties, and must therefore lack at least one kind of emotional trigger. Moreover, this basic sensory/motor distinction between actions and omissions may carry over into the realm of more physically amorphous behaviors, depending on how they are conceptualized. For example, the idea of "firing" someone (active) feels worse than "letting someone go" (passive). This parallels the results of a study by Neeru Paharia, Karim Kassam, Max Bazerman, and myself, showing that jacking up the price of cancer drugs feels less bad if it's done indirectly through another agent, even if the physical action itself is no more indirect.

The hypothesis, then, is that harmful omissions don't push our emotional moral buttons in the same way that harmful actions do. We represent actions in a basic motor and sensory way, but omissions are represented more abstractly. Moreover, this difference in how we represent actions and omissions has nothing to do with morality; it has to do simply with the more general cognitive constraints placed on our brains—brains that couldn't possibly keep track of all the actions we fail to perform and that originally evolved as sensory and motor devices, not as abstract thinking devices. Once again, it seems that a hallowed moral distinction may simply be a cognitive by-product. (But, as I'll explain shortly, there is room for some utilitarian accommodation of the action/omission distinction.)

UTILITARIANISM VERSUS THE GIZMO

Maximizing happiness sounds like a splendid idea, but it could, at least in principle, mean doing terrible, horrible things. What to do? Our understanding of the moral brain tells us that these "in principle" problems may not be such big problems after all. This isn't because "in principle" problems are never worth worrying about. Rather, it's because we now have evidence concerning the reliability of our gut reactions to alarming acts. Given how our dual-process moral brains work, it's all but guaranteed that there will be cases in which doing something that really is good feels terribly, horribly wrong. Likewise, it's all but guaranteed that there will be cases in which doing something that really is bad feels perfectly fine. As we wrap up our account of alarming acts, we'll return to our favorite fruit fly, the *footbridge* dilemma, and then consider the broader significance of what we've learned.

It seems wrong to push the man off the *footbridge*, even though doing so saves more lives. Why is that? We know from chapter 4 that this is the work of an automatic setting. But that is, once again, only a partial answer. A more complete answer comes from our new understanding of this cognitive gizmo's operating characteristics: What does it respond to? And what does it *not* respond to? Let's start with the first question.

First, this automatic setting responds more to harm caused as a *means* to an end (or as an end), rather than as a side effect. (But not if the harmful means is structured like a side effect, a quirky signature of the mechanism.) In other words, it responds to harms that are *specifically intended.** Second, it responds more to harm caused *actively*, rather than passively. And, third, it responds more to harm caused directly by *personal force*, rather than more indirectly. It seems that these are not three separate criteria, employed in checklist fashion. Rather, they appear to be intertwined in the operation of our alarm gizmo, forming an organic whole. Once again, the personal-force factor and the means/side-effect factor *interact*. When the harm is not specifically intended, it doesn't matter if it's caused by personal force. And if the harm is not caused by personal force, then it matters much less that the harm is specifically intended. What's more, the distinction between active and passive harm appears to be intertwined with the other two factors, as demonstrated by a common error that people make when justifying their intuitive trolley judgments.

If you ask people to explain why it's wrong to push the man in the *footbridge* dilemma but acceptable to hit the switch in the *switch* dilemma, they will often appeal to the act/omission distinction, even though it does not apply. People say things like this: "Pushing the guy off the footbridge is *murder*. You're *killing* the guy. But in the other case, you're just allowing him to be killed by the trolley." This explanation doesn't actually work. In both cases, the killing is active. Consider this: If you were to turn a trolley onto someone with the *specific intention* of killing that person, that would be plenty active and plenty murderous. In the *switch* case, the physical action is no less active than in this case of murder by trolley. But because the harm in the *switch* case is caused as a side effect, and without the direct application of personal force, it *feels* less active. Thus, it appears that these three factors are all contributing to the same feeling.

This also fits with our theory about the cognitive mechanics of the means/side-effect distinction. Once again, according to the modular myopia hypothesis, harmful side effects don't set off the alarm because the harmful event is not on the primary chain of the *action plan*. But harms caused passively don't have action plans (at least not usually), because they're not active. There is no body movement, and therefore no chain of events to connect the nonexistent body movement to a goal. Thus, the theory of action plans, which explains our sensitivity to the means/side-effect distinction, gives us, for free, a more detailed explanation of our sensitivity to the action/omission distinction.*

Personal force may play a role in action plans as well. The events in an action plan are arranged not just in a temporal sequence but in a *causal* sequence. Each event causes the next, as we go from the body movement to the goal (hit the switch . . . turn the trolley . . . save the five). There is evidence that we represent causes in terms of *forces*. When you see one billiard ball knock into another, all that you pick up with your retinas is balls in a series of locations, like the frames of a movie. Nevertheless, we intuit, apparently correctly, the delivery of a force from one ball to another. Thus, the kinds of forces represented in an action plan—personal force versus other kinds—may affect the extent to which one feels as if one is personally causing harm. And, of course, the application of personal force is related to the action/omission distinction because omissions, by definition, do not involve the application of personal force.

Putting these three features together, it seems that our alarm gizmo responds to actions that are *prototypically violent*—things like hitting, slapping, punching, beating with a club, and, of course, pushing.* There are behaviors that lead to harm and that lack all three features—behaviors such as not saving people by not giving money to charity—but such actions don't feel at all violent. Likewise, it's hard, maybe impossible, to think of actions that don't feel violent but that involve causing harm actively, through the application of personal force, and with the specific intention of causing harm, if only as a means to an end.* To say that this automatic alarm system responds to violence probably gets things backward. Rather, I suspect that our conception of violence is *defined* by this automatic alarm system.

We've talked about what the gizmo responds to. But what does it ignore? Among other things, it seems to ignore any *benefits* that might be achieved through violence. My collaborators and I gave people a version of the *footbridge* case in which millions of people can be saved by pushing. The trolley, if not stopped, will collide with a box of explosives as it's crossing the top of a large dam, thus bursting the dam, flooding a large city, and killing millions of people. In our sample, 70 percent approved of pushing in this case. That's substantially lower than the 87 percent approval we got for turning the trolley in the *switch* case, despite the fact that, here, the benefits of pushing are roughly a million times larger. Thus, it seems that the gizmo doesn't "care" about what else might be at stake. Of course, many more people approved of pushing here, as compared with the original *footbridge* case, in which only five lives are at stake. Clearly, our *judgments* are affected by the numbers. But this seems to be because the emotional alarm gets *ignored* or *overridden* when the numbers are high, not because the alarm is silent. This is clear enough from introspection: Pushing someone off a footbridge to save a million people doesn't *feel* any better than pushing to save five. And there's experimental evidence for this as well. People who are less willing to trust their intuition (more "cognitively reflective") are more likely to approve of pushing to save millions, indicating that this action is counterintuitive, despite the fact that a majority of people approve of it.

Thus, we now have a pretty good sense for what the gizmo does and does not do. In a nutshell, it responds negatively to prototypically violent acts, independent of whatever benefits those acts may produce. In light of this, how seriously should we take the advice we get from this gizmo?

In general, I think we should take its advice very seriously. Violence is generally bad, and therefore it's very good that we have in our brains a little gizmo that screams at us when we contemplate using violence to achieve our goals. Without the gizmo, we would all be more psychopathic.

This alarm system also provides a good hedge against overconfidence and bias. Even if you're contemplating violence with the best of intentions ("The revolution may be bloody, but think of our glorious future!"), the alarm bell says, "Careful! You're playing with fire!" That's a good voice to have in one's head. (Lenin, Trotsky, Mao . . . take note.) In sum, the alarm system that makes us wince at the thought of pushing an innocent person to his death is, on the whole, a very good thing.

But . . . however indispensable the human antiviolence gizmo may be, it makes little sense to regard it as *infallible* and to elevate its operating characteristics into moral principles. The gizmo may distinguish between means and foreseen side effect not because this distinction has any inherent moral value, but simply because the gizmo lacks the cognitive capacity to keep track of multiple causal chains. Likewise, this gizmo may distinguish between active and passive harm not because actively causing harm is inherently worse than passively causing harm, but because this system is designed to evaluate plans for *action* and because our brains represent actions differently from nonactions. Finally, the gizmo may respond more to harms caused using personal force not because personal force matters per se, but because the most basic nasty things that humans can do to one another (hitting, pushing, etc.) involve the direct application of personal force.

This is not to say that these distinctions bear no meaningful relationship to things that matter morally. For one thing, the fact that most of us draw these distinctions allows us to make inferences about the moral character of people who *don't* draw them. As noted earlier, someone who is willing to cause harm through personal force shows especially strong signs of having a defective antiviolence alarm system: If this person had a normal moral sense, he wouldn't do that. And if one lacks a *normal* moral sense, there's a good chance that one lacks an *adequate* moral sense. (The odds of having a good but abnormal moral sense are slim.) Likewise for people who specifically intend harm and for people who actively cause harm. All of this, however, can be accommodated within a utilitarian framework. In evaluating *people,* it makes sense to take the action/omission distinction, the means/side-effect distinction, and the personal-force/no-personal-force distinction seriously—not because these distinctions reflect deep moral truths, but because people who ignore these distinctions are morally abnormal and therefore especially likely to cause trouble.

The action/omission distinction may also have substantial utilitarian value. Without it, you're responsible for all of the trouble that you can prevent, as if you yourself had caused it all (*napkins, napkins everywhere . . .*). Given that we can't all be superheroes responsible for solving all of the world's problems, it makes sense for each of us to take special responsibility for our own actions (*Just clean up your own napkin!*). Note, by the way, the parallel with our explanation for why the action/omission distinction is intuitive. Both explanations are based on the fact that omissions vastly outnumber actions. For even one moment of behavior, it's impossible for our brains to keep track of all the things that one doesn't do (omissions). Likewise, it's impossible for us to take responsibility for all of the problems that arise from the things we don't do. Still, this doesn't mean that causing harm actively is really inherently worse than harm caused passively. (*A napkin is a napkin is a napkin!*)

Our intuitive distinction between harm caused as a means and harm caused as a side effect may also have important practical benefits. There may be no good reason to distinguish specifically intended harms from harms that are *foreseen* side effects. But it's certainly important to distinguish harms that are specifically intended from harms that are *unforeseen* side effects—that is, *accidents*. Someone who harms people by accident may be dangerous, but someone who specifically intends to harm people as a means to his ends is *really* dangerous. Such people may or may not be more dangerous than people who knowingly cause harm as collateral damage. But at the very least, a moral alarm that responds to specifically intended harms distinguishes those "Machiavellian" harms from accidental harms. That's a good thing, even if the alarm draws the line in the wrong place, treating *foreseen* side effects as if they were accidents.

Thus, as I've said, there are many good reasons to be glad that we have this antiviolence gizmo in our brains. But our key question is this: Should we let the gizmo determine our overarching moral philosophy? Should we be persuaded by the gizmo not to pursue the greater good? From the *footbridge* dilemma and others like it, should we draw the lesson that sometimes it's really, truly wrong to maximize happiness? Or should we conclude instead that the *footbridge* dilemma is *weird* and not worth worrying about?

I've called the *footbridge* dilemma a moral fruit fly, and that analogy is doubly appropriate because, if I'm right, this dilemma is also a moral *pest*. It's a highly contrived situation in which a prototypically violent action is guaranteed (by stipulation) to promote the greater good. The lesson that philosophers have, for the most part, drawn from this dilemma is that it's sometimes deeply wrong to promote the greater good. However, our understanding of the dual-process moral brain suggests a different lesson: Our moral intuitions are generally sensible, but not infallible. As a result, it's all but guaranteed that we can dream up examples that exploit the cognitive inflexibility of our moral intuitions. It's all but guaranteed that we can dream up a hypothetical action that is actually good but that seems terribly, horribly wrong because it pushes our moral buttons. I know of no better candidate for this honor than the *footbridge* dilemma.

Now, you may be wondering—people often do—whether I'm really saying that it's right to push the man off the footbridge. Here's what I'm saying: If you don't *feel* that it's wrong to push the man off the footbridge, there's something wrong with you. I, too, feel that it's wrong, and I doubt that I could actually bring myself to push, and I'm glad that I'm like this. What's more, in the real world, not pushing would almost certainly be the right decision. But if someone with the best of intentions were to muster the will to push the man off the footbridge, knowing for sure that it would save five lives, and knowing for sure that there was no better alternative, I would approve of this action, although I might be suspicious of the person who chose to perform it.

Next question: If dilemmas like the *footbridge* case are weird, contrived, and worth ignoring, then why have I spent so much time studying them? Answer: These dilemmas are worth ignoring for some purposes, but not for others. If we're looking for a guide to a workable metamorality, then we should ignore the *footbridge* dilemma. We should not let the alarm bells that it sets off stop us from pursuing the greater good. But if we're looking for a guide to *moral psychology*, then we should pay it gobs of

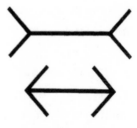

Figure 9.l3. In the famous Müller-Lyer illusion two lines of equal length appear to differ in length.

attention. As I hope is now clear, these weird dilemmas are wonderful tools for figuring out how our moral brains work. Indeed, their scientific role is almost exactly analogous to the role of visual illusions in the science of vision. (Visual illusions are so prized by vision scientists that the Vision Sciences Society awards a prize for the best new illusion each year.) Take, for example, the familiar Müller-Lyer illusion, which shows that the visual system uses converging lines as depth cues:

In figure 9.13 the top horizontal line appears to be longer than the bottom one, but they are actually the same length. Just as visual illusions reveal the structure of visual cognition, bizarre moral dilemmas reveal the structure of moral cognition. They are moral illusions—revealing for the manner in which they mislead us.

Trolley dilemmas may be weird, but there are real-world moral problems that mirror their weirdness, with life-and-death consequences. Once again, the *footbridge* dilemma is weird because it's a case in which a proto-typically violent act promotes the greater good. This rarely happens in ordinary life, but it happens regularly in the domain of bioethics, where modern knowledge and technology give us the opportunity to do pseudo-violent things that promote the greater good.

Consider, once again, the American Medical Association's stance on physician-assisted suicide. The AMA has essentially endorsed both the Doctrine of Double Effect and the Doctrine of Doing and Allowing. If I'm right, the AMA is essentially endorsing the operating characteristics of a myopic module. As a result, chronically ill patients may suffer simply be-cause we lack the fortitude to actively, intentionally, and personally do what's best for them—and what they want for themselves. (This is not to say, of course, that there aren't utilitarian reasons for being extremely cautious when it comes to ending patients' lives. But the AMA is categorically opposed to physician-assisted suicide as a matter of "principle.") The gizmo's operating characteristics may also influence people's attitudes toward mandatory vaccination and policies surrounding organ donation and abortion. Indeed, trolleyology was born as part of a discussion about abortion and the Doctrine of Double Effect.

Beyond medicine, the gizmo's operating characteristics may influence our attitudes toward capital punishment, torture, and war. In all of these cases, violent or quasi-violent actions that push our moral buttons may serve the greater good, and we may reject these actions not because we've worked through all of the relevant moral considerations, but because of how they feel. And just as our moral alarm bells may overreact by being blind to the benefits of certain alarming actions, our alarm bells may underreact by being blind to the costs of actions that are not alarming. For example, when we harm people (including future people) by harming the environment, it's almost always as a side effect, often passive, and never through the direct application of personal force to another person. If harming the environment felt like pushing someone off a footbridge, our planet might be in much better shape.*

It's worth noting that skepticism concerning the wisdom of the gizmo cuts both ways politically. Our alarm bells may fuel opposition to physician-assisted suicide and abortion, but they may also fuel opposition to torture and capital punishment. I am not, here, coming out for or against any of these policies. Rather, I'm suggesting that we train ourselves to think about them differently.

It's good that we're alarmed by acts of violence. But the automatic emotional gizmos in our brains are not infinitely wise. Our moral alarm systems think that the difference between pushing and hitting a switch is of great moral significance. More important, our moral alarm systems see no difference between self-serving murder and saving a million lives at the cost of one. It's a mistake to grant these gizmos veto power in our search for a universal moral philosophy.

10.

Justice and Fairness

In the last chapter, we focused on actions that use alarming means in pursuit of good ends. Here we'll focus on ends: What should we be aiming for? Some say that the greater good is incompatible with justice, forcing us to do unfair things to others and even ourselves. We'll consider this objection and the psychology behind it.

As before, we'll employ our two strategies. Sometimes we'll *accommodate*, arguing that maximizing happiness in the real world doesn't have the absurd conclusions that some think it has. Other times we'll argue for *reform*, using our cognitive and evolutionary understanding of moral psychology to cast doubt on our intuitive sense of justice.

IS UTILITARIANISM TOO DEMANDING?

As explained in chapter 8, being a happiness maximizer is tough because the world is full of avoidable unhappiness. Once again, if you want to save someone's life, you can probably do it for about $2,500, which you can think of as $500 a year for five years, or $500 each from you and four friends. And short of saving a life, you can still relieve a lot of misery with a modest donation—less than you might spend on dinner in a restaurant. Simply put, a dollar spent in the right way on others buys a lot more happiness than it buys you, your family, or your friends.

Perhaps you're wondering if this is really true. And let's be honest. Part of you may be *hoping* that it's not true. Because if there's nothing you can do to help the world's most unfortunate people, then you're off the hook. Sorry, but you're on the hook. And so am I. It's true that some charitable projects backfire, doing more harm than good. And it's true that money sent with the best intentions sometimes falls into the wrong hands, filling the coffers of nasty dictators. But today, you can't excuse yourself from helping on the grounds that helping is impossible. International aid organizations are more effective and more accountable than ever. And even if some of them are bad, you only need one good one to be on the hook. There are plenty of good ones, and even if the world's best humanitarian organizations were to frivolously waste half of their money (which they don't), we'd still be on the hook, because merely doubling the cost of helping doesn't fundamentally change the math. Today, there's no denying that you can—if you want to—use your money to help people who desperately need help.

Ultimately, that's good news, but it does put people like us—people with at least some disposable income—in a tough moral position. A hundred dollars from you can feed a poor child for months. How, then, can you spend that money on things you don't really need? And what about the next $100, and the next $100? Can you take a vacation? Take someone out on a date? Have hobbies? Choose a career that is not maximally lucrative? Have a birthday party for your child? *Have a child in the first place?* Can you order toppings on your pizza? Are you even allowed to have pizza? For a perfect utilitarian, the answer to all of these questions is the same: Such indulgences are allowed insofar as they are absolutely necessary to keep yourself minimally happy and to maximize your ability to increase the happiness of others, mostly strangers you will never meet. In short, being a perfect utilitarian requires forsaking almost everything you want out of life and turning yourself into a happiness pump.

A few years ago, a philosopher gave a talk at a conference defending this utilitarian ideal. During the question period, another philosopher stood up and pointed to the speaker's laptop: "That thing cost at least a thousand dollars," he said. "How can you justify *that* when there are people starving in the world?" To which the speaker replied, "I can't! But at least I have the decency to admit that I'm a hypocrite!" This answer, I think, is not just funny, but enlightened. (It's also probably incorrect. As I explain below, he almost certainly *can* justify owning a laptop.)

The worry that utilitarianism is too demanding is a devastating worry only if we expect ourselves to be *perfect* utilitarians, and trying to be a perfect utilitarian is, in fact, a very un-utilitarian thing to do. Consider, for example, the analogous dilemma we face as we try to eat a healthy diet. As a perfectly healthy eater, you would identify the healthiest possible set of foods and consume only those foods, in precisely optimal quantities. If you were to maintain a perfect diet, chances are you would never eat your favorite foods, not even on your birthday. You would travel with bundles of optimal food, because, odds are, optimal food will not be available wherever you're going. Upon receiving a dinner invitation from a friend, you would either decline, eat before, eat after, or bring your own optimal food. (BYOOF!) You would never take a date to a restaurant, or you would only take someone to restaurants that serve optimal food. And so on.

If you were a food-consuming computer, maintaining an optimal diet might be a realistic goal. But as a real person with limited time, money, and willpower, trying to maintain a physiologically optimal diet is not, in fact, optimal. Instead, the optimal strategy is to eat as well as you can, given your real-world constraints, including your own psychological limitations and including limitations imposed on you as a social being. This is challenging because there's no magic formula, no bright line between the extremes of perfectionism and unbridled gluttony. To be the healthiest that you can actually be—not in principle, but in practice—you have to set reasonable goals, which will inevitably be somewhat arbitrary, and then work reasonably hard to attain them.

The same goes for being a real-world, flesh-and-blood utilitarian, only more so. The ideal utilitarian "moral diet" is simply incompatible with the life for which our brains were designed. Our brains were not designed to care deeply about the happiness of strangers. Indeed, our brains might even be designed for indifference or malevolence toward strangers. Thus, a real-world, flesh-and-blood utilitarian must cut herself a lot of slack, even more than a real-world healthy eater.

What does this mean in practice? Here, too, there's no magic formula, just an ill-defined Goldilocks zone between two extremes. Being a flesh-and-blood utilitarian does not mean trying to turn yourself into a happiness pump. To see why, one need only consider what would happen if you were to try: First, you wouldn't even try. Second, if you *were* to try, you would be miserable, depriving yourself of nearly all of the things that motivate you to get out of bed in the morning (that is, if you still have a bed). As a halfhearted happiness pump, you would quickly rationalize your way out of your philosophy, or simply resign yourself to hypocrisy, at which point you'd be back where you started, trying to figure out how much of a hypocrite, and how much of a hero, you're willing to be.

At the same time, being a flesh-and-blood utilitarian doesn't mean being a *complete* hypocrite, giving yourself a free pass. Your inability to be a perfect utilitarian doesn't get you off the hook any more than your inability to eat a perfect diet justifies gorging yourself at every meal. There clearly are things you can do that relieve a lot of suffering while requiring comparably little sacrifice on your part. How much sacrifice should you make? Again, there's no magic formula, and it all depends on your personal circumstances and limitations. There's a social dimension to the problem that may, in the long run, favor strong efforts over heroic ones. Your life is a model for others, especially your children (if you have children). If you improve the lives of hundreds of people every year through your charitable donations, but your life remains happy and comfortable, you're a model that others can emulate. If, instead, you push yourself to just shy of your breaking point, you may do more good directly with your personal donation dollars, but you may undermine the larger cause by making an unappealing example of yourself. Promoting a moderate, sustainable culture of altruism may, in the long run, do more good than pushing yourself to your personal limit. Heroes who make enormous sacrifices for others are "inspiring," but when it comes to motivating real-world behavior, research shows that the best way to get people to do something good is to tell them that their neighbors are already doing it.

The more general point is this: If what utilitarianism asks of you seems absurd, then it's not what utilitarianism actually asks of you. Utilitarianism is, once again, an inherently practical philosophy, and there's nothing more impractical than commanding free people to do things that strike them as absurd and that run counter to their most basic motivations. Thus, in the real world, utilitarianism is demanding, but not overly demanding. It can *accommodate* our basic human needs and motivations, but it nonetheless calls for substantial *reform* of our selfish habits.

Still, one might object to the moderate reform that real-world utilitarianism demands. Helping strangers, one might say, is admirable but entirely optional. Is that a defensible moral stance or just a comfortable rationalization? With this in mind, let's consider the moral problem as Peter Singer originally posed it* and the psychology behind our intuitive sense of duty to help others.

THE DUTY TO HELP

Suppose that you're out for a stroll in the park when you encounter a young child drowning in a shallow pond. You could easily wade in and save this child's life, but if you do, you will ruin your new Italian suit, which cost more than $500. Is it morally acceptable to let this child drown in order to save your suit? Clearly not, we say. That would be morally monstrous. But why, asks Singer, is it morally acceptable for you to spend $500 on a suit if you could instead use that money to save a child's life by making a donation to an international aid organization? In other words, if you think that saving the drowning child is morally obligatory, then why is saving poor, faraway children morally optional?

(Again, $500 might not be enough to save a life, but according to a well-dressed colleague, it's also not enough for a truly snazzy suit. In any case, you can imagine that you have four friends who will donate if you will, or that this situation plays out four more times over the coming years.)

First, let's take a moment to acknowledge the value of nice suits. Suppose you're a corporate lawyer who orchestrates deals worth hundreds of millions of dollars. For you, shopping at J.C. Penney would be penny-wise and pound-foolish. Your elegant clothing projects confidence and competence and is a sound financial investment. The same goes for the handsome oak furniture in your office, your country club membership, your fine home, suitable for entertaining, and much more besides. (And the same goes for your laptop computer, if you're an academic who makes his living by reading and writing.) The more general point is this: Many apparently unnecessary luxuries can, despite appearances, be justified on utilitarian grounds. That's a valid point, and a congenial one for those of us, myself included, who are reluctant to radically change our lifestyles. But it doesn't make Singer's problem go away. That's because there's still no denying that we in the affluent world have some income that is truly disposable. If you let a child drown to save your financially functional suit, you're still a moral monster. Why? Because if you can afford a suit like that, you can also afford to replace it. And if you can afford to replace your suit after saving a drowning child's life, then you can afford to save a faraway child's life before buying your next suit. Once we've taken care of our own needs—broadly construed—we must face up to our moral opportunities.

Perhaps we're allowed to ignore the plight of faraway children because they are (in our version of the story, at least) citizens of foreign nations. But then, is it acceptable to let a foreign child drown in a foreign pond while traveling abroad? Perhaps your obligation to distant children is diminished by the existence of many other people who could help them. In the case of Singer's drowning child, you're the only one there who can help. But how important is that? Suppose there are other people standing around the pond, aware of the drowning child and doing nothing to help. Now is it okay to let the child die? Lesson: It's surprisingly hard to justify treating the nearby drowning child and the faraway starving child differently.

Nevertheless, it seems obvious that we must save the nearby drowning child, and it seems obvious that donating to foreign aid is, for the most part, morally optional. In other words, these two cases are *intuitively* very different. Should we trust our intuitions? When we see these two cases as very different, are we having a moral insight? Or does this simply reflect the inflexibility of our automatic settings?

To help us answer this question, Jay Musen and I conducted a series of experiments aimed at identifying the factors that affect our judgments in cases like these, essentially doing the "trolleyology" of Peter Singer's problem.* In these experiments, we didn't look at all of the potentially relevant factors. Obviously it matters, for example, if one of the children is your child or your niece. But this difference in relatedness plays no role in Singer's original problem.* We want to know why we say "You must!" to saving the drowning child while insisting that fighting global poverty is admirable but optional.

In our experiments, the factor that had the biggest effect by far was *physical distance.* For example, in one of our scenarios you're vacationing in a developing country that has been hit by a devastating typhoon. Fortunately for you, you've not been affected by the storm. You've got a cozy little cottage in the hills overlooking the coast, stocked with everything you need. But you can help by giving money to the relief effort, which is already under way. In a different version of this scenario, everything is the same except that, instead of being there yourself, it's your friend who is there. You're at home, sitting in front of your computer. Your friend describes the situation in detail and, using the camera and microphone in his smartphone, gives you a live audiovisual tour of the devastated area, re-creating the experience of being there. You can help by donating online.

In response to the versions of this scenario in which you are physically present, 68 percent of our subjects said that you have a moral obligation to help. By contrast, when responding to the versions in which you're far away, only 34 percent of our respondents said that you have a moral obligation to help. We observed this big difference despite the fact that, in the faraway versions, you have all of the same information and you are just as capable of helping.

It's worth emphasizing that this study controlled for many of the factors that people typically cite when they resist Singer's utilitarian conclusion. In none of our scenarios does one have a unique ability to help, as in the case of Singer's drowning child. In all of our scenarios, the aid is delivered in the same way: through a reputable organization that accepts donations. Our experiments control for whether the aid is sought in response to a specific emergency (drowning child) or to an ongoing problem (poverty). In all of our scenarios, the victims are citizens of foreign countries, thus removing a patriotic reason for helping some more than others. Finally, our experiments control for whether the unfortunate circumstances were brought about by accident rather than by the actions of other people who might bear more responsibility for helping, thus relieving you of yours. In short, there are very few differences between the near and far versions of our helping dilemmas, suggesting that our sense of moral obligation is heavily influenced by mere physical distance or other factors along these lines.*

Should mere physical distance matter? As in Trolleyland, one can argue that physical distance matters when it comes to assessing people's *character*. Someone who allows a child to drown right in front of her because she's worried about her suit is a moral monster. But clearly, we're not all moral monsters for buying things like suits instead of making donations. Still, that doesn't mean that distance *really* matters. It just means that people who are insensitive to distance are morally abnormal, and being morally abnormal matters. Consider, once again, a friend who calls for moral advice: "Should I help those poor typhoon victims or not?" It would be rather strange to respond, "Well, it depends. How many feet away are they?"

It seems that we are, once again, being led around by our inflexible automatic settings: Nearby drowning children push our moral buttons; faraway starving children don't,* but the differences between them are morally irrelevant, just like personal force. Why would our moral buttons work this way? A better question is: Why *wouldn't* they? Once again, as far as we know, our capacity for empathy evolved to facilitate cooperation—not universally, but with specific individuals or with members of a specific tribe. If you help an in-group member in need, odds are good that your "friend in need" will, at some point, be a "friend indeed" (reciprocity). And it's possible that by helping members of one's tribe, one helps oneself indirectly, by making our tribe stronger than competing ones. (The drowning child you save today might someday lead your tribe into battle.) By contrast, there is no biological advantage—and I emphasize *biological*—to being universally empathetic. Traits prevail by natural selection because they confer a competitive advantage, either at the individual level or, perhaps, at the group level. Thus, it's not hard to explain why we are generally unmoved by the plight of faraway people. The harder question, from a biological point of view, is why we sometimes *are* moved by the plight of nearby strangers. That explanation may come from cultural evolution rather than biological evolution. As explained in chapter 3, some cultures have evolved norms whereby strangers are expected to behave altruistically toward one another, at least if the costs are not too high.

There is another salient difference between the nearby drowning child and the faraway poor children. The drowning child is a specific, identifiable person, whereas the children you might save with your donation dollars are, from your perspective, unidentified "statistical" people.* The economist Thomas Schelling observed that people tend to respond with greater urgency to identifiable victims, as compared with indeterminate, "statistical" victims. This is known as the "identifiable victim effect," which is illustrated by the case of Jessica McClure, a.k.a. "Baby Jessica."

In 1987, eighteen-month-old Baby Jessica fell down a well in Midland, Texas, and remained trapped there for nearly sixty hours. Strangers sent more than $700,000 to her family in support of the rescue effort, a sum that could have saved the lives of many children if spent on preventive healthcare. To let Baby Jessica die in the well would be unthinkable, morally monstrous. But failing to increase the state's budget for preventive healthcare for children is not so unthinkable. As Schelling observed in his seminal article on the identifiable-victim effect, the death of a particular person evokes "anxiety and sentiment, guilt and awe, responsibility and religion, [but] . . . most of this awesomeness disappears when we deal with statistical death."

Inspired by Schelling's observation, Deborah Small and George Loewenstein conducted a series of experiments examining our reactions to identifiable, as opposed to "statistical," victims. They gave each of ten participants an "endowment" of ten dollars. Some of them randomly drew cards that said "KEEP" and were allowed to retain their endowments, while others drew cards that said "LOSE" and subsequently had their endowments taken away, thus rendering them "victims." Each nonvictim drew a victim's number, and thus each nonvictim was paired with a victim. Critically, the nonvictims didn't know the identities of the people with whom they were paired. As a nonvictim, you might know that you're paired with "person #4," but you have no idea who that person is, and you never will. The nonvictims were allowed to give a portion of their endowments to the victims with whom they were paired, and each could choose how much to give. However—the crucial manipulation—some of the nonvictims were paired with (but didn't meet) specific victims *before* deciding how much to give, while others were paired up *after* deciding how much to give. The ones who decided how much to give before being paired up knew at the time that they would be paired up later. Thus, some people faced this question: "How much do I want to give to person #4 [determined victim]?" Others faced this question: "How much do I want to give to the person whose number I will draw [undetermined victim]?" Once again, at no point did the decision makers know who would receive their money.

The results: The median donation to the determined victims was more than double the median donation to the undetermined victims. In other words, people were more inclined to give money to "randomly determined person #4" than to "person #? to be randomly determined." This makes no sense. Surely it's irrelevant whether you choose the recipient first and then choose the amount, or vice versa, so long as you know nothing about the recipient in either case.

In a follow-up study, Small and Loewenstein measured self-reports of sympathy (effectively the same as "empathy") and found, as expected, that sympathy for the victims predicted donation levels. They also conducted a field experiment in which they gave people the opportunity to donate money to Habitat for Humanity. Each donation would help provide a home for a needy family. In some cases, the recipient family had been determined in advance, while in other cases the recipient family was to-be-determined. As in the lab experiment, none of the decision makers knew who would benefit from their donations. As predicted, the median donation to the determined families was twice as high as the median donation to the undetermined families. A more recent study shows, not surprisingly, that people are more inclined to give money to charity when donations are directed toward a specific needy person—in this case, a poor seven-year-old Malian girl named Rokia—rather than to the larger cause of poverty in Africa. No surprise, either, that people's donations to Rokia were correlated with their self-reported levels of sympathy for her. What is surprising, however, is that people were *less* willing to help Rokia when, in addition to presenting Rokia's personal story, the researchers also presented statistics describing the broader problem of poverty in Africa. This turned Rokia into a "drop in the bucket." The numbers, however, need not be mind-numbingly large to dull our sympathies. Tehila Kogut and Ilana Ritov solicited donations to help either one sick child in need of a costly medical treatment or eight such children. People felt more distressed about the single child and gave more money to the single child than to the group of eight. A more recent study shows that our sympathies are dulled with numbers as small as *two*.*

So . . . Is there really a moral difference between the nearby drowning child and the faraway children who need food and medicine? These cases certainly *feel* different, but we now know that our intuitive sense of moral obligation is at least somewhat unreliable, sensitive to things that don't really matter, such as mere physical distance and whether we know, in a trivially minimal way, whom we are helping. This doesn't mean that our automatic empathy programs are, on the whole, bad. On the contrary, without our natural empathic feelings, we'd be moral monsters. Our capacity for empathy may be the most quintessentially moral feature of our brains. Nevertheless, here, too, it would be foolish to let the inflexible operating characteristics of our empathy gizmos serve as foundational moral principles.

PERSONAL COMMITMENTS

Let's agree, then, that we are unduly indifferent to the plights of faraway, "statistical" people. Still, does our duty to help distant strangers swallow everything? What about everything else that we care about?

We humans are not just resource allocators. We are mothers and fathers, sons and daughters, brothers and sisters, lovers and friends, fellow citizens, keepers of our faiths, and champions of myriad worthy causes, from the arts to the pursuit of knowledge to the life well lived. These commitments, it seems, give us legitimate moral obligations and options. If you never buy your children birthday presents because your money is better spent on poor, anonymous, faraway children, you may be admirable in a way, but you're a bad parent. If your utilitarian commitments leave you no money for socializing, then you can't be a good friend. Likewise, supporting the arts or the local high school's sports teams hardly seems like a moral mistake. Must maximizing global happiness crowd out the rest of life's worthy causes?

Here, too, utilitarianism can do plenty of accommodating. If it seems absurd to ask real humans to abandon their families, friends, and other passions for the betterment of anonymous strangers, then that can't be what utilitarianism actually asks of real humans. Trying to do this would be a disaster, and disasters don't maximize happiness. Humans evolved to live lives defined by relationships with people and communities, and if our goal is to make the world as happy as possible, we must take this defining feature of human nature into account.

However, along with this agreeable utilitarian accommodation comes some challenging reform. *Of course* you should get your child a birthday present. But does your child need three birthday presents? Five? Ten? At some point, spending money on your own child instead of children who badly need food and medicine may indeed be a moral mistake. Supporting the arts is wonderful, and likely better than spending money on yourself. But maybe it isn't morally defensible to give a million dollars to the Metropolitan Museum of Art—enough to acquire one moderately priced piece of world-class art—if that money could feed, heal, clothe, and educate a thousand poor children. As a practical matter, it's probably counterproductive to pooh-pooh philanthropists for doing good in the "wrong" way; better to spend money on the Met than on a fourth vacation home. But helping the people who need help most would be better still. As with selfish desires and personal relationships, when it comes to noble causes, there's no formula for drawing the line between reasonable and indulgent uses of our resources. But in the real world, we can't draw the line in a place that seems absurd, because if we do, the line won't be respected.

Thus, here, as always, utilitarianism is firm but reasonable in practice, accommodating our needs and limitations. Still, it may seem that utilitarianism has missed something deeply important about human values.

HUMAN VALUES VERSUS IDEAL VALUES

Utilitarianism will forgive you for nurturing personal relationships and interests. But is this something for which you need to be *forgiven*? A morally ideal human, you say, is not a happiness pump. Someone who invests in friends and family isn't exhibiting an acceptable human frailty. She's being a *good person*—plain and simple. Isn't there something wrong with utilitarianism if it says that we ought, ideally, to care more about unfortunate strangers than anything else?

Maybe not. Perhaps if we step back far enough from our human values we can see that they are not ideal, even as we continue to embrace them. Here a thought experiment may help.

Imagine that you're in charge of the universe, and you've decided to create a new species of intelligent, sentient beings. This species will live in a world, like ours, in which resources are scarce and in which allocating resources to the "have-nots" eliminates more suffering and produces more happiness than allocating those resources to the "haves." You get to design the minds of your new creatures, and thus choose how they will treat one another. You've narrowed down your choice of species to three options.

Species 1, *Homo selfishus*: These creatures don't care about each other at all. They do whatever they can to make themselves as happy as possible and care nothing about the happiness of others. The world of *Homo selfishus* is a rather miserable world in which no one trusts anyone else and everyone is constantly fighting over scarce resources.

Species 2, *Homo justlikeus*: The members of this species are rather selfish, but they also care deeply about a relatively small number of specific individuals and, to a lesser extent, individuals who belong to certain groups. If all else is equal, they prefer that others be happy rather than unhappy. But they are, for the most part, not willing to lift much more than a finger for strangers, especially strangers who belong to other groups. This is a loving species, but their love is very limited. Many members of this species are very happy, but the species as a whole is far less happy than it could be. This is because members of *Homo justlikeus* tend to amass as many resources as they can for themselves and their close associates. This leaves many members of *Homo justlikeus*—a bit less than half of them—without the resources they need to be happy.

Species 3, *Homo utilitus*: The members of this species value the happiness of all members equally. This species is as happy as it could possibly be, because its members care about one another as much as they care about themselves. This species is infused with a spirit of universal love. That is, the members of this species love one another with the same passionate intensity that members of *Homo justlikeus* love their family members and close friends. Consequently, they're a very happy lot.

Were I in charge of the universe, I'd go for *Homo utilitus*, the much happier species suffused with universal love. You may disagree. You might insist that members of *Homo utilitus* are mindless drones whose indiscriminate devotion pales in comparison with the rich, partisan love we humans have on earth—Romeo and Juliet versus the Borg. This, however, is just a failure of imagination. To aid the imagination, consider some of earth's real-life heroes. Some people donate their kidneys to strangers, asking nothing in return. Amazingly, these people don't view themselves as heroic. They insist, with heartwarming optimism, that others would do the same if only they knew about the opportunities to help. And then there's Wesley Autrey, who dove in front of a subway car to save the life of a man who had stumbled onto the tracks during an epileptic seizure. Autrey held the man down with his body as the train passed over them, grazing Autrey's hair. When we imagine *Homo utilitus*, we should imagine heroes, not drones, people who are like us but who are willing to do more for others than the vast majority of us are willing to do.

My point is this: It's not reasonable to expect actual humans to put aside nearly everything they love for the sake of the greater good. Speaking for myself, I spend money on my children that would be better spent on distant starving children, and I have no intention of stopping. After all, I'm only human! But I'd rather be a human who knows that he's a hypocrite, and who tries to be less so, than one who mistakes his species-typical moral limitations for ideal values.

JUST DESSERTS

There's a straightforward utilitarian rationale for punishing people who break the rules: Without the threat of punishment, people won't behave. Others, however, say that punishment isn't, or shouldn't be, primarily about encouraging good behavior. They say that we should punish transgressors simply because they *deserve* it, independent of the practical benefits of punishing. This approach to punishment, known as *retributivism*, is favored by many moral and legal theorists, among them Immanuel Kant. Kant, in fact, once said that if an island community were to abandon its home, the islanders' departing "to do" list should include the execution of any murderers remaining in prison, just to squeeze in a little extra justice before pushing off.

Retributivists make some compelling objections to utilitarianism. First, it seems that utilitarians will sometimes punish when they shouldn't. You'll recall, from chapter 3, the *Magistrates and the Mob* case (page 79), in which the magistrates can stave off a violent riot by imprisoning an innocent person. It seems just plain wrong to punish an innocent person, even if doing so will produce better overall consequences. Second, it seems that utilitarians will sometimes punish too little. For the retributivist, the ideal world is one in which the good people are rewarded and the bad people suffer. But for a utilitarian, the ideal world is one in which everyone is maximally happy, including the baddies. In fact, the ideal utilitarian punishment system is one in which punishments are convincingly faked rather than actually delivered. In an ideal utilitarian world, convicts would be sent to a happy place where they can't bother anyone, while the rest us believe that they're suffering, the better to keep us on our best behavior.

Punishing the innocent? Rewarding the guilty? It seems that utilitarianism turns a deaf ear to justice. And that, some say, gives us good reason to reject the utilitarian ideal of maximizing happiness. As ever, we begin by accommodating as much common sense as we can, bearing in mind how things work in the real world.

This worry about punishing the innocent and rewarding the guilty, at least at the policy level, has no real-world merit. We can dream up cases like the *Magistrates and the Mob*, in which punishing the innocent makes things go better, but in the real world it would be disastrous to adopt such a policy. Likewise for a policy of faking punishments. For such policies to fulfill their utilitarian aims, government officials would have to maintain, indefinitely, an enormous conspiracy of Orwellian proportions while forgoing daily opportunities to abuse their power. This cannot be expected to lead to a happier world.

Utilitarianism naturally accommodates other features of common-sense justice. For example, we punish people far less (or not at all) if they harm others *accidentally*, rather than intentionally. As explained in the last chapter, there's a perfectly good utilitarian justification for this common-sense policy: People who cause harm intentionally are, in general, far more dangerous than people who cause harm accidentally, and therefore it's more important to deter them. Moreover, intentional acts are under conscious control, and therefore more likely to be deterred by the threat of punishment. Of course, we do punish people for accidentally causing harm if the harm was caused through *negligence*, and that, too, makes utilitarian sense. We wish to deter dangerous carelessness as well as intentional harm.

Likewise, there are utilitarian justifications for all of the standard excuses and justifications recognized by the law and by common sense. For example, the law recognizes "infancy" (being a child) as a legitimate legal excuse, and this, too, makes utilitarian sense. A ten-year-old who commits a crime is far more likely than, say, a thirty-year-old to respond to gentler incentives to behave, and to be irreparably damaged by harsher treatment. People who are mentally ill are less likely to be deterred by the threat of punishment, giving us less of a utilitarian reason to punish them. Finally, there are natural utilitarian justifications for breaking the law in self-defense and out of "necessity" (e.g., stealing a boat to save one's own life or another's): We don't want to deter people from doing these things.

Thus, in the real world, a legal system that maximizes happiness will not involve precarious Orwellian schemes in which innocent people are intentionally punished and guilty people are intentionally let off or rewarded. Likewise, punishment that promotes the greater good in the real world will acknowledge all of the standard excuses and justifications, distinguishing intentional crimes from accidents, children from adults, and so on. That said, punishment aimed only at the greater good will almost certainly involve some controversial reforms.

Consider, for example, the safety and well-being of prisoners, a cause that garners little public support and may be a liability for politicians. Prisoners are frequently sexually abused by other prisoners. We find such abuse regrettable—at least many of us do—but we're not sufficiently bothered by it to demand better protections for incarcerated victims. But consider this: Would you support a policy by which prisoners are, as an official part of their punishment, raped by official state rapists? Prison rape as a *foreseen side effect* of incarceration, with victims *to be determined* by fortune, strikes many people as regrettable but tolerable. But rape as a *means* of state punishment, delivered intentionally to *specific individuals*, is barbaric, we say.* From a utilitarian perspective, however, these two forms of sexual abuse are not so morally different—*a rape is a rape is a rape*—and we ought to do more to prevent prisoner-on-prisoner violence.

Reducing sexual violence among prisoners is just one example highlighting the tension between what feels right to society and what does good for society when it comes to criminal law. A more general question concerns the nature and consequences of the prison experience: Does being in prison make one more likely to lead a productive, law-abiding life upon release? Does the misery of prison encourage others to behave? There's little doubt that the general threat of punishment has an important deterrent effect. But it's an open question whether the unusually harsh and frequent punishments delivered in the United States are a necessary response to an unusual set of social problems, or just bad policy. Do would-be criminals know what the local statutes are, and do they care? The more general point is this: A criminal justice system aimed at the greater good would not be an absurd Orwellian machine, but it would likely look different from our current criminal justice system, which is highly retributive.

IDEAL JUSTICE

Utilitarian justice is reasonable in practice, but this still leaves us with the perennial "in principle" problem. Suppose that punishing an innocent person really would promote the greater good. Would that then be the right thing to do? And suppose that we really could convincingly fake punishments at low cost. Would it really be better to give murderers and rapists cushy lives rather than actually punishing them, provided that we could do so without losing the usual benefits of punishment (deterrence, etc.)? For utilitarians, punishment is just a necessary evil. But isn't there something inherently *right* about punishing people who do bad things? The utilitarian conception of justice, however unabsurd it may be in practice, seems to miss the deeper truths of justice.

That's what our critics say. Another possibility is that our intuitive sense of justice is a set of heuristics: moral machinery that's very useful but far from infallible. We have a *taste* for punishment. This taste, like all tastes, is subtle and complicated, shaped by a complex mix of genetic, cultural, and idiosyncratic factors. But our taste for punishment is still a taste, implemented by automatic settings and thus limited by its inflexibility. All tastes can be fooled. We fool our taste buds with artificial sweeteners. We fool our sexual appetites with birth control and pornography, both of which supply sexual gratification while doing nothing to spread our genes. Sometimes, however, our tastes make fools of us. Our tastes for fat and sugar make us obese in a world of abundance. Drugs of abuse hijack our reward circuits and destroy people's lives. To know whether we're fooling our tastes or whether our tastes are fooling us, we have to step outside the limited perspective of our tastes: To what extent is this thing—diet soda, porn, Nutella, heroin—really serving our best interests? We should ask the same question about our taste for punishment.

As I said, our intuitive sense of justice is extremely useful, and we'd be lost without it. As explained in chapter 2, punishment promotes cooperation, encouraging people to behave in ways that are good for Us, instead of merely good for Me. In other words, the natural function of punishment is quasi-utilitarian: We're natural punishers because punishment serves a social function.*

If you ask people why we ought to punish transgressors, people give the obvious utilitarian answer: Without the threat of punishment, people will misbehave. But this is manual-mode talking. If you look at people's punishment judgments in response to specific cases, it's clear that they're not thinking primarily about deterrence. Rather, punishment is, as explained in chapter 2, motivated primarily by feelings of anger, disgust, et cetera. These feelings are triggered by the transgressions themselves and the people who commit them, not by the prospect of deterring future transgressions. When people assign punishments for transgressions, they tend to ignore factors that are specifically related to deterrence, instead punishing based solely on how they *feel* about the transgressions. For example, it makes utilitarian sense to punish crimes more severely when the crimes are harder to detect: More deterrent incentive is needed when the odds of getting caught are low. (California, for example, imposes fines of up to $1,000 for littering—not because dropping a paper cup on the ground does such horrible damage, but because it's so easy to get away with it.) People tend to ignore this utilitarian consideration. Crimes with lower detection rates don't make us angrier, and therefore we don't intuitively assign more punishment. As explained in chapter 2, this disregard for the costs and benefits of punishing is very likely a design feature: If you punish only when it's "worth it," then you're not a reliable punisher, making you an attractive mark. But if you're a hothead with a taste for vengeance, and you're known as such, then you can deter more effectively.

In some cases, our punishment judgments are clearly irrational. Small and Loewenstein, the researchers who documented the preference for helping determined victims, documented similar behavior in a study of punishment. People played a game in which they could play cooperatively or selfishly. After the game, cooperators were given the opportunity to punish selfish players. Some cooperators were given the opportunity to anonymously punish a determined individual: "How much would you like to punish selfish person #4?" Others were given the opportunity to anonymously punish an undetermined individual: "How much would you like to punish the selfish person whose number you will draw?" As expected, people punished "determined" transgressors about twice as much, and their punishments were proportional to their emotional responses.

Our emotions also influence our judgments about who should and should not be held morally responsible. Shaun Nichols and Joshua Knobe presented people with the following description of a "deterministic" universe:

Imagine a universe (Universe A) in which everything that happens is completely caused by whatever happened before it. This is true from the very beginning of the universe, so what happened in the beginning of the universe caused what happened next, and so on right up until the present. For example one day John decided to have French Fries at lunch. Like everything else, this decision was completely caused by what happened before it. So, if everything in this universe was exactly the same up until John made his decision, then it *had to happen* that John would decide to have French Fries.

Nichols and Knobe asked their subjects whether people in this universe are fully morally responsible for their actions. Fewer than 5 percent of the respondents said yes. A different group of subjects read the same description of Universe A, but instead of responding to a general question about responsibility in this universe, they got a more specific question, one designed to arouse their emotions.

> In Universe A, a man named Bill has become attracted to his secretary, and he decides that the only way to be with her is to kill his wife and 3 children. He knows that it is impossible to escape from his house in the event of a fire. Before he leaves on a business trip, he sets up a device in his basement that burns down the house and kills his family.

Here 72 percent of respondents said that Bill is fully morally responsible for his actions. This is an amazing about-face. If you ask people about responsibility in a deterministic universe in the *abstract*, nearly everyone says there is no such thing. But if you present people with a particular instance of emotionally arousing malfeasance, that abstract judgment goes out the window.

Perhaps, as Kant thought, making transgressors suffer is a truly worthy goal, just for its own sake. But if that's right, it's a remarkable coincidence. How strange if the true principles of justice just happen to coincide with the feelings produced by our punishment gizmos, installed in our brains by natural selection to help us stabilize cooperation and thus make more copies of our genes. Knowing how our brains work and how they got here, it's more reasonable to suppose that our taste for justice is a useful illusion. We see punishment as inherently worthy and not just a means to better behavior, much as we experience food as inherently tasty and not just a means to nutrition. The enjoyment we get from food is typically harmless, but making people suffer is never harmless. Thus, we should be wary of punishment that tastes good but does more harm than good. And we shouldn't fault utilitarianism for seeing beyond the limitations of our taste for punishment.

THE JUST SOCIETY

Utilitarianism is a very egalitarian philosophy, asking the haves to do a lot for the have-nots. Were you to wake up tomorrow as a born-again utilitarian, the biggest change in your life would be your newfound devotion to helping unfortunate others. Despite this, one of the most persistent objections to utilitarianism is that it's not egalitarian enough, that it fails—or could fail—to respect the interests of the downtrodden.

According to John Rawls, who is widely regarded as the most important moral philosopher of the twentieth century, maximizing happiness can lead to gross injustice. Utilitarians say that it's okay to reduce some people's happiness if doing so affords greater gains in happiness for others. This is the principle behind progressive taxation: Asking the wealthy to pay more does little to cramp their style, but the revenue generated can do a lot of good for the rest of society. Nevertheless, says Rawls, distributing resources so as to maximize happiness will sometimes be unjust. Recall Rawls's example from chapter 8: Imagine a society in which the majority enslaves a minority. If the majority is happy with this arrangement—happy enough to offset the unhappiness of those enslaved—does that make it right? According to Rawls, a well-ordered society begins with certain basic rights and freedoms, and not with the overarching goal of maximizing happiness.

This is, on the face of it, a very compelling argument. Certainly, slavery is unjust, and any moral standard that endorses slavery is a bad moral standard. The question, then, is whether utilitarianism really endorses slavery. To see whether it does, we need to divide the problem into two parts: "in principle" and "in practice." I'm going to focus on "in practice," because that's what matters for the case I'm making in this book. I am not claiming that utilitarianism is the absolute moral truth. Instead I'm claiming that it's a good metamorality, a good standard for resolving moral disagreements *in the real world*. As long as utilitarianism doesn't endorse things like slavery *in the real world*, that's good enough.

I do not believe that, in the real world, maximizing happiness could ever lead to anything like slavery. And I say this as a committed empiricist, as someone who is very reluctant to make bold pronouncements from the armchair about how the world works. But in this case I will be bold. Utilitarianism could endorse slavery in principle, but only if human nature were radically different from what it is. To find a world in which maximizing happiness leads to slavery, we have to enter the world of science fiction. (Which is a world in which our moral intuitions cannot necessarily be trusted.)

It is exceptionally difficult to think clearly about the relationship between utilitarianism and social justice. More specifically, it is very hard to think clearly about "utility," because we naturally confuse utility with *wealth*. We'll consider this "wealthitarian" fallacy in the next section. For now, I want to show you, in a different way, just how hard it would be for slavery (or other forms of oppression) to make the world happier.

Slavery has generated enormous wealth for some people. It has also generated enormous misery. When we think about these enormous gains and losses in the abstract, and in aggregate, it's not obvious that the losses *must* outweigh the gains in all realistic circumstances. Instead, this may seem like an open empirical question. I don't think it is. To think more clearly about how slavery affects human happiness, it will help to zoom in on the happiness of representative individuals.

In a society with slavery, there are, of course, slaves and slave owners. To make things more concrete, let's imagine a representative slave society in which half of the people are slave owners and half are slaves. In other words, each free person owns exactly one slave. (Note that this one-to-one ratio is a *conservative* assumption** with respect to the argument I'm about to make.) For slavery to maximize happiness, each slaveholder must, on average, gain more happiness from *having* a slave than his slave loses by *being* a slave. Is that at all plausible?

Let me put this question to you in stages. Right now you are neither a slave nor a slaveholder. Your first question: How much would your happiness go up by having a slave? Of course, because you're a decent person, having a slave wouldn't make you at all happier, but we're trying to imagine your life as a happy slave owner, with no such moral reservations. To make this easier, we can imagine your having a high-tech robot slave. Your robot slave can do anything that an able-bodied, uneducated human can do. But your robot slave, like your laptop and your toaster, feels nothing. Thus, you are not more bothered by owning your robot slave than slaveholders in the past were by owning their human slaves.

What would you do with your robot slave? If you're like nearly all slave owners of the past, you'll try to get as much economic value out of your slave as possible. You'll put your slave to work. Let's suppose that your slave can get you an extra $50,000 per year. (This, too, is a conservative assumption, erring on the side of higher utility gained from slave ownership. Even with a lot of overtime, $50,000 per year is a lot to expect from an unskilled laborer. And we're not even considering the costs of providing for your slave's basic needs.) So how good is it to get an extra $50,000 per year? Good, perhaps. But not as good as you might think. If you're in a position to own a slave, you're already doing pretty well financially. And if there's one thing that we've learned from research on happiness, it's that additional income (above a fairly modest level) adds relatively little to one's happiness.* Some research suggests that additional income above a modest level adds nothing at all. The happiness gained from additional income varies from person to person, but we know that on average, a unit of additional income for wealthy people does very little compared with what it does for the poor, and that's what matters here. I emphasize that this is not a tentative finding.* After decades of research, the weak relationship between wealth (above a modest level) and happiness is more like a law of human nature. Past a certain point, wealth simply doesn't buy (much) happiness.

Thus, we may conclude, conservatively, that owning a slave would give you a substantial boost in *wealth* and a modest gain in happiness. Now for your second question: How much happiness would you lose by becoming a slave? The answer, of course, is *a lot*, and for all of the obvious reasons. I will not dwell on the horrors of slavery, which historically have involved beatings, rape, backbreaking labor, a complete absence of personal freedom, and the dissolution of one's family. Suffice it to say that being another person's property is very bad under the best of circumstances, and unthinkably miserable under typical circumstances. Going from being the free person you are today to being a slave would, needless to say, mean an enormous loss in your happiness.

Having considered how individuals gain and lose happiness from slavery, you're ready to answer your third and final question: Could the gains really outweigh the losses? Perhaps the best way to think about this question is to translate it into an equivalent personal choice: Would you be willing to spend half of your life as a slave so that, in the other half, you could make an extra $50,000 per year? Or would you rather just live your life the way it is now? I hope the answer is obvious. And if the answer is obvious, then it should be equally obvious—now that you've thought about it in more concrete terms—that slavery comes nowhere close to maximizing happiness in the real world. And what goes for slavery goes for oppression more generally. Gross injustice is *gross* injustice because it involves horrible outcomes for some people. Outside the realm of fantasy—inhabited, for example, by "utility monsters" who derive incomprehensible quantities of happiness from eating humans—there are no goods to be extracted from people that are so good as to outweigh the horrors of oppression.**

What almost certainly *does* exist in the real world are social inequalities and restrictions on freedom that promote the greater good. Having a free market leads to economic inequality, raising the question of how much, if at all, we ought to redistribute wealth. In a regime of maximal redistribution (communism), the inequality is eliminated, along with any economic incentive to be productive. Understanding this, nearly all of us, like the Northern herders, believe that some amount of economic inequality is justified by greater productivity (if not also as a matter of fairness). Likewise for inequalities of freedom. In the United States and many other nations, it can be illegal for someone who is HIV-positive to have unprotected sex without informing his or her partner of the risk. Such laws restrict the freedom of people who are HIV-positive, a downtrodden group to begin with, but most of us believe that this is justified by the greater good. The greater good justifies restricting our freedoms in other ways, as in the old case of shouting "Fire!" in a crowded theater. Lesson: It is certainly possible to have inequalities and restrictions on freedom that promote the greater good, but there's no reason to think that, in the real world, such inequalities and restrictions are grossly unjust. Such things may seem unjust to some people. But the point for now is that real-world utilitarianism does not, as its critics claim, lead to social arrangements that are clearly unjust, such as slavery. Say what you will about utilitarianism *in principle*; as a practical matter, making the world as happy as possible does not lead to oppression.

Why, then, have so many thoughtful people concluded that utilitarianism underwrites gross injustice? Part of the answer, as I've said, lies in a confusion about the meaning of the word *utility*. People confuse utility with *wealth*, and this makes maximizing utility seem less attractive, and perhaps unjust. Jonathan Baron and I documented this confusion in an experiment, which I'll describe in the next section. It's a bit complicated, and if you want to skip it, feel free. For the skippers, the take-home message is this: If you think that oppression can maximize happiness in the real world, you're almost certainly imagining the wrong thing. You're imagining oppression that maximizes wealth, not happiness.

THE "WEALTHITARIAN" FALLACY

"Utilitarianism" is, once again, an awful name for a splendid idea. Utilitarianism is not what we think about when we think about "utility."

As explained in chapter 7, the first key utilitarian idea is the primacy of experience: All good things are good, and all bad things are bad, because of their effects on experience. (The second idea, once again, is that everyone's experience counts equally.) Experience—happiness and unhappiness, broadly construed—is the utilitarian currency. But the word *utility* suggests something more like "useful stuff." To have lots of useful stuff is to be wealthy. Thus, it's easy to mistake utilitarianism for "wealthitarianism," the idea that we should, above all else, maximize wealth. This is not a splendid idea.

The confusion between utilitarianism and wealthitarianism runs deeper than a misleading word, however. Whether one calls it "utility," "happiness," or "the quality of experience," it's hard to think about it properly. We're used to quantifying *stuff*, things out in the world, or features of things in the world: *How many apples? How much water? How long is the meeting? How many square feet? How much money?* But we don't ordinarily quantify the quality of our experiences. And thus, when we imagine possible distributions of "utility," it's very hard not to think of distributions of stuff, rather than distributions of experiential quality.

Utility is closely related to stuff, but it is not itself stuff. First, utility need not come from market goods. The positive experience you get from friendship, sunny days, proving mathematical theorems, and being respected by your neighbors is all "utility." Second, utility isn't equivalent to stuff because the amount of utility we get from a given quantity of stuff varies from person to person and from situation to situation. To a poor Cambodian farmer, an extra $2,000 can easily be life changing, but to a rich businessperson, $2,000 is just an upgrade to first class on the way to Singapore. Thus, when we talk about distributions of utility, it's important to remember that we're talking about distributions of *utility*, elusive mental stuff, not stuff out in the world or in a bank account.

The question, then, is whether social inequalities that maximize *utility*—not stuff, not wealth—could be deeply unjust in the real world. When Rawls imagines utilitarian inequality, he imagines things like slavery. Slavery is certainly unjust, but why would anyone think that slavery (or something like it), could make the world happier? Because slavery that maximizes utility sounds plausible if you confuse utility with stuff—with *wealth*.

As noted above, it's not plausible that slave owners gain more happiness (i.e., utility) from having slaves than slaves lose by being slaves. Again, you wouldn't choose to be a slave for half your life in order to boost your income during the other half. Likewise, it's not plausible that $2,000 means as much to a rich businesswoman as it does to a poor farmer. But the math could work out in favor of the haves over the have-nots if we do our accounting in terms of wealth rather than happiness. The slave owner might make gobs of money by exploiting his slaves, and a better night's sleep for a traveling vice president might make or break a multimillion-dollar deal. In financial terms, the greater gains to the well off might outweigh the lesser losses to the poor. This is what Rawls and others have in mind when they envision utilitarian injustice: "Sorry," says the happy exploiter, "but my big, big gains justify your big losses." But the gains outweigh the losses only if we're counting dollars, rather than happiness.**

The experiment mentioned above shows that people readily confuse utility with wealth, and that this leads them to Rawlsian conclusions. Jonathan Baron and I presented people with hypothetical societies that had varying distributions of annual income (wealth). For example, in Country A, the people in the bottom third might get $25,000, while the people in the middle third get $45,000, and the people in the top third get $70,000. In Country B, everything is the same except that the bottom third gets only $15,000. We then asked people to compare Countries A and B. More specifically, we asked them to rate how much they would like to live in each country, knowing that they would have an equal chance of landing in the top, middle, or bottom economic tiers. The difference in the ratings for Countries A and B tells us how much people value a jump of $10,000 at the bottom of the scale, going from $15,000 to $25,000. We used similar comparisons to determine how much people value other income jumps, such as the $10,000 jump from $40,000 to $50,000. As you might expect, not all $10,000 jumps are equal. People value going from $15,000 to $25,000 more than they value going from $40,000 to $50,000, reflecting the declining marginal utility of wealth. The more dollars you have, the less each additional dollar means to you, and the less happiness each additional dollar brings.

In the next phase of the experiment, people rated the values of the various incomes used in the first phase of the experiment. In other words, we had people assign utility levels to each income level. We told them to give the lowest income a rating of 0 and the highest income a rating of

100. We then had them rate the other income levels using numbers between 0 and 100, being careful to make sure that every increment had the same value; for example, the improvement from 0 to 50 should be just as desirable as the improvement from 50 to 100. The subjects gave their ratings, and, as expected, they gave more weight to increases at the bottom of the income scale, consistent with the declining marginal utility of wealth. For example, the difference in ratings between $15,000 and $25,000 was typically bigger than the difference in ratings between $40,000 and $50,000.

In the third phase of the experiment, our subjects once again rated the desirability of living in various countries, but this time the countries were described not in terms of income distributions but in terms of *utility* distributions. Instead of telling people how much money they would make in the bottom, middle, and top third of each society, we gave them a *rating* of the value (the utility) of the income in each third of the society. These ratings were presented as "RATINGS OF INCOMES given by someone just like you, as you gave them in [the last part of the experiment]." We could then compare the values people place on different jumps, but here the jumps are jumps in utility rather than jumps in income. For example, we can determine how much people value jumping from an income rated 0 to an income rated 25. Likewise, we can determine how much people value jumping from an income rated 75 to an income rated 100. And, most critically, we can find out whether people think that the first 25-point jump in utility is more desirable than the second 25-point jump in utility.

If people are consistent, these two jumps will be equally desirable. This is because we are now dealing with utility rather than income levels. Remember that subjects in the second part of the experiment made their ratings so that each point on the scale had the same additional value. That is, a 25-point increase in utility should be the same, whether it's going from 0 to 25 or from 75 to 100.

Results: The responses in the third part of the experiment (rating countries based on their utility distributions) were inconsistent with their responses in the second part of the experiment (assigning utilities to different levels of income). Instead of treating jumps of equal utility as adding equal value, they treated the utilities *exactly as they treated the incomes in the first part of the experiment.* That is, they (like Rawls) gave more weight to improvements at the bottom of the scale, saying that the jump from 0 to 25 is more desirable than the jump from 75 to 100. In fact, they put as much weight on the bottom of the scale here as they had when they were rating income levels. This pattern of evaluation is internally inconsistent, but it's exactly what Baron and I predicted people would do, based on reading Rawls.*

This experiment shows that people have a very hard time thinking clearly about utility. On the one hand, people understand that stuff and utility are different. This is demonstrated by their giving different ratings for a $10,000 jump depending on where that jump starts (e.g., starting from $15,000 as opposed to $40,000). On the other hand, if you ask people to evaluate distributions of utility, they just treat utility as if it were wealth, rather than the peculiar abstraction that it is. In other words, people look at the jump from an income that gets a utility rating of 0 to one that gets a rating of 25 and they think, "With so little utility to start with, that increase would make a very big difference." Then they look at the difference between incomes with ratings of 75 and 100 and they think, "That would be a nice improvement, but, starting with a decent amount of utility already, it's not quite as big an improvement." This thinking is simply erroneous. You can get more or less utility out of a given quantity of money, depending on your situation, but you can't get *different levels of utility out of your utility. Utility is utility is utility.* And it's not that our subjects were just a little bit off. When they switched over from thinking about income distributions to thinking about utility distributions, they didn't modify their thinking *at all.* They treated utility *exactly* as if it were stuff. (And the handful of professional philosophers we tested did the same thing.)

So what does this mean? It means that the Rawlsian critique of utilitarianism, the idea that utilitarianism underwrites gross injustice, is empirically debunkable. The Rawlsian objection to utilitarianism derives its force from a misunderstanding of utility, which we can easily demonstrate in the lab. People *sort of* understand the difference between wealth and utility. They understand that, as one's wealth grows, additional wealth matters less and less. But when people evaluate distributions of utility, they forget this entirely and instead treat utility exactly as if it were wealth. They confuse utilitarianism with wealthitarianism. Thus, countless philosophers have convicted poor, innocent utilitarianism of crimes against humanity.

(What about the "in principle" version of this objection?*** And what about Rawls's argument from the "original position"?**)

JUSTICE AND THE GREATER GOOD

Is utilitarianism unjust? Let's review.

Does utilitarianism require us to turn ourselves into happiness pumps? To enslave ourselves to the greater good? No. Because this is not a realistic goal for flesh-and-blood humans, whose brains were not designed for moral heroism. Instead, utilitarianism asks only that we push ourselves to be morally better, to care more than we do about people beyond our immediate circles. Utilitarianism doesn't ask us to be morally perfect. It asks us to face up to our moral limitations and do as much as we humanly can to overcome them. Here science can help, showing us just how fickle and irrational our sense of duty can be.

Does utilitarianism endorse perversions of criminal justice, punishing the innocent and rewarding the guilty? In imaginary worlds, yes. But in the real world, these are disastrous ideas, and nothing that a wise happiness maximizer would endorse. As ever, this accommodation comes with reform. Our taste for punishment is useful, but it's not infallible. Just as our taste for fat and sugar can make us obese in a world full of milkshakes, our taste for retribution can create a criminal justice system that satisfies our taste for punishment while undermining our social health.

Does utilitarianism endorse slavery and other forms of oppression? Not in the real world. In the real world, oppression offers only modest gains in happiness to the oppressors while heaping misery upon the oppressed. The idea that utilitarianism underwrites social injustice is based on the wealthitarian fallacy, a subtle confusion between maximizing wealth and maximizing happiness. "In principle," one can maximize happiness by oppressing people, but in the real world, with human nature as it is, oppression does not make the world a happier place.

Thus, in the real world, there is no fundamental tension between happiness and justice. However, we can refine our sense of justice through a better understanding of where its cognitive gizmos come from and how they work.

PART V

Moral Solutions

11.

Deep Pragmatism

Now it's time to put it all together, to translate what we've learned from biology, psychology, and abstract philosophy into something we can use. Knowing what we now know, how should we think about the problems that divide us? How should we think about the foolish man who refused to buy health insurance? Is healthcare a right? (Even for foolish people?) Or is it just another product for which one saves one's pennies? Ten percent of Americans control 70 percent of America's wealth. Is that unfair? Or is that just how it goes in the land of opportunity? Is the threat of global warming real, or just a do-gooder hoax? And if, as the experts say, the threat is real, who pays to stop it, and how much? Does Iran have the right to develop nuclear technology? Does Israel have the right to stop them? Is Amnesty International correct to call the death penalty a "fundamental violation of human rights"? Or is Judge Alex Kozinski (United States Court of Appeals, Ninth Circuit) correct when he says that murderers "forfeit their right to their own life"? Is gay marriage a civil right in the making, or an abomination before God?

Should doctors help terminally ill patients end their lives if they want to? Or should we trust the American Medical Association, which calls physician-assisted suicide "fundamentally incompatible with the physician's role as healer"?

Back in chapter 6, we introduced a splendid idea: We, the modern herders of the new pastures, should put aside our respective ideologies and instead do whatever works best. Presented like so, this prescription sounds very reasonable and not especially helpful. But as we've seen, and as we'll see more shortly, this is in fact an extremely powerful and challenging idea. To take it seriously is to fundamentally change the way we think about our moral problems.

TWO COMPASSES

This philosophy, which I've called deep pragmatism, comes across as agreeably bland because we believe that we've already adopted it. We all believe that what we want is for the best. But, of course, we can't all be right about that. To give this philosophy some teeth, we need to get specific about what counts as "best." We need a shared moral standard, what I've called a *metamorality*. Once again, a metamorality's job is to help us make tough choices, to make *trade-offs* among competing tribal values. Can this be done in a principled way?

The proverbial "relativist" says that it can't be done. There are different tribes with different values, and that's all there is to say. The relativist might be right in some ultimate metaphysical sense. Perhaps our moral questions have no objectively correct answers. But even if that's true, knowing that it's true is not much help. Our laws have to say *something*. We have to choose, and unless we're content to flip coins, or allow that might makes right, we must choose for *reasons*. We must appeal to some moral standard or another.

Forging ahead, there are two general strategies. The first appeals to an independent moral authority of some kind—God or Reason or Science.

As we saw in chapter 7, none of these has delivered the kind of non-question-begging moral truth that can resolve our disagreements. Thus, we're thrown back on "the morass," the tangled web of values and beliefs that simultaneously holds us together and pulls us apart.

The second strategy, the deep pragmatist's strategy, is to seek agreement in shared values. Rather than appeal to an independent moral authority (God/Reason/Science says: "The right to life trumps the right to choose"), we aim instead to establish a *common currency* for weighing competing values. This is, once again, the genius of utilitarianism, which establishes a common currency based on *experience*. As revealed by the buttons we will and will not push, we all care about experience, both our own and others'. We all want to be happy. None of us wants to suffer. And our concern for happiness and suffering lies behind nearly everything else that we value, though to see this requires some reflection. We can take this kernel of personal value and turn it into a *moral* value by valuing it *impartially*, thus injecting the essence of the Golden Rule: Your happiness and your suffering matter no more, and no less, than anyone else's. Finally, we can turn this moral value into a moral *system* by running it through the outcome-optimizing apparatus of the human prefrontal cortex. This yields a moral philosophy that no one loves but that everyone "gets"—a second moral language that members of all tribes can speak. Our respective tribes have different moral intuitions, different automatic settings, and therein lies our strife. But fortunately for us, we all have flexible manual modes. With a little perspective, we can use manual-mode thinking to reach agreements with our "heads" despite the irreconcilable differences in our "hearts." This is the essence of deep pragmatism: to seek common ground not where we think it ought to be, but where it actually is.

The pragmatist is a familiar character, an object of both admiration and suspicion. We admire the pragmatist for getting "results," for practicing the "art of the possible," and for bridging the gap between Us and Them. But the worry about pragmatists is that they, in their zeal to move things along, may lose their sense of direction. Suppose two children are fighting over a cake. One wants to split it down the middle. The other wants it all for himself. Then along comes the "pragmatist," ever the catalyst of compromise: "There, there, children. Let's be reasonable now. You get three-quarters, and you get one-quarter." An indiscriminate willingness to compromise is no virtue. Some compromises are bad, and some uncompromising feelings are good. But if, in the spirit of compromise, we've put aside our uncompromising moral feelings, what's left to guide us? Where is our moral compass?

This is why the "deep" in deep pragmatism is essential, and why it's not enough for modern herders simply to say, "Let's be reasonable and open to compromise." A pragmatist needs an explicit and coherent moral philosophy, a second moral compass* that provides direction when gut feelings can't be trusted. This is why I've spent precious pages explaining, clarifying, and defending utilitarianism. I understand that this is not everyone's idea of a good time. But if we acknowledge that our tribal feelings can't all be right, and yet aspire to resolve our differences in a principled way, then we need some kind of "ism," an explicit moral standard to guide us when our emotional compasses fail.

Utilitarianism isn't love at first sight. As I explained back in chapter 6, this philosophy is very easily misunderstood. It's not about "utility"—valuing mundane functionality over the things that give life meaning. Nor is it the shallow pursuit of our "favorite things." Nor is it egoistic, hedonistic, or blindly utopian. Utilitarianism doesn't require magical or high-tech abilities to measure happiness with great precision, and it doesn't require us to be constantly "calculating." On the contrary, there are utilitarian reasons to reject all of these naive, pseudo-utilitarian ways. Utilitarianism, properly understood, bears little resemblance to its many caricatures. Properly understood, and wisely applied, utilitarianism is deep pragmatism. It's our second moral compass, and our best guide to life on the new pastures. In this chapter we'll consider what it means to be a deep pragmatist and contrast this philosophy with some of its tempting alternatives.

WHEN TO POINT AND SHOOT?:
(ME VS. US) VERSUS (US VS. THEM)

I've said two apparently contradictory things. On the one hand, I've said that we should put our gut reactions aside, shift into manual mode, and rely on our utilitarian moral compass for direction. (My apologies for the mixed mechanical metaphors.) On the other hand, I've said that we, deep pragmatists, shouldn't be constantly making utilitarian calculations. So which is it?

It depends on the kind of problem we're facing. Throughout this book, we've relied on three guiding metaphors, two of which come together here. The first is the Parable of the New Pastures, illustrating the Tragedy of Commonsense Morality. The second is the camera metaphor, illustrating the strengths and weaknesses of our gut reactions (automatic settings) and explicit reasoning (manual mode). To answer this question— *When to point and shoot?*—we need to put our first two guiding metaphors together. (The third is *common currency*, which we'll return to soon.)

As explained in part 1, we face two fundamentally different kinds of moral problems. The first problem is Me versus Us. This is, once again, the basic problem of cooperation, the Tragedy of the Commons. Our moral brains solve this problem primarily with emotion. Feelings of empathy, love, friendship, gratitude, honor, shame, guilt, loyalty, humility, awe, and embarrassment impel us to (sometimes) put the interests of others ahead of our own. Likewise, feelings of anger and disgust impel us to shun or punish people who overvalue Me relative to Us. Thanks to these automatic settings, we do far less lying, cheating, stealing, and killing than we otherwise could, and that enables Us to succeed.

Complex moral problems are about Us versus Them. It's our interests versus theirs, or our values versus theirs, or both. This is the modern moral tragedy—the Tragedy of Commonsense Morality—and the source of strife on the new pastures. Here our disparate feelings and beliefs make it hard to get along. First, we are tribalistic, unapologetically valuing Us over Them. Second, different tribes cooperate on different terms. Some are more collectivist, some more individualist. Some respond aggressively to threats. Others emphasize harmony. And so on. Third, tribes differ in their "proper nouns"—in the leaders, texts, institutions, and practices that they invest with moral authority. Finally, all of these differences lead to biased perceptions of what's true and what's fair.

Our second guiding metaphor, the camera, illustrates our two modes of moral thinking: We have automatic settings: emotional gut reactions that are efficient but inflexible. And we have manual mode: a general capacity for explicit, practical reasoning, which is inefficient but flexible.

Thus, we have two kinds of moral problems and two kinds of moral thinking. And now we can answer our question: The key to using our moral brains wisely is to *match the right kind of thinking with the right kind of problem*. Our moral emotions—our automatic settings—are generally good at restraining simple selfishness, at averting the Tragedy of the Commons. That's what they were designed to do, both biologically and culturally. Thus, when the problem is Me versus Us (or Me versus You), we should trust our moral gut reactions, also known as *conscience*: Don't lie or steal, even when your manual mode thinks it can justify it. Cheat on neither your taxes nor your spouse. Don't "borrow" money from the office cash drawer. Don't badmouth the competition. Don't park in handicapped spots. Don't drink and drive. And *do* express your contempt for people who do such things. When it's Me versus Us, trust your automatic settings. (The moral ones, not the greedy ones!)

But . . . when it's the Tragedy of Commonsense Morality—when it's Us versus Them—it's time to stop trusting your gut feelings and shift into manual mode. How do we know which situation we're in? This question has a surprisingly simple answer: *controversy*. When someone commits a straightforward moral transgression, such as fraud or murder, there is a moral *problem*, but there is no moral *controversy*. There are no protesters outside the courthouse standing up for Bernie Madoff's "right" to defraud investors. It's Him versus Us. Here our instincts about what's right or wrong are likely to serve us well.

But when there's controversy, when whole tribes disagree, then you know that you're on the new pastures, dealing with Us versus Them. And that's when it's time to shift into manual mode. Why? Because when tribes disagree, it's almost always because their automatic settings say different things, because their emotional moral compasses point in opposite directions. Here we can't get by with common sense, because our common sense is not as common as we think.

As it happens, the decision strategy advocated here—when intuitions conflict, shift into manual mode—is one that our brains already use in other contexts. The camera analogy leaves us with a general mystery about human decision making: In photography, it's the photographer who decides when to point and shoot and when to shift into manual mode. What, then, plays the role of the photographer in human decision making? How do we decide how to decide? Here we're threatened with an infinite regress. Before we decide, don't we have to decide how to decide? And before we decide how to decide, don't we have to . . . ?

The pioneering research of Matthew Botvinick, Jonathan Cohen, and colleagues shows how the brain gets out of this pickle.* You may recall from chapter 4 the color-naming Stroop task. Here the challenge is to name the color in which the word is printed, even when the word names the wrong color. For example, if it shows the word "red" written in blue, you're supposed to say "blue." This is hard, because reading is more automatic than color naming, and to do this quickly and accurately requires cognitive control—that is, manual mode. So how does manual mode know when to kick in? Must one ask oneself each time, "Is this a tricky one?" and then decide how to think?

Botvinick and Cohen argue that the brain solves this problem using a *conflict monitor*, based in a part of the brain called the anterior cingulate cortex (ACC). The ACC perks up whenever incompatible responses are activated simultaneously. For example, when you see the world "red" written in blue, one population of neurons starts firing, initiating your saying "red," while a different population of neurons initiates your saying "blue." According to the conflict-monitoring theory, the ACC detects that the brain has been firing up two incompatible behaviors and then sends a wake-up signal to the DLPFC, the seat of manual mode, which can, like a higher court, resolve the conflict. Consistent with this, my research with Cohen and others shows that difficult moral dilemmas, which by their nature evoke conflicting responses, engage the ACC and DLPFC.

In the Stroop task, and in some moral dilemmas, the conflict is *within* a single brain. But when we herders disagree, the conflict is *between* human brains. What I'm suggesting, then, is that we take the strategy that our brains automatically apply to *intra*cranial disagreement and deliberately apply it to *inter*cranial disagreement: In the face of conflict, shift into manual mode.

OUT OF OUR DEPTH

Let's agree, then, that when we herders disagree, we'll stop and think. Hard. *Really* hard. That's a splendid idea, but there's a big danger here. When we think about divisive moral problems, our first instinct is to think of all the ways in which We are right and They are wrong.

Recall, once again, the experiment in which opponents and proponents of the death penalty were shown mixed evidence concerning its efficacy as a deterrent. Instead of becoming more moderate ("the evidence is mixed"), people became more polarized. People latched on to the evidence that suited them and dismissed the rest. Likewise, when it comes to climate change, Americans who are more scientifically literate and "numerate"—not climate experts, but ordinary people who like to use their manual modes—are especially polarized. And recall that when we evaluate evidence, our biases creep in unconsciously. Embattled negotiators lose money when they bet on what unbiased third parties will say.

Knowing all this, you might think that evidence-based, manual-mode morality is hopeless, that thinking hard about divisive problems can only make things worse. Perhaps. Alternatively, manual-mode thinking may bring us together, provided that we use our manual modes in the right way. Most controversial real-world moral problems, such as global warming and healthcare reform, are very complicated. Nevertheless, people without expertise on these topics have strong opinions about them. In an ideal world, we'd all transform ourselves into experts and make judgments based on extensive knowledge. Given that this will never happen, our next best option is to emulate the wisdom of Socrates: We become wiser when we acknowledge our ignorance.

Psychologist Frank Keil and colleagues have documented what he calls "the illusion of explanatory depth." In short, people think they understand how things work even when they don't. For example, people typically think they understand how a zipper or a flush toilet works, but when they actually try to explain how these things work, they fail miserably. But—and this is key—when people try to explain how these things work and they fail, they recognize that they've failed and then revise their estimates of how much they understand.

In a brilliant set of experiments, Philip Fernbach, Todd Rogers, Craig Fox, and Steven Sloman applied this idea to politics. They asked Americans to consider six controversial policy proposals, such as a single-payer healthcare system and the cap-and-trade system for reducing carbon emissions. In one version of the experiment, they asked people to offer their opinions about these policies and to indicate how well they understood them. They then asked people to explain in detail how these policies are supposed to work. Finally, they asked people to once again offer their opinions and rate their understanding. They found that people, after being forced to explain the mechanics of these policies, downgraded their estimates of their own understanding and became more moderate in their opinions. The experimenters ran a control version of this experiment in which people, instead of explaining how the policies are supposed to work, offered reasons for their opinions. For most people, offering reasons left their strong opinions intact.

What these studies elegantly demonstrate, then, is that the right kind of manual-mode thinking can bring us closer together. Simply forcing people to justify their opinions with explicit reasons does very little to make people more reasonable, and may even do the opposite. But forcing people to confront their ignorance of essential facts does make people more moderate. As these researchers note, their findings suggest an alternative approach to public debate: Instead of simply asking politicians and pundits *why* they favor the policies they favor, first ask them to explain *how* their favored (and disfavored) policies are supposed to work. And what goes for *Meet the Press* goes for *Meet the Relatives*. When your opinionated, turkey-stuffed uncle insists that national health insurance is a historic step forward/the end of civilization as we know it, you may yet shift his opinion in your direction without overtly challenging him: "That's very interesting, Jim. So how exactly does national health insurance work?"

THE SECRET JOKES OF OUR SOULS: RATIONALIZATION AND THE DUAL-PROCESS BRAIN

In the early 1970s, Donald Dutton and Arthur Aron sent an attractive female experimenter to intercept men crossing two different bridges in a park in British Columbia. One of the bridges was a frightening, wobbly suspension bridge spanning a deep gorge. The other was a sturdy wooden bridge, closer to the ground. The attractive confederate interviewed these men (one at a time) about their experiences in the park and then gave each of them her phone number, in case they wanted to learn more about the study—*wink, wink*. The men she'd met on the wobbly bridge were far more likely to call her back, and far more likely to ask her out. Why? As Dutton and Aron predicted, the men on the wobbly bridge mistook their thumping hearts and sweaty palms for feelings of intense attraction. The lesson: When we don't know why we feel as we do, we make up a plausible-sounding story and go with it.

This is not an isolated phenomenon. In another classic experiment, Richard Nisbett and Timothy Wilson asked people to choose one of several pairs of panty hose displayed in a row. When asked to explain their preferences, people gave sensible enough answers, referring to the relevant features of the items chosen—superior knit, sheerness, elasticity, et cetera. However, their choices had nothing to do with such features, because the items on display were in fact identical. People simply had a preference for items on the right-hand side of the display. In a similar experiment, the same duo presented people with word pairs, one of which was "ocean-moon." Later, those people had to choose among different laundry detergents. The subjects who had earlier read the words "ocean-moon" were twice as likely to choose the laundry detergent Tide over other brands, but when subjects explained their preferences, they said things like "Tide is the best-known detergent" or "My mother uses Tide" or "I like the Tide box."

This tendency to make up stories about why we do what we do is dramatically illustrated in neurological patients who have a hard time making sense of their behavior. Patients with Korsakoff's amnesia, for example, will often attempt to paper over their memory deficits with elaborate stories, typically delivered with great confidence and no awareness that they are making stuff up. Neurologists call this "confabulation." In one study, for example, an amnesic patient seated near an air conditioner was asked if he knew where he was. He replied that he was in an air-conditioning plant. When it was pointed out that he was wearing pajamas, he said, "I keep them in my car and will soon change into my work clothes." One sees similar effects in "split-brain" patients, people whose cerebral hemispheres have been surgically disconnected to prevent the spread of seizures. With the two cerebral hemispheres disconnected, each half of the brain is denied its usual inside information about what the other half is up to. In one study, a patient's right hemisphere was shown a snow scene and instructed to select a matching picture. Using his left hand, the hand controlled by the right hemisphere, he selected a picture of a shovel. At the same time, the patient's left hemisphere, the hemisphere that controls language, was shown a picture of a chicken claw. The patient was asked verbally why he had chosen the shovel with his left hand. The patient (i.e., the patient's left hemisphere, seeing the chicken claw but not the snow scene) answered, "I saw a claw and picked a shovel, and you have to clean out the chicken shed with a shovel."

Confabulation is strange, but the lesson that cognitive neuroscientists have drawn from it is even stranger: It's not that this brain damage somehow creates or unleashes a capacity for confabulation. After all, damaging the brain is unlikely to endow it with a new ability or motive. Rather, the lesson is that we're *all* confabulators, and those of us with healthy brains are just better at it. We're constantly interpreting our own behavior, fashioning it into a plausible narrative about what we are doing and why. The critical difference between confabulating neurological patients and the rest of us is that they, thanks to their deficits, are forced to construct their narratives from more meager raw material. To catch healthy people in the act of confabulation, you have to set up a controlled experiment, such as the bridge experiment or the Tide experiment.

The moral equivalent of confabulation is *rationalization*. The confabulator perceives himself doing something and makes up a rational-sounding story about what he's doing and why. The moral rationalizer *feels* a certain way about a moral issue and then makes up a rational-sounding justification for that feeling. According to Jonathan Haidt, we're all consummate moral rationalizers, and this makes perfect sense, given our dual-process brains. Our automatic settings gives us emotionally compelling moral answers, and then our manual modes go to work generating plausible justifications for those answers, just like the manual modes of amnesiac patients trying to explain what they're up to. Here, for example, is Immanuel Kant, explaining why masturbation is a violation of the categorical imperative, the supreme moral law, in a passage entitled "Concerning Wanton Self-Abuse":

> That such an unnatural use (and so misuse) of one's sexual attributes is a violation of one's duty to himself and is certainly in the highest degree opposed to morality strikes everyone upon his thinking of it. . . . However, it is not so easy to produce a rational demonstration of the inadmissibility of that unnatural use . . . of one's sexual attributes as being a violation of one's duty to himself. . . . The ground of proof surely lies in the fact that a man gives up his personality (throws it away) when he uses himself merely as a means for the gratification of an animal drive.

Recall the Doctrine of Double Effect, which distinguishes between harming someone as a means and harming someone as a side effect. Kant, like Aquinas, endorses the idea that certain actions are wrong because they involve using someone as a means. Here Kant takes this idea and applies it to the sin of autoeroticism: Masturbation is wrong because it involves using *yourself* as a means.

This is very clever. It's also kind of funny. We who don't share Kant's sexually repressed mores can have a good chuckle over his earnest attempt to deduce the immorality of masturbation from abstract principles. The nineteenth-century German philosopher Friedrich Nietzsche found Kant's rationalistic moralism amusing as well:

> *Kant's Joke*—Kant wanted to prove, in a way that would dumbfound the common man, that the common man was right: that was the secret joke of this soul. He wrote against the scholars in support of popular prejudice, but for scholars and not for the people.

In other words, Kant has the same automatic settings as his surrounding tribespeople. But Kant, unlike them, felt the need to provide esoteric justifications for their "popular prejudices." Kant also developed an elaborate theory to explain the superiority of whites and the inferiority of blacks, whom he regarded as "born slaves."

Rationalization is the great enemy of moral progress, and thus of deep pragmatism.* If moral tribes fight because their members have different gut feelings, then we'll get nowhere by using our manual modes to rationalize our feelings. We need to shift into manual mode, but we need to use our manual modes wisely. We've seen some of this already (explaining *how* in addition to *why*), but we can do more. We can learn to recognize rationalization, and we can establish ground rules that make it harder to fool ourselves—and each other.

"HEADS I WIN, TAILS YOU LOSE": RIGHTS AS RATIONALIZATION

As deep pragmatists, we want to focus on the hard, empirical work of figuring out what works best in the real world. But tribal loyalists, with their infallible gut reactions, have every reason to resist our call for wonkery. Death penalty opponents will gladly tell you that the death penalty doesn't reduce crime, citing the best evidence they can find. And death penalty proponents will gladly do the opposite. But for tribal loyalists, these pragmatic, utilitarian arguments are just window dressing. If, as Amnesty says, the death penalty is a "fundamental violation of human rights," then the policy debate is "Heads I win, tails you lose." If the facts come out against the death penalty, Amnesty will cheer. But if they don't, the death penalty is still wrong on "principle." And, of course, the same goes for death penalty proponents, who will, when empirical push comes to rhetorical shove, simply insist that the death penalty is an aggrieved society's moral right.

Thus, appeals to "rights" function as an intellectual free pass, a trump card that renders evidence irrelevant. Whatever you and your fellow tribespeople feel, you can always posit the existence of a right that corresponds to your feelings. If you feel that abortion is wrong, you can talk about a "right to life." If you feel that outlawing abortion is wrong, you can talk about a "right to choose." If you're Iran, you can talk about your "nuclear rights," and if you're Israel you can talk about your "right to self-defense." "Rights" are nothing short of brilliant. They allow us to rationalize our gut feelings without doing any additional work.

Rights and their mirror images, duties, are the perfect rhetorical weapons for modern moral debate. As we saw in the preceding chapters, our automatic settings issue moral commands, telling us that certain things are *not to be done* while other things are *to be done*. These feelings correspond more or less perfectly to the concepts of *rights* and *duties*. If we feel that an action is simply *not to be done*, we can express this feeling by saying that it violates people's *rights*. And likewise, if we feel that an action is simply *to be done*, we can express that feeling by appealing to a corresponding *duty*. Pushing the man off the *footbridge* feels very, very wrong, and therefore we say that it's a gross violation of his rights, whether or not it saves five lives. But hitting the switch doesn't feel nearly so bad, and thus we say that it's not a violation of the victim's rights, or that his rights are "outweighed" by the rights of the five.* Likewise, we have a duty to save the nearby drowning child, but faraway "statistical" children don't tug on our heartstrings nearly as hard, and thus we have no duty to save them. The rights and the duties follow the emotions.*

Talk of rights and duties aptly expresses our moral emotions in two ways. First, when our gut reactions tell us what we must and must not do, these commands come across as *nonnegotiable*, reflecting the inflexibility of our automatic settings. Once again, the feeling that tells us not to push the man off the footbridge doesn't "care" whether there are zero, five, or a million other lives at stake. Such feelings can be overridden, but the feeling itself is, so to speak, unwilling to negotiate. It is, as experimental psychologists say, "cognitively impenetrable." This nonnegotiability is built into the concepts of *right* and *duty*. Rights and duties can be overridden, but doing so involves more than tipping the balance of considerations. Rights and duties are *absolute*—except when they're not.

Second, we embattled moralists love the language of rights and duties because it presents our subjective feelings as perceptions of objective facts. We like this because our subjective feelings often feel like perceptions of things that are "out there," even when they are not. Consider, for example, the experience of sexual attraction. When you find someone sexy, you don't feel as though your mind is projecting an aura of sexiness onto the object of your desire. And yet we know that this is what's happening. We humans find other humans sexy (some of them), but we don't find baboons sexy (most of us). And baboons, of course, are similarly interested in one another and not in us. As our interspecies disagreement reminds us, sexiness is in the mind of the beholder.* Nevertheless, that's not how it feels when one is in the grip of sexual attraction. A sexy person's sexiness strikes us not as a subjective projection, but as something no less "out there" than that individual's height and weight. And thus, it's natural to describe someone as "sexy," rather than as "provoking sexual desire in people like me." In the same way, talk of rights and duties presents subjective feelings as objective facts that are "out there," whether or not they really are. When you say that someone has a right, you appear to be stating an objective fact about what this person *has*, like the fact that she has ten fingers.

If I'm correct, rights and duties are the manual mode's attempt to translate elusive feelings into more object-like things that it can understand

and manipulate. Manual mode exists primarily to deal with physical things out in the world: actions and events and the causal relationships that connect them. Thus, the manual mode's native ontology is one of concrete "nouns" and "verbs." How, then, can it make sense of the outputs of automatic settings, mysterious feelings that come out of nowhere, protesting actions that otherwise seem perfectly sensible? (Or commanding actions that otherwise seem optional.) Answer: It represents such feelings as perceptions of external *things*. The feelings get *nounified*. An amorphous feeling of *not-to-be-doneness* is conceived as a perception of a thing called a "right," an abstract but nonetheless real thing that can be gained, lost, relinquished, transferred, expanded, restricted, outweighed, suspended, threatened, traded, violated, and defended. By conceptualizing our moral emotions as perceptions of rights and duties, we give ourselves the ability to think about them explicitly, using the cognitive apparatus that we ordinarily use to think about concrete objects and events.

Thus, for all of these reasons, rights and duties are the modern moralist's weapons of choice, allowing us to present our feelings as nonnegotiable facts. By appealing to rights, we excuse ourselves from the hard work of providing real, non-question-begging justifications for what we want. As long as we allow ourselves to play the rights card, evidence is secondary, because it's "heads I win, tails you lose."

At this point you might think that I'm being too hard on rights: Is this an argument against appeals to rights? Or is this just an argument against unsubstantiated assertions? Sure, we can rationalize our gut reactions by appealing to rights, but we can also make utilitarian rationalizations: Whatever we want in our hearts, we can say that it's for the greater good. So what's the difference?

The difference, as suggested above, is that claims about what will or won't promote the greater good, unlike claims about rights, are ultimately accountable to *evidence*. Whether or not a given policy will increase or decrease happiness is ultimately an empirical question. One can *say* that national health insurance will improve/destroy American healthcare, but if one is going to say this, and say it with confidence, one had better have some evidence. First, one had better understand how national health insurance is actually supposed to work (see above). Then, as a seeker of evidence, one must understand how different healthcare systems work and how different systems have fared in various states and nations: Who lives longest? Who has the best quality of life following care? Which citizens are most satisfied overall with their healthcare? These are, of course, just the sorts of questions that policy wonks try to answer, and not just about healthcare but about all significant social issues: When nations abolish the death penalty, do murder rates go up? Do nations that redistribute wealth more widely encourage laziness? Are the citizens of such nations less happy overall? Figuring out what makes societies happier is challenging and prone to bias. But in the end, with ten steps forward and nine steps back, such questions can be answered with evidence.

The same cannot be said for questions about rights. As explained in chapter 7, we have, at present, no non-question-begging way to figure out who has which rights. If, someday, philosophers produce a theory of rights that is demonstrably true, then everything I'm saying here will go out the window. But for now, at least, arguing about rights is a dead end. When you appeal to rights, you're not helping to resolve the issue. Instead you're pretending that the issue has already been resolved in some abstract realm to which you and your tribespeople have special access.

Now, at this point you, as a longtime believer in rights, may still be torn. You agree that much of our rights talk is empty rationalization. But still, it seems that the idea of rights captures something deeply important, something that can't be captured with utilitarian balance sheets. What about selling little girls into prostitution? What about torturing people for expressing their beliefs? Do these things not violate people's rights? *Have you no moral compass?*

For you, I have good news. As deep pragmatists, we can appreciate the vital role that thinking about rights has played, and continues to play, in our moral lives. Arguing about rights may be pointless, but sometimes arguing is pointless. Sometimes what you need is not arguments, but weapons.** And that's when it's time to stand up for rights.

RIGHTS AS WEAPONS AND SHIELDS

The law professor Alan Dershowitz once told a handful of students the following story. There was a Holocaust denier who insisted on having a public debate with Dershowitz, who refused. The man hounded the professor with angry letters, challenging his intellectual integrity. *You call yourself a champion of free speech, and yet you try to silence me! Why are you opposed to an open exchange of ideas? You're afraid to debate me because you know that I will win!* Finally, Dershowitz agreed. "I'll debate you," he said. "But on one condition: Our debate must be part of a three-part series. First, we'll debate whether the Earth is flat. Then we'll debate the existence of Santa Claus. And then we'll debate whether the holocaust really happened." His would-be opponent declined.

Dershowitz's clever response illustrates a valuable, pragmatist lesson: Moral debate is not just about seeking truth. Deciding whether and how to engage with one's opponents is a pragmatic decision like any other, involving costs and benefits. In Dershowitz's case, it was the benefits of having an open exchange of ideas versus the costs of lavishing time and attention on a pernicious crank.* Some issues are not worth debating. In Dershowitz's case, the issue was a matter of historical fact, but the same goes for matters of value.

As a bit of Googling reveals, there are still people out there who think that blacks deserve to be enslaved, that some women deserve to be raped, and that it's a shame Hitler didn't finish off the Jews. These people, too, are not worth debating. We modern herders have agreed that slavery, rape, and genocide are simply unacceptable. We offer different reasons for this. Some of us appeal to God's will. Some appeal to human rights. Some, like me, oppose these things because of the overwhelming and unnecessary suffering that they cause. And some people—probably most people—are just simply opposed, as a matter of moral common sense, without any specific justification in mind. But we all agree that these things are completely unacceptable. In other words, some moral judgments really are *common* sense. Common doesn't mean *universal*. It means common enough for practical, political purposes. The matter has been settled.

When dealing with moral matters that truly have been settled, it makes sense to talk about rights. Why? Because the language of rights aptly expresses our firmest moral commitments. It's good to have some firm commitments and to reject some ideas out of hand,* not because we're guaranteed to be correct in all such cases, but because the risk of being incorrect is smaller than the risk of being insufficiently firm. We want our children to understand—not just intellectually, but emotionally—that some things are simply beyond the pale. And we want the extremists in our midst—the Klansmen, the neo-Nazis, the misogynists—to understand clearly that they are not welcome.

Above I said that I'm opposed to slavery because the costs overwhelmingly outweigh the benefits. But doesn't it make you a little uncomfortable to hear me put it that way? Me, too. Thus stated, it sounds like maybe, just maybe, if someone were to come along with the right kind of argument, I would consider changing my mind about slavery. Well, rest assured that on this particular matter, my mind is closed. If you send me an e-mail with a subject heading "why slavery might be justified in some cases," I'll simply hit DELETE, thank you very much. I still believe, as stated above, that the only non-question-begging argument against slavery is the utilitarian one offered long ago by Bentham and Mill. But now, in this new millennium, the slavery question is one that I'm more than happy to "beg." In my estimation, the costs of talking about slavery as if it's an open question, to be settled by the available evidence, outweigh the benefits. And thus, as a deep pragmatist, I'm happy to join the chorus: *Slavery violates fundamental human rights!*

"But," you object, "you don't really mean it!" Yes, I do. To a deep pragmatist, declarations about human rights are, when properly deployed, like wedding vows. When you tell your beloved, "Till death do us part," you are not, if you're a reasonable adult with an active manual mode, saying that there are absolutely no circumstances under which you would seek a divorce. You're not saying that the odds of your marriage ending by choice are 0.00000 percent. You are expressing a *feeling*, a deep commit-ment. And it would be a very poor expression of that feeling and that com-mitment to declare at the altar: "My love, the odds of our staying together are, in my estimation, very, very high." Likewise, it's a poor expression of your opposition to slavery to say that, in your estimation, slavery very clearly fails to maximize happiness. When someone asks you, "Do you believe that slavery violates fundamental human rights?" the correct an-swer is "I do."

As deep pragmatists, we can appeal to rights when moral matters have been settled. In other words, our appeals to rights may serve as *shields*, pro-tecting our moral progress from the threats that remain. Likewise, there are times when it makes sense to use "rights" as weapons, as rhetorical tools for making moral progress when arguments have failed. Consider, for example, the moral struggles of the American civil rights movement. There are utili-tarian arguments for allowing blacks to vote and to eat alongside whites in restaurants. These are good arguments. But these arguments, like all utili-tarian arguments, depend on a premise of impartiality, on the Golden Rule, on the idea that no one's happiness is inherently more valuable than anyone else's. It was precisely this premise that the opponents of the civil rights movement rejected. Thus, arguing about explicit racial discrimination is not like arguing about higher versus lower taxes, capital punishment, or physician-assisted suicide. From an impartial moral perspective, there's nothing to debate. Jim Crow was a simple matter of one tribe's dominating another,* and by the 1950s it was clear that moral reasoning alone was not going to get the job done. What was needed was force, and an emotional commitment on the part of third parties to using force. Thus, during this important moral and political struggle, the emotionally salient language of rights was the right language to use. The issue might not have been settled, but there was, at the same time, no more room for rational debate.

Thus, there are times when a deep pragmatist should feel free to speak of rights—and not just legal rights but moral rights. These times, how-ever, are rarer than we think. If we are truly interested in persuading our opponents with reason, then we should eschew the language of rights. This is, once again, because we have no non-question-begging (and nonutilitarian) way of figuring out which rights really exist and which rights take precedence over others. But when it's not worth arguing—either because the question has been settled or because our opponents can't be reasoned with—then it's time to stop arguing and rally the troops. It's time to affirm our moral commitments, not with wonky estimates of probabilities but with words that stir our souls.

But *please* do not take this as license to ignore everything else that I've said about "rights." Most moral controversies are not simple cases of one tribe's dominating another. In nearly all moral controversies, there are truly moral considerations on both sides.* There is something to be said for individualist systems that encourage people to take care of themselves. And there is something to be said for collective systems in which everyone gets the help they need. There is something to be said for not killing any human fetuses, and there is something to be said for letting people make their own tough bioethical choices. Here the solution is not for us to bludgeon one another with heartfelt assertions about rights, however tempting this may be. The solution is, once again, to put our automatic settings aside and shift into manual mode, seeking bargains brokered with the common currency.

ABORTION: A CASE STUDY

The abortion debate is both bitter and enduring. Thus, it's a good test case for deep pragmatist thinking. If the deep pragmatist's approach can help us here, it can probably help us elsewhere. (And here I must emphasize that I am by no means the first to take this approach. Many of the ideas presented in this section and the next follow closely those of Peter Singer, among others.)

The moral peacemakers say that we should all be more reasonable, flexible, open-minded. But what does that mean? If you believe that abortion is murder—killing an innocent human being—should you be "reasonable" and allow people to commit murder? If you believe that outlawing abortion violates women's fundamental rights, should you be "reasonable" and relinquish a woman's right to choose? Simply urging people to be reasonable does little to solve the problem because each of us believes that we're already reasonable. To make real progress, we have to put our gut reactions aside and shift into manual mode. As it turns out, almost no one, on the left or the right, takes a coherent moral stand on abortion, one that stands up well to manual-mode scrutiny.

Let's start with the pro-choicers. As you well know, liberals tend to view abortion as a matter of "rights," and of women's rights, more specifically. But almost no one believes that a woman has the right to abort a nine-month-old fetus. Why not? The fetus is still inside the body—does a woman not have a right to control her body? Do elderly Christian fundamentalist congressmen from the Deep South now have the right to tell young women in San Francisco that they can't choose abortion? At some point, apparently, they do.

To be a coherent pro-choicer, one must explain why early-term abortions are morally acceptable but late-term abortions are not.* Both first- and third-trimester fetuses have the potential to become fully developed humans. Thus, the moral difference can't be a matter of *potential*. Both early- and late-term abortions prevent a human life from being lived.* If it's not a matter of potential, then the key difference must be *actual*: a matter of what the fetus *is* in early versus late pregnancy. There are plenty of candidate differences.

The most influential distinction, famously drawn by the U.S. Supreme Court in *Roe v. Wade*, concerns the fetus's viability outside the womb, a distinction that separates early-term and late-term fetuses. But is viability what really matters? Viability is as much a function of technology as it is of the fetus itself.* Today, infants born as early as twenty-two weeks can survive, and that figure will almost certainly change as technology develops further. It's possible that at some point, perhaps within our lifetimes, struggling fetuses could develop in artificial wombs outside the mother, beginning in early pregnancy. Will pro-choice advocates then say that, thanks to new technology, first-trimester abortions have become immoral?* And what about late-term fetuses that aren't viable outside the womb? Suppose that a fetus just shy of nine months has a rare condition that would prevent it from living outside the womb at the moment. And suppose that this condition will resolve just before birth. Is it acceptable to abort this nearly nine-month-old fetus because it's not (yet) viable outside the womb?

Viability outside the womb seems to be a convenient proxy for whatever it is that really matters. What, then, really matters? What is the special something that only late-term fetuses have, endowing them with a right to life? Finding that special something will be tough, because whatever it is, it's almost certainly shared by animals that we (most of us) eat. Is it the ability to feel pain? Pigs feel pain. (In any case, we're at least as certain of adult-pig pain as we are of late-term human-fetus pain.) Likewise, pigs, as compared with all human fetuses, are at least as likely to be conscious, at least as likely to have a robust sense of self, at least as likely to have complex emotions, and more likely to have meaningful relationships with others. The morally significant features that late-term fetuses have, but that early-term fetuses lack, are almost certainly going to be shared with adult pigs and other animals that we kill for food.

The pro-choice position isn't dead, but options are running out. One option is to say that certain features of late-term fetuses (e.g., rudimentary consciousness) make late-term abortions wrong, and also make it wrong to eat certain animals. But this is no easy way out. Consistency requires more than being a moral vegetarian.* It requires being a *militant* vegetarian. Many vegetarians, including those with moral motivations, choose not to eat meat themselves yet remain "pro-choice" about eating meat. They don't regard their meat-eating friends as murderers, and they don't believe that eating meat should be illegal. (Some do, but most don't.) If you're not going to be pro-choice about late-term abortions because you think that late-term fetuses have rudimentary consciousness (or whatever), then you shouldn't be pro-choice about eating pigs. This position is an option, but the vast majority of pro-choicers are unwilling to go that far.*

Another pro-choice option: You might say that late-term fetuses have a magic combination that endows them with a right to life. Like pigs, they have rudimentary consciousness (or whatever), but unlike pigs, they are *human*. And like early-term fetuses, they are human, but unlike early-term fetuses, they have rudimentary consciousness. Neither of these two things alone is enough to grant one a right to life, you say, but put them together and—*pow!*—you've got a creature with rights. The first thing to notice about this theory is that it's completely ad hoc. Second, what's especially ad hoc is the idea that humanness per se is a critical factor. Few liberals would say that being a member of *Homo sapiens* is a necessary ingredient for having a right to life. For example, most liberals believe that nonhuman animals such as chimpanzees have a right to life, that we can't just kill chimps if it serves our interests. To make this point more sharply, consider the moral rights of nonhuman aliens who think and feel just like us. Take, for example, the lovely Deanna Troi from *Star Trek: The Next Generation*. Surely it's not okay to kill her simply because she's not human.* To the chagrin of countless Trekkies, Troi is not real, but her character is enough to make the point: What endows us with rights is not our being human, per se, but rather our having features that members of others species could, or do, have.

The idea that something like "human consciousness" is what really matters suggests a more familiar idea, the idea of the *soul*. We'll talk more about souls shortly when we consider the pro-lifer's predicament. But first, let's consider what a pro-choice appeal to the soul might look like. We'll suppose that humans have souls and that other animals, such as pigs, either don't have souls or have qualitatively different souls—pig souls, et cetera. And we'll suppose that having (or being) a human soul is what grants one an unequivocal right to life. If you're a soul-minded pro-choicer, then you'll say that late-term fetuses have souls and that early-term fetuses don't. The problem with this claim is that there's simply no reason to believe that it's true. Early-term fetuses can move their bodies. They are *animate*. And if it's not a human soul that's animating them, then what is it? A temporary fetal soul? In any case, we surely can't be *confident* that "ensoulment" occurs sometime after the first trimester, if it occurs at all. If we think that humans have souls, and we think that early-term fetuses might have souls, then this is hardly a strong case for being pro-choice.

In sum, to construct a coherent justification for the pro-choice position on abortion is actually very hard. I'm not saying that it can't be done.

I'm saying that if it can be done at all, it will require complex, manual-mode philosophical maneuvering of a rather esoteric sort. In our popular moral discourse, it's perfectly acceptable to say, "I believe in a woman's right to choose," without further explanation. But without further explanation, appealing to this "right" is just a bluff, a bald assertion to the effect that, somewhere out there, there is a coherent, pro-choice theory of reproductive rights.

What about the pro-lifers? Can they do any better? One kind of pro-life argument focuses on the human life that never gets lived because of an abortion. The problem with this argument is that it applies too widely for its proponents' tastes. Abortion denies a person an existence, but so does contraception, and most pro-lifers (at least in the United States) are not ready to outlaw contraception. Of course, many pro-lifers, most notably devout Catholics, are opposed to contraception, but the problem doesn't end there. The life-denial argument also applies to *abstinence*. Couples who choose not to have children, or to have fewer children than they otherwise could, are also preventing human lives from being lived. And this is true even for couples who can't afford to support more children, so long as there are others willing to adopt them. Unless you think that morality requires us to make as many happy babies as possible, you can't argue that abortion is wrong on the grounds that it prevents human lives from getting lived.

This, however, is not the kind of argument that most pro-lifers want to make. They want to draw a distinction between lives that are merely possible and lives that are, in some sense, already under way. For most pro-lifers, the critical moment is conception. (I will use the term "conception" interchangeably with "fertilization," to refer to the joining of sperm and egg.) It's often said that "life" begins at conception, but that is not literally true. The sperm and egg that form a zygote (the single cell from which a fetus develops) are both undeniably alive. The idea, then, is not that *life* begins at conception, but that *someone's* life begins at conception. Could that be correct?

This brings us back to the topic of souls. But before we go there, let's see if we can make sense of this idea in a more metaphysically modest way. You might say that conception is special because, once sperm hits egg, the identity of the individual has been determined. There is now an answer to the question "Whose life is at stake?" Answering this question, however, does not necessarily require the physical joining of sperm and egg. Consider what happens in fertility clinics, in which fertilization occurs outside the body. Typically, the fertilization container (usually a petri dish, rather than the proverbial "test tube") holds several eggs and many sperm, one of which will be the lucky one, should fertilization succeed. But a fertility clinician could select a single sperm and a single egg and let them have at it. Before they meet, they might be held in separate containers. At that moment, when the lucky sperm is "on deck," it's been determined which human shall be, if anyone is to be.* The would-be zygote's genetic identity has been determined. But is there, at that moment, a person with a right to life, divided between those two containers? If the woman backs out of the procedure while the sperm is still on deck, is that murder? Will she have robbed an innocent person of his life?**

Most pro-lifers, I suspect, will not call a woman who backs out of in vitro fertilization a murderer, even if the (un)lucky sperm and egg have already been selected, thus determining the identity of the would-be child. And that means that it's not really about the determination of the would-be child's genetic identity. Instead, the idea is that something morally significant happens when the sperm and egg physically join, that "life" begins at conception. This raises the all-important question "What happens at conception?"

Well . . . A lot of interesting things happen. I'll spare you the full biology lesson, which I'm not qualified to give, in any case. The critical point for us is that fertilization and the processes surrounding it are fairly well understood on a mechanical, molecular level. We understand the chemical processes that allow sperm to move: The mitochondria in the midpiece of the sperm produce ATP, the fuel that is used to power the movements of the sperm's tail (flagellum). Dynein proteins in the flagellum convert ATP's chemical energy into movement, and ultimately into the movements of the sperm's tail, which propels the sperm forward. We understand how the sperm finds the egg: The sperm is sensitive to a combination of chemical and thermal signals emanating from the egg. We know what happens when the lucky sperm hits the surface of the egg: The egg is surrounded by a glycoprotein membrane called the zona pellucida, which contains chemical receptors that match chemical receptors on the head of the sperm. This chemical interaction causes the sperm to release digestive enzymes that enable it to burrow through the zona pellucida toward the egg cell membrane. The sperm's membrane fuses with the egg's. This triggers a set of chemical reactions that prevent other sperm from entering the egg. The genetic material from the sperm is released into the egg, and a new membrane forms around the male genetic materials, creating the male pronucleus. Meanwhile, the fusion of the sperm with the egg causes the female genetic material to finish dividing and to form the female pronucleus. Thin polymer structures called microtubules pull the two pronuclei together. They fuse, and the two sets of genetic material are now contained within a single nucleus, the nucleus of the zygote, and fertilization is complete. The zygote will then divide into two cells, four cells, eight cells, and so on, until it forms a ball of cells called a morula, which then hollows out to form an empty ball of cells called a blastula. The blastula develops into a gastrula, which consists of three distinct cell layers (ectoderm, mesoderm, and endoderm), each of which go on to form different bodily tissues. For example, the ectoderm goes on to form the nervous system (brain and spinal cord) as well as the tooth enamel and the outer layer of skin (epidermis).

I've told you all of this not to impress you with my knowledge of developmental biology—I had to look most of this up—but to impress upon you the marvelous extent to which we understand the mechanics of life at the earliest stages of development. Indeed, my little summary hardly does justice to the step-by-step, molecule-by-molecule understanding that biologists now have. There are gaps in this understanding, to be sure. But there are no big, gaping mysteries—just little holes waiting to be filled by the next research article describing the next protein in a long chain of chemical reactions.

Our mechanistic understanding of human development poses a serious problem for most pro-lifers. They want to say that fertilization creates, in an instant, a new person with a right to life. Fertilization is amazing, a pivotal moment in the development of a new human being. But so far as we can tell, it's not magic. What's more, the fertilization of a human egg appears to be no more or less magical than the fertilization of a mouse egg or a frog egg. There is no evidence whatsoever for the occurrence of "ensoulment" at fertilization, or at any other point in development. So far as we can tell, it's all just organic molecules operating according to the laws of physics.

What's a pro-lifer to do? One can insist that something magical must happen at fertilization and that if scientists haven't found it yet, this simply reflects their ignorance or, worse yet, their godless, materialistic biases. This, however, is just a hope, a bald assertion with no evidence behind it. The pro-lifer who says this is no different from a pro-choicer who declares, without a shred of evidence, that the moral magic happens later, during the third trimester, rather than at conception.

A more modest pro-lifer might admit that we don't know when "ensoulment" occurs but argue that we, in light of our ignorance, should be on the safe side. Because we don't know when it happens, we should not allow abortions of any kind. But if that's right, why stop at fertilization? Why not suppose that God attaches a soul to each unfertilized egg, and that the sperm just supplies some useful molecules? Or why not assume that God attaches souls to sperm? (Cue Monty Python.) How can we be sure that contraception does not kill souls? To be safe, shouldn't we outlaw contraception? And how do we know that abstinence doesn't kill souls? To be really safe, shouldn't we require women to take on as many (potentially soul-laden) sperm as their wombs can handle?

The pro-lifer's troubles multiply when faced with possible exceptions to a ban on abortions. In 2012, Republican Senate candidate Richard Mourdock ignited a firestorm when he explained why he is opposed to abortion even in cases of rape:

> I think, even if life begins in that horrible situation of rape, that it is something that God intended to happen.

With that remark, his campaign went up in flames. *Mourdock says that God wants women to be raped!* The problem that Mourdock stumbled upon is actually a much bigger problem, one that goes well beyond abortion. It's the age-old "problem of evil," which has dogged theologians for centuries: If God is all-knowing and all-powerful, why does he allow things like rape (and child abuse and mass school shootings and deadly earthquakes) to happen? This isn't just Mourdock's problem. It's a problem for anyone who believes in an omniscient, omnipotent, and benevolent deity. In any case, Mourdock's remarks did not go over well with voters, especially women, and he lost the election. But I don't think Mourdock hates women, or more specifically rape victims. I think he was just trying to be a consistent pro-lifer. As he said at the time:

> I believe that life begins at conception. The only exception I have
> for . . . to have an abortion, is in that case of the life of the mother. I
> just struggled with it myself for a long time but I came to realize life
> is a gift from God . . .

If you really believe that "life" begins at conception, and that God attaches souls to biological matter at that moment, then, really, what business do we have undoing God's metaphysical injections? From a pro-life perspective, the only questionable part of Mourdock's position should be his willingness to allow abortion to save a mother's life. Would it be okay to kill a three-year-old if, somehow, that were the only way to save her mother?

In the end, the pro-lifers may be right. We may have souls, and God may attach human souls to biological matter at the moment of fertilization. But we have absolutely no evidence that this is true, and we have no more evidence for this than we have for other theories of ensoulment, including theories that place ensoulment late in pregnancy or before fertilization. When pro-lifers declare with confidence that a fetus has a "right to life," they, like their pro-choice counterparts, are just bluffing, pretending that they have a coherent argument when in fact they have only strong feelings and unsubstantiated assumptions.

―――――――

Some ideas about abortion resonate deeply with people, and some don't. And some ideas resonate with people on both sides of the debate. If most people's attitudes about abortion are not backed up by a coherent philosophy, then where do these attitudes come from? As ever, a dose of psychological understanding can go a long way.

You may recall from chapter 2 (pages 46–48) the experiment in which babies chose to play with the nice triangle with googly eyes, the one that helped the googly-eyed circle get up the hill. And you may recall that this preference disappeared when the circle's googly eyes were removed and when the children never saw the circle move on its own. Without eyes (the proverbial "windows of the soul") and without the appearance of spontaneous movement, the circle was just a shape. Likewise, you may recall how mere images of eyes can set off alarm bells in our amygdalas (figure 5.2, page 142) and cause us to be more generous (figure 2.3, page 45). Eyes turn on our social brains.

Although eyes do a lot, it turns out that movement alone is enough to turn faceless entities into creatures with hearts and minds. In the 1940s, the pioneering social psychologists Fritz Heider and Marianne Simmel created a famous film in which three shapes enacted a silent drama. A big, mean triangle tormented two smaller shapes, chasing after them as they tried to escape. The film involved nothing more than moving shapes, but people automatically attributed to them intentions ("The big triangle is trying to get them," "The smaller shapes are trying to get away"), emotions ("The big triangle is angry that they got away," "The little shapes are happy because they escaped"), and even moral character traits ("The big triangle is a bully"). These attributions happen so automatically that people can't stop themselves from making them. We see social drama as automatically as we see color and shape.

Fetuses move, and fetuses have eyes. Before the advent of medical imaging, many who contemplated the ethics of abortion located the moral turning point at "quickening," the point at which a fetus begins producing detectable movements. Medical imaging has allowed us to see not only fetal movement but also fetal features, such as eyes, and at stages prior to quickening. For many, this pushed the magic moment further back in time.

Movement and eyes have a powerful effect on us, but that can't explain everything. The animals that most of us eat without a second thought move and have eyes.* But fetuses, unlike the animals we eat, at some point start to look *human*. They have little human hands, little human feet, and little human faces, and they move in very human ways. This is undoubtedly why pro-life advocates are so keen to present images of fetuses, especially close-up images of hands, feet, and faces. It also explains why the 1984 pro-life film *The Silent Scream* was such a stunning success. The film, which is eerily similar to Heider and Simmel's film, features within it an ultrasound recording of an abortion. We see the fetus moving inside the womb. The narrator explains that the fetus is "purposefully" moving away from the device, describing its movements as "agitated" and "violent." In the film's defining moment, the fetus's mouth opens as the suction device approaches. Later, its head is crushed, allowing it to pass through the cervix. Whether you're pro-life or pro-choice, the film is very hard to watch, and that's precisely the point. *The Silent Scream* engages one's automatic settings, providing an "argument" against abortion that's more powerful than any actual (manual-mode) argument.

The Silent Scream works because the fetus looks rather human. Had the film depicted an abortion early in the first trimester, when the developing human is just a cluster of cells, there would have been no show. Destroying a cluster of cells does not feel like a horrible thing to do. And therein lies the intuitive moralist's dilemma. Abortion doesn't *feel* wrong until the fetus starts to look human, and it doesn't feel *horribly* wrong until the fetus truly looks like a baby. But there is no bright line between when the fetus looks like a baby and when it looks merely humanoid. Nor is there a bright line between the humanoid stage and the earlier stages during which a developing human is, to the untrained eye, indistinguishable from a developing mouse or frog. The only sharply discontinuous event in the whole process is fertilization. But at that point, at the zygote stage, the developing human has none of the features that engage our automatic settings. It is transparently a bag of organic molecules. If we place no restrictions on abortion, then we're allowed to kill something (*someone!*) that looks just like a baby. But if we outlaw all abortion, then we force otherwise free people to severely disrupt their lives for the sake of a bag of molecules. And yet there is no emotionally comfortable resting point in between these two extremes.

What, then, do we do? To a large extent, we do whatever the other members of our tribe do. Most tribes believe in souls—a very natural belief, for a variety reasons. If you're committed to the idea that people have souls, you have to believe that ensoulment happens *at some point*, and conception seems like the most likely point. It's true that a single cell doesn't look much like a creature with a soul, but what's the alternative? Before conception, you've got two distinct bodies, and after conception there is no discrete event. Fertilization is by far the least implausible moment for ensoulment. And so, if your tribal elders tell you that this is when "life" begins, and you have no better theory of your own, then you go with it. And besides, to say otherwise would make you sound like one of *Them*.

If, by contrast, your tribe doesn't believe in souls, or allows people to draw their own conclusions about them, then what do you do? Deciding the abortion issue based on someone's speculation about the timing of ensoulment is not appealing. This is especially true if your tribe values personal choice—not just in the context of abortion, but more generally. Still, not everything goes. Your tribe might not have a party line on ensoulment, but it's sure of at least this: Killing babies is definitely not allowed. Thus, to be safe, you can't allow people to kill anything that looks like a baby or could be a baby right now—that is, anything that is viable outside the womb. Unfortunately, looking babylike is a matter of degree. From early in development, fetuses move spontaneously and have human-looking hands, feet, and faces. What to do?

The pro-choice position requires an awkward, emotional balancing act. Few pro-choicers are completely comfortable with killing human-looking things, and many are uncomfortable with killing things that look like animals. But pro-choicers are also uncomfortable with telling others what to do, especially women. Thus, pro-choicers must strike an uncomfortable, but apparently unavoidable, balance between "Don't tell other people what to do!" and "You can't kill a thing that looks like that!"

What does all of this mean for the abortion debate? It means that nearly all of us are bluffing. All of our confident talk about a "right to life" and a "right to choose" is just so much manual-mode confabulation, our attempts to put a rational face on our half-baked intuitive theories, driven by cognitive gizmos that we barely understand. When you strip away the high-minded talk of rights, there's really not much left. An honest pro-lifer sounds like this:

> I believe that a person is a soul inhabiting a physical body. I have no real evidence for this, but it seems right to me, and this is what all of the people I trust believe. I don't know how souls get into bodies, but the people I trust say that new souls arrive when the sperm hits the egg. I don't know exactly how this works, but I don't have any better ideas of my own. My best guess, then, is that there's a human soul in there starting at conception. You can't rightly kill an innocent human soul. I know that this is partly a matter of faith. And I understand that we're supposed to respect each other's beliefs. But I just can't see letting people kill something, even if it's small, so long as there might be a human soul in there. I know that's hard on a lot of people who don't want to be pregnant. But those people made a choice to have sex (except in the case of rape, which is different), and killing something that maybe has a human soul is not a legitimate way to undo that choice. That's how I feel.

And an honest pro-choicer sounds like this:

> I believe that people should be free to think for themselves and make their own choices, and that's how I feel about abortion. At least during the early stages of pregnancy. A first-trimester fetus looks kinda like a person, but it also looks kinda like a frog. And while I don't like the idea of killing a froggy little human, I think that forcing a woman

to go through with an unwanted pregnancy is even worse. I know that there are people who want to adopt babies, but giving birth to a baby and giving it away must be agonizing. Forcing a woman to do that seems worse to me than killing a froggy little human. Third-trimester fetuses, however, don't look froggy. They look like babies. And killing babies is clearly wrong. So if the fetus you're carrying looks kinda froggy, then I think it's okay for you to kill it, if that's your choice. But if your fetus looks like a real baby, and not a little froggy thing, then I think you have to let it live, even if you don't want to. That's how I feel.

This is what the abortion debate really comes down to: strong, but complicated, feelings that we can neither justify nor ignore. What, then, are modern moral herders to do?

ABORTION: THE PRAGMATIC APPROACH

Having called the "rights" bluff on both sides, we're ready to think like deep pragmatists. Rather than try to figure out when "life" begins, we start with a different set of questions: What happens if we restrict legal access to abortion? What happens if we don't? And what impact would these policies have on our lives? These are complex empirical questions, difficult to answer, but we can begin with some educated guesses.

If abortion were outlawed, people would adjust their behavior in one of three general ways. First, some people would change their sexual behavior. Some would abstain from sex completely, at least for a time. Others would have sex less frequently, and others would take further measures to reduce the likelihood of pregnancy. Second, some people would seek abortions by other means, illegally or abroad. Third, some people would give birth to babies who would otherwise not be born. Of these people, some would give their babies up for adoption and some would choose to raise these children themselves.

How does all of this add up? Let's start with people who change their sexual behavior. For most adults, nonprocreative sex is a highly enjoyable and fulfilling part of life. Nonprocreative sex is a major source of happiness, not only for the young and the restless but for couples in stable monogamous relationships. For fertile couples, nonprocreative sex is made possible by contraception, but as we all know, contraception provides no guarantee, even when used responsibly. Thus, for millions of sexually active adults, the option to have an abortion provides an important safeguard against unwanted pregnancy.

On the other side of the pragmatic ledger, some sex is harmful, and there might be less harmful sex if abortion were outlawed. Examples of harmful sex include emotionally damaging sex between consenting adults, sex between teenagers who are not emotionally ready to have sex, incest, and rape. Avoiding sex for fear of pregnancy might also have the beneficial side effect of reducing the spread of sexually transmitted disease. What's less clear is whether outlawing abortion would substantially reduce the amount of harmful sex. It seems unlikely that rapists would be deterred by the knowledge that their victims couldn't get abortions. Certainly, outlawing abortion would prevent some teenagers from having sex, though it's not clear whether this would be, on balance, good or bad. The teenagers who are most mindful of the consequences of their choices are, presumably, the ones who are most ready to be sexually active.

In sum, when it comes to changing people's sexual behavior, making abortion illegal would take a big toll on millions of sexually active adults without any clear compensating benefit, as measured in terms of happiness.

Next, let's consider alternative routes to abortion. For people of means, making abortion illegal would simply make obtaining an abortion more expensive and inconvenient. Less fortunate women would turn to a domestic market that caters to the desperate for an illegal abortion. I'll not recount here the horrors of illegal abortion. From a utilitarian perspective, causing people to seek alternative routes to abortion leaves them with options that range from bad to horrible.

Finally, let's consider the effects of increased birth. Forcing women into unwanted pregnancies is horrible. Pregnancy is an enormous emotional strain under the best of conditions, and women carrying unwanted fetuses may, consciously or unconsciously, take less good care of them. Carrying a fetus/baby to term not only is a great emotional strain, but can severely disrupt one's life. In sum, forcing women to have babies against their will is very, very bad.

Nevertheless, one could argue that the benefits of forcing women to go through with unwanted pregnancies are even greater. By giving birth, a woman allows a new person to live. If the woman does not want to keep her baby, she can give the baby up for adoption. In the best case, the baby will go to a loving home with plenty of resources. Here it's hard to argue that the costs endured by the birth mother, however high, are so high that they outweigh her biological child's entire existence. Of course, and unfortunately, not all adoptive children find nurturing homes, and if abortion were illegal, there might be fewer good adoptive homes available. Still, even when an adopted child's conditions are far from ideal, it's hard to argue that the mother's pain and suffering should take precedence. So long as the adopted child's life is overall worth living, it's hard to say that the biological mother's suffering outweighs the value of her biological child's entire life.

In some cases, the mother, and perhaps the father, too (or instead), will choose to raise the child. In many cases—perhaps most cases—things will turn out well. Many happy families include children born of unplanned, and initially unwanted, pregnancies. In other cases, the unwanted child's life may not go so well as we would like, but for abortion to be preferable, it would have to be the case that the child's life is, overall, not worth living—either that or that the child's existence would have to make the world worse off overall. Or, more realistically, the child's existence would have to preclude the existence of another child who would go on to live a happier life, or make the world happier overall.

It's here, with this awkward utilitarian accounting, that the pro-lifer makes her strongest case. If abortion were illegal, some additional people would exist. In some cases, their existence would result in a net loss of happiness. But overall, it's hard to claim with any degree of confidence that the additional people created by outlawing abortion would, on balance, be unhappy, or make the world less happy overall. This is, of course, a complicated empirical question. A lot depends on the availability of good adoptive homes. To the extent that good adoptive homes are available, it's hard to argue that the fetuses/babies in question are better off being aborted, or that the world is better off if they're aborted.

Where does this leave us? My informal tally looks like this: On the one hand, outlawing abortion would pull an important safety net out from under millions of people, cause some wealthy people to seek abortions at great expense, and cause some desperate women and girls to seek horribly dangerous illegal abortions. Outlawing abortion would also disrupt many people's life plans, causing them to have children when they are not yet ready to have children, or not interested in having children at all. These are very, very high costs. On the other hand, outlawing abortion would grant life to many people who would otherwise not get to exist. And, depending on the availability of good adoptive homes, among other things, their existence would likely be good. So where does *that* leave us? Are we deadlocked once again?

I don't think so. The pro-lifer's life-saving utilitarian argument is a good one. The problem is that it's *too good*. You may recall from our earlier discussion that lives are lost not only from the restricting of abortion, but also from the restricting of contraception and abstinence. If we're opposed to abortion because it denies people their existence, then we should be opposed to contraception and abstinence, too, since both of these practices have the same effect. This, however, is an argument that almost no pro-lifers want to make.

This deeply pro-life argument is, in fact, analogous to the utilitarian argument in favor of extreme altruism, or turning ourselves into happiness pumps. One way to pump out happiness is to allocate resources more efficiently, helping the have-nots at the expense of the haves. Yet another way is to pump out more happy people. (Better still is to breed happy little utilitarians who are willing to work hard for the happiness of others.) It's not that this argument doesn't make sense. It is, instead, simply too much to ask of nonheroic people. Were I a god choosing between two species I could create—one that makes as many happy members as it can, and one that holds back—I would, if all else were equal, go for greater happiness.

Our resistance to making more happy people is not, I think, a moral one. Rather, we living humans are engaged in a conspiracy against the most underrepresented people of all, an underrepresented *majority*, in fact: the helpless hypothetical masses who, thanks to our selfish choices, never get the chance to even protest their nonexistence.

Oh, well. Too bad for them. For better or worse, we can't take the pro-lifer's life-saving utilitarian argument seriously. But the pro-choicer's utilitarian arguments are not *too* good. They're just plain good. Disrupting people's sex lives, disrupting people's life plans, and forcing people to seek international or illegal abortions are all very bad things that would make many people's lives much worse, and in some cases much shorter. And that's why, in the end, I believe that deep pragmatists should be pro-choice. I make no appeal to "rights," just to a realistic consideration of the consequences.

If you're an honest pro-lifer, unwilling to bluff with "rights," you have two choices. First, you can be upfront about your tribal metaphysical beliefs and insist, with a straight face, that the rest of the world live by them. But if you do that, expect your pro-choice opponents to ask questions like this: "Does God attach the soul when the head of the sperm makes contact with the zona pellucida? Or does ensoulment occur when the sperm hits the cell membrane? Is it enough for all of the sperm's genetic material to enter the egg cell? Or does God wait until the male pronucleus and the female pronucleus have fused? All the way fused, or partway?" At this point, pro-lifers will have to admit that they have no evidence-based answers to these questions, while nevertheless insisting that their faith-based answers dictate the law of the land.

The pro-choice position, by contrast, need not rely on unsubstantiated metaphysical assertions or on arguments that we can't take seriously when applied consistently. It's true that pro-choicers still haven't found a principled place to draw the line. But that may be inevitable. Whatever it is that makes people worthy of moral consideration, these things don't all appear in one magic moment. Without a magic moment to believe in, pro-choicers simply have to draw the line somewhere, while acknowledging that the line they've drawn is somewhat arbitrary. There may be no better place to draw the abortion line than where we currently draw it. But if there are good arguments, based on common currency, for drawing the line elsewhere, then deep pragmatists should listen.

WAITING FOR GODOT

Perhaps you've found this pragmatic, utilitarian "solution" to the problem of abortion unsatisfying. Indeed, it doesn't feel as though we've found the *right* answer. Our tentative pro-choice conclusion feels less like a victory and more like an indefinite cease-fire, albeit with terms that heavily favor one side. Thus dissatisfied, you may yet seek a true moral victory. You may hold out for a theory of abortion that is rationally defensible and that also feels right. Indeed, this is what we want whenever we engage in moral inquiry. Have we given up too soon?

Many moral thinkers will say yes. There are many bioethicists who try to do what I failed to do above: to make intuitively satisfying, non-utilitarian arguments about what's right and wrong when it comes to life and death. Beyond abortion and bioethics, moral philosophers have been busy devising sophisticated moral theories that purport to do a better job than good old-fashioned nineteenth-century utilitarianism. Are all of these people barking up the wrong tree? I believe that they are. I can't prove that they are, and I won't attempt to. Instead, in this section, I want to explain why I'm less optimistic than others about the prospects for sophisticated moral theory.* (And if you don't care about why sophisticated moral theories are unlikely to succeed, please feel free to skip this section.)

It all comes back to our dual-process moral brains. What we want is a manual-mode moral theory—an explicit theory that we can write out in words—that always (or as often as possible) gives the same answers as our automatic settings. If our automatic settings tell us that it's wrong to kill a third-trimester fetus but that it's okay to kill a first-trimester embryo, then we want a moral theory that will tells us *that* these intuitions are correct and *why* they're correct. And so on. In short, we want a moral theory that *organizes* and *justifies* our gut reactions. In Rawls's terms, we want to find a "reflective equilibrium" in which our moral theory matches our "considered judgments."*

But our gut reactions were not designed to be organized, and they weren't necessarily designed to serve truly moral ends. Automatic settings are heuristics—efficient algorithms that get the "right" answers most of the time, but not always. I put "right" in scare quotes because our automatic settings, even when functioning as they were designed to, need not be "right" in any truly moral sense. Some of our gut reactions may simply reflect the biological imperative to spread our genes, causing us, for example, to favor ourselves and our tribes over others. With this in mind, we might attempt to clean house. Before organizing our moral intuitions, we might first attempt to jettison all of our biased intuitions. If we use our scientific self-knowledge to debunk our biased intuitions, where will we end up?

I believe that we'll end up with something like utilitarianism. Why? First, as I explained in chapter 8, utilitarianism makes a whole lot of sense—not just to me and you, but to every nonpsychopath with a manual mode. The only truly compelling objection to utilitarianism is that it gets the intuitively wrong answers in certain cases, especially hypothetical cases. In chapters 9 and 10, we examined many of these cases and grew suspicious of our anti-utilitarian moral intuitions. These anti-utilitarian intuitions seem to be sensitive to morally irrelevant things, such as the distinction between pushing with one's hands and hitting a switch. I expect that we'll find more of this.

Second, I wonder: What would it mean for our anti-utilitarian moral intuitions to be sensitive to morally *relevant* things? One possibility is that our intuitive judgments serve to promote good consequences: We have negative gut reactions to things that tend to produce bad results (such as violence) and positive reactions to things that tend to do good (such as helping people). (In other words, our nondebunkable moral intuitions are "rule utilitarians.") If that's what we find, it only strengthens the case for utilitarianism, suggesting that our automatic settings are just imperfect utilitarian devices. What, then, would it mean for our automatic settings to be sensitive to morally relevant things that are *not* about producing good consequences? A natural thought is that our gut reactions track things like rights. For example, our sense that it's wrong to push the man off the footbridge might reflect the fact that pushing him violates his rights. But how could we know that this is true without an independent nonutilitarian theory of rights (one derived from self-evident moral axioms)? How could we know whether our gut reactions are tracking people's rights, or whether "rights" are just phantoms of our gut reactions? What would a complete (non-utilitarian) vindication of rights even look like?

At some point, it dawns on you: Morality is not what generations of philosophers and theologians have thought it to be. Morality is not a set of freestanding abstract truths that we can somehow access with our limited human minds. Moral psychology is not something that occasionally intrudes into the abstract realm of moral philosophy. Moral philosophy is a manifestation of moral psychology. Moral philosophies are, once again, just the intellectual tips of much bigger and deeper psychological and biological icebergs. Once you've understood this, your whole view of morality changes. Figure and ground reverse, and you see competing moral philosophies not just as points in an abstract philosophical space but as the predictable products of our dual-process brains.

There are three major schools of thought in Western moral philosophy: utilitarianism/consequentialism (à la Bentham and Mill), deontology (à la Kant), and virtue ethics (à la Aristotle). These three schools of thought are, essentially, three different ways for a manual mode to make sense of the automatic settings with which it is housed. We can use manual-mode thinking to explicitly *describe* our automatic settings (Aristotle). We can use manual-mode thinking to *justify* our automatic settings (Kant). And we can use manual-mode thinking to *transcend* the limitations of our automatic settings (Bentham and Mill). With this in mind, let's take a quick psychological tour of Western moral philosophy.

What happens if you're the chief philosopher of a single tribe? Within one's tribe, there are moral disagreements, but they are primarily about Me versus Us (or Me versus You)—the Tragedy of the Commons. Within a tribe, there are no full-blown moral controversies, clashes between the values of Us and Them, because within a tribe there is only Us. Thus, as your tribe's chief philosopher, it's not your job to resolve tensions between competing moral worldviews. It's not your job to question your tribe's common sense, but rather to *codify* it, to serve as a repository for your tribe's accumulated wisdom. Your job is to reflect back to your tribe what it already knows but sometimes forgets.

Among Western philosophers, Aristotle is the great champion of common sense. Unlike his mentor, Plato, Aristotle offers no radical moral ideas. Nor does Aristotle offer a formula. For Aristotle, being good—morally good, and good more generally—is a complex balancing act that is best described in terms of *virtues*, durable habits and skills that enable one to flourish. For example, in the face of danger, says Aristotle, one must be neither *rash* nor *cowardly*. Instead, one must be *brave*, exhibiting a virtuous balance between the vicious extremes. The virtues associated with love, friendship, work, play, conflict, leadership, and so on require their own balancing acts. For Aristotle, there is no explicit set of principles that tells one how to achieve a good balance. It's just a matter of practice.

As an ethicist, Aristotle is essentially a tribal philosopher. Read Aristotle and you will learn what it means to be a wise and temperate ancient Macedonian-Athenian aristocratic man. And you will also learn things about how to be a better human, because some lessons for ancient Macedonian-Athenian aristocratic men apply more widely. But Aristotle will not help you figure out whether abortion is wrong, whether you should give more of your money to distant strangers, or whether developed nations should have single-payer healthcare systems. Aristotle's virtue-based philosophy, with its grandfatherly advice, simply isn't designed to answer these kinds of questions. One can't resolve tribal disagreements by appeal to virtues, because one tribe's virtues are another tribe's vices—if not in general, then at least when tribes disagree.

Among contemporary moral philosophers, Aristotelian virtue theory has undergone a revival.* Why? The great hope of the Enlightenment was that philosophers would construct a systematic, universal moral theory—a metamorality. But as we've seen, philosophers have failed to find a metamorality that *feels* right. (Because our dual-process brains make this impossible.) Faced with this failure, one option is to keep trying. (See above.) Another option is to give up—not on finding a metamorality, but on finding a metamorality that feels right. (My suggestion.) And a third option is to just give up entirely on the Enlightenment project, to say that morality is complicated, that it can't be codified in any explicit set of principles, and that the best one can do is hone one's moral sensibilities through practice, modeling oneself on others who seem to be doing a good job. When confronted with the morass of human values, modern Aristotelians simply bless the mess.

In sum, Aristotle and the like do a very nice job of describing what it means to be a good member of a specific tribe, with some lessons for members of all tribes. But when it comes to modern moral problems defined by intertribal disagreement, Aristotelians have little to offer, because, once again, one tribe's virtue is another tribe's vice. (Yet another way of giving up on the Enlightenment project is to be a "relativist" or a "nihilist," one who, rather than embracing the common sense of any particular tribe, muddles along while denying that any tribe has got things right.)

If you're more ambitious than Aristotle (or modern "relativists"), you can use your manual mode not only to describe your tribe's morality, but to *justify* it. You may attempt to demonstrate that your tribal moral principles are universally true, like mathematical theorems. Enter Immanuel Kant.

Mathematicians try to prove theorems, but they don't try to prove just any theorems. They take interesting mathematical statements that seem to be true, or possibly true, and attempt to derive those statements from first principles, from axioms. At this, mathematicians have been wildly successful, from the Pythagorean theorem to Andrew Wiles's proof of Fermat's Last Theorem. So why not for ethics? Why can't philosophers derive interesting moral truths from first principles?

This is Kant's hope, and it's the hope of many ambitious philosophers since. Nietzsche, once again, called it the "secret joke" of Kant's soul, this ambition to prove that his tribe's morality is correct. But is this an embarrassment for Kant? It's no shame on mathematicians that they try to prove things that seem to them true. Why, then, is this a secret joke rather than a noble ambition?

Kant's problem is not so much with his ambition but with his unwillingness to admit failure. Mathematicians have successfully resolved countless mathematical controversies with proofs, but not one moral controversy has ever been resolved with a proof from first principles. Kant wants so badly to prove that his moral opinions are correct that he's blind to the flaws in his arguments. In other words, Kant crosses the line from *reasoning* to *rationalization*, and that's why Nietzsche is chuckling.

It's easy to see that Kant's arguments don't work when we're unsympathetic to his conclusions. Take, for example, his argument that masturbation is wrong because it involves using oneself as a means. (Really? And is it also wrong if one massages one's own arm, just because it feels good?) Kant's argument against masturbation is not considered his best work, but his more famous arguments don't, as far as I can tell, work any better. For example, Kant famously argues that lying, promise breaking, stealing, and killing are wrong because the "maxims" of lying, promise breaking, stealing, and killing cannot be "universalized": If everyone were to lie (or break promises), then the institution of telling the truth (or keeping one's word) would be undermined, and it would be impossible to lie or break a promise. Likewise, if everyone were to steal, then the institution of personal property would be undermined, and it would be impossible to steal. And if everyone were to kill people, there'd be no one left to kill.

These arguments are very clever, but they fall far short of proof. For one thing, it doesn't follow, either logically or intuitively, that an action must be wrong if it can't be universalized in Kant's sense. Take, for example, being fashionable: If everyone is fashionable, then no one is fashionable. Universal fashionableness is self-undermining. Nevertheless, we don't think that being fashionable is immoral. Likewise, there are nasty behaviors, such as beating people up, that are not self-undermining: No reason why we can't beat on one another until the end of time. This would be bad, but it wouldn't be impossible, and Kant's argument requires impossibility.*

Kant's fans are well aware of the flaws in Kant's arguments, and they have their replies, but we can at least say this: After nearly two and a half centuries, no one has ever managed to transform Kant's flawed arguments into rigorous moral proofs, and it's not for lack of effort. Nor has anyone else managed to prove any substantive moral claims true. By this I mean, once again, that no moral controversy has ever been resolved with a proof. Of course, many very smart people have tried. Most famously, John Rawls, in *A Theory of Justice*, attempts to show that, given a few minimal assumptions, one can derive the kind of egalitarian liberal political theory that he favors. I don't know whether anyone believes that Rawls's argument counts as a bona fide proof of his conclusions, but many people believe that Rawls makes a fine case for a nonutilitarian moral and political philosophy. I'm skeptical. In fact, I think that Rawls's central argument in *A Theory of Justice*, like Kant's before him, is essentially a rationalization.***

We can describe our tribal automatic settings (Aristotle), and we can attempt to prove that they are correct (Kant). However, neither of these philosophical approaches does much to solve our modern moral problems, because it's our tribal intuitions that are causing the trouble in the first place. The only way forward, then, is to *transcend* the limitations of our automatic settings by turning the problem (almost entirely) over to manual mode. Instead of trusting our tribal moral sensibilities, or rationalizing them, we can instead seek agreement in shared values, using a system of common currency.

Perhaps Aristotle, or someone like him, is right. Perhaps there is a single set of moral virtues to which we all should aspire. Or perhaps Kant, or someone like him, is right: Perhaps there is a true moral theory waiting to be proved from first principles. Or perhaps, more modestly, we can organize the morass of human values into something more coherent, a sophisticated moral theory that better captures our intuitive sense(s) of right and wrong. Perhaps. But while we're waiting for Godot, I recommend a more pragmatic approach: We should simply try to make the world as happy as possible. This philosophy doesn't give us everything we want, but for now, it's the best that modern herders can do.

WHY I'M A LIBERAL, AND WHAT IT WOULD TAKE TO CHANGE MY MIND

I'm a university professor. I live in Cambridge, Massachusetts. It requires no sociological sleuthing to guess that I'm a liberal. (And by "liberal" I mean liberal in the American sense: left of center, and less averse to active government than libertarians and some "classical liberals.") My liberalism is predictable enough, but is it justified? Is my liberal tribe just another tribe, with its own gut reactions and well-rehearsed rationalizations?

Yes, to some extent. As we saw in our discussion of abortion, liberals make their share of unsubstantiated assertions and incoherent arguments. But the liberal tribe is not, in my view, just another tribe. The world has many traditionally tribal tribes, people with a shared history, bound together by a set of "proper nouns" (Gods, leaders, texts, holy places, etc.). Today, however, there are two global meta-tribes—post-tribal tribes—bound together not by a shared history, and not by proper nouns, but by a set of abstract ideals. One of these meta-tribes is my liberal tribe. I wasn't always a liberal,* and it's conceivable that I might not be again someday. I'm a liberal because I believe that, in the real world, my tribe's policies tend to make the world happier. But I'm not a liberal to my core. I'm a deep pragmatist first, and a liberal second. With the right kind of evidence, you could talk me out of my liberalism.

To understand why the liberal tribe is special, it will help to contrast my understanding of morality and politics with that of Jonathan Haidt, whose work we've discussed throughout this book, and who's been a major influence on my own thinking. Haidt and I agree on the general evolutionary and psychological picture of morality presented in chapters 1, 2, and 3. The key ideas are as follows:

> Morality is a suite of psychological capacities designed by biological and cultural evolution to promote cooperation. (chapter 1)

At the psychological level, morality is implemented primarily through emotional moral intuitions, gut reactions that cause us to value the interests of (some) others and encourage others to do the same. (chapter 2)

Different human groups have different moral intuitions, and this is a source of great conflict. Conflicts arise in part from different groups' emphasizing different values and in part from self-serving bias, including unconscious bias. When people disagree, they use their powers of reasoning to rationalize their intuitive judgments. (chapter 3)

In addition to these scientific descriptions, Haidt and I agree on at least one normative prescription. This prescription is the central message of Haidt's wonderful book *The Righteous Mind*, which I summarize as follows:

> To get along better, we should all be less self-righteous. We should recognize that nearly all of us are good people, and that our conflicts arise from our belonging to different cultural groups with different moral intuitions. We're very good at seeing through our opponents' moral rationalizations, but we need to get better at seeing our own. More specifically, liberals and conservatives should try to understand one another, be less hypocritical, and be more open to compromise.

These are important lessons. But unfortunately, they get us only so far. Being more open-minded and less self-righteous should facilitate moral problem solving, but it's not itself a solution.

My first important disagreement with Haidt concerns the role of reason, of manual mode, in moral psychology. I believe that manual-mode thinking has played an enormously important role in moral life, that it is, once again, our second moral compass. Haidt disagrees. He thinks that moral reasoning plays a minor role in moral life, a conclusion neatly expressed by the title of his famous paper "The Emotional Dog and Its Rational Tail." (For the record, Haidt does not accept this characterization of his view.**) We'll return to the topic of moral reasoning shortly. For now, let's consider why reducing self-righteousness and hypocrisy is not enough to solve our moral problems (a conclusion with which Haidt agrees).

Consider once more the problem of abortion. Some liberals say that pro-lifers are misogynists who want to control women's bodies. And some social conservatives believe that pro-choicers are irresponsible moral nihilists who lack respect for human life, who are part of a "culture of death." For such strident tribal moralists—and they are all too common—Haidt's prescription is right on time. But what then? Suppose you're a liberal, but a grown-up liberal. You understand that pro-lifers are motivated by genuine moral concern, that they are neither evil nor crazy. Should you now, in the spirit of compromise, agree to additional restrictions on abortion? Likewise, should grown-up liberals, in the spirit of compromise, favor *more* civil rights for gay couples but not *full* civil rights? Should open-minded liberals fight for environmental regulations that are strong but not quite strong enough to stave off global warming? Grown-up social conservatives, of course, face parallel questions: Should they be "reasonable" and relax their position on early-term abortions, even though they think it's murder?

It's one thing to acknowledge that one's opponents are not evil. It's another thing to concede that they're right, or half right, or no less justified in their beliefs and values than you are in yours. Agreeing to be less self-righteous is an important first step, but it doesn't answer the all-important questions: *What should we believe?* and *What should we do?*

Haidt has a more specific theory about why liberals and conservatives disagree. According to this theory, called Moral Foundations Theory, liberals have impoverished moral sensibilities. Haidt identifies six "moral foundations," which can be labeled in positive or negative terms: care/harm, fairness/cheating, loyalty/betrayal, authority/subversion, sanctity/degradation, and the recently added liberty/oppression. Each foundation has a corresponding set of moral emotions. For example, the value of care is associated with feelings of compassion. The value of sanctity is associated with feelings of awe (for that which is sanctified) and disgust (for that which is defiling, the opposite of sanctifying). Haidt compares these moral-emotional dispositions to the tongue's five chemical taste receptors. Just as our tongues have distinct receptors for sweet, salty, sour, bitter, and savory foods, our moral minds have six distinct moral receptors, capacities to respond emotionally to actions and events that are related to the six moral foundations. For example, a suffering child engages the moral mind's care/harm receptor, producing feelings of compassion. Critically, different cultural groups (tribes, in my parlance) have different moral palates emphasizing different moral tastes. And liberals, according to Haidt, have exceptionally bad tongues. They can easily "taste" caring, fairness, and liberty, but they can barely taste loyalty, authority, and sanctity.

I have my doubts about this six-part theory, but there is an important aspect of Haidt's theory that rings true and is well supported by the evidence:** Some moral values are shared more or less equally by liberals and conservatives, while others are not. Haidt asks his survey participants: Would you stick a sterile hypodermic needle into a child's arm in order to get money (care/harm)? Would you accept a stolen television as a gift (fairness/cheating)? Both conservatives and liberals reliably answer no to questions like these.* But on the following questions, liberals and conservatives tend to differ: Would you anonymously criticize your country as a caller on a foreign radio show (loyalty/betrayal)? Would you slap your father in the face (with his permission) as part of comedy skit (authority/subversion)? Would you attend a short avant-garde play in which the actors acted like animals, crawling around naked and grunting like chimpanzees (sanctity/degradation)? Here social conservatives are far more likely than liberals to say no (or "No!!!"). Why is that?

Haidt's answer, once again, is that liberals have impoverished moral tongues, with half of their moral taste receptors severely weakened. How did this happen? The culprits, according to Haidt, are Western moral philosophers and other children of the Enlightenment. Very smart people with autistic tendencies, most notably Bentham and Kant,* decided that avoiding harm and being fair are the only things that matter. These ideas caught on, and before long a new cultural breed was born: the WEIRD* (Western, Educated, Industrialized, Rich, Democratic) modern liberal, with her enfeebled moral palate. As predicted by Haidt's theory, social conservatives (with all six moral taste receptors switched on) do a better job of predicting what liberals will say in response to moral questions than vice versa.

What, then, should we make of liberals' narrow moral tastes? Is this a deficiency that liberals need to correct? In some ways, yes. If you're a liberal social scientist, your impoverished palate puts you at a disadvantage. If you think that morality is just about avoiding harm and being fair, then you're likely to miss—and misunderstand—a lot of human behavior. Likewise, if you're a political operative trying to sway swing voters, you'll lose votes if your ads engage only one moral taste receptor while your opponents' ads engage them all. Finally, as noted above, if you're a liberal who wants to understand conservatives, it's helpful to know that they have broader moral tastes. But none of this answers the most critical question: *Are liberals morally deficient?* I think the answer is no. Quite the opposite, in fact.

I take a different view of modern moral history, illustrated by the Parable of the New Pastures: The modern world is a confluence of different tribes with different moral values and traditions. The great philosophers of the Enlightenment wrote at a time when the world was rapidly shrinking, forcing them to wonder whether their own laws, their own traditions, and their own God(s) were any better than anyone else's. They wrote at a time when technology (e.g., ships) and consequent economic productivity (e.g., global trade) put wealth and power into the hands of a rising educated class, with incentives to question the traditional authorities of king and church. Finally, at this time, natural science was making the world comprehensible in secular terms, revealing universal natural laws and overturning ancient religious doctrines. Philosophers wondered whether there might also be universal *moral* laws, ones that, like Newton's law of gravitation, applied to members of all tribes, whether or not they knew it. Thus, the Enlightenment philosophers were not arbitrarily shedding moral taste buds. They were looking for deeper, universal moral truths, and for good reason. They were looking for moral truths beyond the teachings of any particular religion and beyond the will of any earthly king. They were looking for what I've called a metamorality: a pan-tribal, or post-tribal, philosophy to govern life on the new pastures.

One might say, as Haidt does, that liberals have narrow moral tastes. But when it comes to moral foundations, less may be more. Liberals' moral tastes, rather than being narrow, may instead be more refined.

For the most part, American social conservatives belong to a specific tribe—a European American, white, Christian tribe that remains lamentably tribal. This tribe dismisses the knowledge gained from science when it conflicts with tribal teachings. Moreover, this tribe regards its own members as the "real" Americans (implicitly, if not explicitly) and regards residents who challenge their tribal beliefs as foreign invaders. According to Haidt, American social conservatives place greater value on respect for authority, and that's true in a sense. Social conservatives feel less comfortable slapping their fathers, even as a joke, and so on. But social conservatives do not respect authority in a *general* way. Rather, they have great respect for the authorities recognized *by their tribe* (from the Christian God to various religious and political leaders to parents). American social conservatives are not especially respectful of Barack Hussein Obama, whose status as a native-born American, and thus a legitimate president, they have persistently challenged. Such conspiracy theories ought to be relegated to the right-wing fringe, but according to a 2011 CBS/*New York Times* poll, 45 percent of Republicans believe that President Obama has lied about his origins. Likewise, Republicans, as compared with Democrats and independents, have little respect for the authority of the United Nations, and a majority of Republicans say that a Muslim American with a position of authority in the U.S. government should not be trusted. In other words, social conservatives' respect for authority is deeply tribal, as is their concern for sanctity. (If the Prophet Muhammad is sacred to you, you shouldn't be in power.) Finally, and most transparently, American social conservatives' concern for loyalty is also tribal. They don't think that *everyone* should be loyal to their respective countries. If Iranians, for example, want to protest against their government, that is to be encouraged.

In sum, American social conservatives are not best described as people who place special value on authority, sanctity, and loyalty, but rather as tribal loyalists—loyal to their own authorities, their own religion, and themselves. This doesn't make them evil, but it does make them parochial, tribal. In this they're akin to the world's other socially conservative tribes, from the Taliban in Afghanistan to European nationalists. According to Haidt, liberals should be more open to compromise with social conservatives. I disagree. In the short term, compromise may be necessary, but in the long term, our strategy should not be to compromise with tribal moralists, but rather to persuade them to be less tribalistic.

I'm not a social conservative because I do not believe that tribalism, which is essentially selfishness at the group level, serves the greater good. And I think the evidence is on my side. If liberals are eroding the moral fabric of American society, then decidedly liberal nations such as Denmark, Norway, and Sweden, where only a minority of citizens report believing in God, should be descending into hell. Instead, they have some of the lowest crime rates, highest-achieving students, and highest levels of quality of life and happiness in the world. According to Haidt, American politics needs the "yang" of conservatism to balance out the "yin" of liberalism. If so, does the same lesson apply to Scandinavia? Should the Danes be importing Christian fundamentalists from rural America in order to balance out their lopsided politics? Here in the "People's Republic of Cambridge," no Republicans hold elected office, and yet Cambridge is one of the few municipalities in the United States whose bonds are rated AAA by all three major credit-rating agencies.

This is not to say that liberals have nothing to learn from social conservatives. As Haidt points out, social conservatives are very good at making each other happy. They are good neighbors, more willing than typical liberals to invest in their communities, with time and with money. They know how to build social capital, to create social networks and institutions that build trust and make collective action possible. In other words, social conservatives are very good at averting the original Tragedy of the Commons. Nevertheless, they are very bad at averting the modern tragedy, the Tragedy of Commonsense Morality. As a liberal, I can admire the social capital invested in a local church and wish that we liberals had equally dense and supportive social networks. But it's quite another thing to acquiesce to that church's teachings on abortion, homosexuality, and how the world got made.

Tribal loyalists are not the only conservatives around. The individualist Northern herders have gone global, forming the world's other meta-tribe. They are the libertarians, the free marketeers, and the "classical liberals" who favor minimal government intervention on both social and economic issues. They want lower taxes, fewer social programs, fewer regulations, and less redistribution of wealth. But they also want the right to an abortion, the right to smoke pot, and the right to marry whomever they want. The libertarians (as I'll call them) are the least tribal people of all, eschewing the moderate collectivism of their modern liberal counterparts.

So why shouldn't a deep pragmatist be a libertarian? To a great extent, one should. Considering the full range of political options, from unfettered free-market capitalism to communism, liberals like me are closer to today's libertarians than to the communists of yore (aspersions from the right not withstanding). The full-blown collectivism of the Southern herders is dead, and the question today is not whether to endorse free-market capitalism, but whether and to what extent it should be moderated by collectivist institutions such as assistance for the poor, free public education, national health insurance, and progressive taxation.

For some libertarians, their politics is a matter of fundamental rights: It's simply *wrong*, they say, to take one person's hard-earned money and give it to someone else. The government has no *right* to tell people what they can or can't do. And so on. I reject this view, for reasons already given: We have no non-question-begging way of knowing who has which rights. With respect to economic matters, this view also presupposes that the world is fair: If government interference in the marketplace is unfair, it must be because the marketplace is itself fair, with winners who deserve all of their winnings and losers who deserve all of their losings. I don't believe that the world is fair. Many people, myself included, begin life with enormous advantages over others. Some people succeed despite great disadvantages, but this doesn't mean that disadvantages don't matter. Ron Paul says that the government shouldn't take care of a man too foolish to buy health insurance. But what about that man's child? Or the child whose family is too poor to afford health insurance? Should the government let those children die? These are familiar liberal points, and I won't belabor them. Unless you believe that the world is fair, or that, with a little gumption, all socioeconomic disadvantages can be sloughed off, the rights-based fundamentalist argument for libertarian policies is a nonstarter.

The pragmatic argument for libertarian policies is that they serve the greater good. As the Northern herders say, punishing the wise and industrious and rewarding the foolish and lazy isn't good for anyone in the long run. It's too bad, says Ron Paul, that some people make foolish choices. But a society that promises to care for people who refuse to care for themselves is bound for ruin. "Spread my work ethic, not my wealth," says a conservative protester's sign.

I believe that libertarians are probably right—righter than many liberals—in some cases. Introducing more competition into public schools sounds to me like a good idea. I am not, on first principles, opposed to having a legal market for human organs, though I worry that the costs of exploitation and organ-related violence will outweigh the benefits of making organs more available. Over the protests of some liberals, I would rather see prostitution legalized and regulated. Refusing to buy products produced in overseas sweatshops may do workers in poor nations more harm than good. The establishment of the Euro as a common European currency may be a brave and brilliant step forward, or it may be a misguided adventure in hypercollectivism. Time will tell. Where exactly the ideal balance between individualism and collectivism lies, I don't know, and I don't pretend to know. But I do know something about moral psychology, which makes me lean to the left more often than I lean to the right. I suspect that many supposedly utilitarian arguments against "big government" are actually rationalizations. In saying this, I'm not claiming liberals make no rationalizations of their own (see above). Nor am I claiming that there are no honest and self-aware advocates of minimal government. What I'm saying—hypothesizing—is that a lot of anti-government sentiment is not what it purports to be. Such sentiment comes in two flavors.

First, why are social conservatives opposed to "big government"? It's not because social conservatives are staunch individualists, like their libertarian allies. I suspect that social conservatives are wary of the U.S. federal government for the same reason that they're wary of the United Nations. Both are trans-tribal power structures, willing and able to take from Us to give to Them. (Or to impose the values of Them on Us.) Social conservatives are perfectly happy to give money to their churches and other local institutions that serve their fellow tribespeople. But when the federal government takes their money, they think, it goes not to hardworking people who just need a helping hand, but to "welfare queens"—to *Them*. It's no accident, I think, that the former slave states are also the states that (in the eastern United States) most reliably vote Republican. For many, what appears to be a philosophical opposition to "big government" is, I suspect, largely just tribalism. Government programs such as Medicare, which very visibly and directly help Us, are not only tolerated, but sacrosanct among social conservatives. (As an angry conservative at a town hall meeting once said, "Keep your government hands off my Medicare!")

Other staunch opponents of "big government" are wealthy people who favor lower taxes, fewer regulations, and the minimization of social programs. They are the proverbial "1 percent," who cast very few votes but wield great power. As Warren Buffett, the voice of the un–1 percent, famously observed, something's wrong when billionaires pay taxes at a lower rate than their secretaries. However, such policies might be justified if you think that the world's wealthiest people deserve extra rewards for being wise and industrious. I'll be neither the first nor the last liberal to observe that such beliefs are self-serving. But let me suggest that such beliefs are, rather amazingly, sincere. Mitt Romney famously pleased a roomful of wealthy donors when he dismissed 47 percent of Americans as irresponsible freeloaders. But what is not often remarked upon is that this was a roomful of *donors*. Mitt Romney's favorite audience is not selfish in any straightforward way. A psychopath would not spend $50,000 on a campaign dinner. One can get more reliable returns,* or have a lot more fun, elsewhere. I believe that Mitt Romney and his wealthy friends sincerely believe that what they want is for the greater good. It's not simple selfishness. It's biased fairness.

Some people earn three million dollars in a year. More typical American workers have an annual income of thirty thousand dollars. Such is the way of the free market. I'm prepared to believe that, on average, people who earn millions work harder than typical workers and deserve to be rewarded. But I don't believe that they work one hundred times harder. I don't believe that the super-rich do more hard work in one week than typical workers do all year. Rich people may deserve to be rich, but they are also beneficiaries of good fortune. I see no reason why the world's luckiest people should keep all of that good fortune for themselves, especially when public schools can't afford to pay teachers competitive professional salaries and billions of children worldwide are born into poverty through no fault of their own. Taking a bit of money from the haves hurts them very little, whereas providing resources and opportunities to the have-nots, when done wisely, goes a long way. That's not socialism. That's deep pragmatism.

I began by comparing my understanding of political psychology to Jonathan Haidt's. Based on what I said above, you might think that Haidt is a staunch conservative, but he's not. He's a centrist, a sometime ambivalent liberal, who in the end endorses, of all things, utilitarianism.* Haidt's ultimate endorsement of utilitarianism is both paradoxical and instructive.

According to Haidt, liberals have narrow moral palates, and utilitarians have the narrowest palates of all. He cites research posthumously diagnosing Jeremy Bentham with Asperger syndrome, a mild form of autism that disconnects people from the social world. Haidt argues that

Bentham's psychopathology shows in his philosophy, which systematically reduces all of morality to a single value. Building on his culinary analogy, Haidt imagines a "utilitarian diner." Like a restaurant that serves only sugar, Bentham's kitchen stimulates only one moral taste receptor—an impoverished philosophy indeed. But later in his book, Haidt says this:

> I don't know what the best normative ethical theory is for individuals in their private lives. But when we talk about making laws and implementing public policies in Western democracies that contain some degree of ethnic and moral diversity, then I think there is no compelling alternative to utilitarianism.

What's going on? When faced with the ultimate question—*What should we do?*—it seems that the autistic philosopher was right all along. What's happening, I think, is that Haidt is now using his other moral compass.

We modern herders have strong moral feelings, and sometimes very different feelings. Unfortunately, we can't all get our way. What to do? The first step, as Haidt tells us, is to understand each other better, to understand that we come from different moral tribes, each sincere in its own way. But that's not enough. We need a common moral standard, a metamorality, to help us get along. The idea that we should aim for maximum happiness is not the arbitrary glorification of a single moral flavor, or the elevation of one tribe's values over others. It's the implementation of a common currency, a metric of value against which other values can be measured, enabling not just compromise but *principled* compromise. According to Haidt "human beings are 90 percent chimp and 10 percent bee," meaning that we are mostly selfish, but also and partly tribal—guardians of our respective hives. I think that this accounting of human nature is incomplete. Which part of us believes that we should maximize global happiness? This is neither chimp nor bee. This metamoral ideal is a distinctively human invention, a product of abstract reasoning. Were we limited to our selfish and tribal instincts, we'd be stuck. But fortunately, we all have the capacity, if not the will, to shift into manual mode.

In the short term, moral reasoning is rather ineffective, though not completely.* This is why, I think, Haidt underestimates its importance. If a herder feels in his heart that something is right or wrong, the odds that your good argument will change his mind, right then and there, are slim. But like the wind and rain, washing over the land year after year, a good argument can change the shape of things.** It begins with a willingness to question one's tribal beliefs. And here, being a little autistic might help. This is Bentham writing circa 1785, when gay sex was punishable by death:

> I have been tormenting myself for years to find if possible a sufficient ground for treating [gays] with the severity with which they are treated at this time of day by all European nations: but upon the principle of utility I can find none.

Manual-mode morality requires courage and persistence. Here is Mill in the introduction to his classic defense of women's rights, *The Subjection of Women*, published in 1869 and possibly co-authored with his wife, Harriet Taylor Mill:

> But it would be a mistake to suppose that the difficulty of the case must lie in the insufficiency or obscurity of the grounds of reason on which my conviction rests. The difficulty is that which exists in all cases in which there is a mass of feeling to be contended against. . . . And while the feeling remains, it is always throwing up fresh intrenchments of argument to repair any breach made in the old.

Today we, some of us, defend the rights of gays and women with great conviction. But before we could do it with feeling, before our feelings felt like "rights," someone had to do it with *thinking*. I'm a deep pragmatist, and a liberal, because I believe in this kind of progress and that our work is not yet done.

12.

Beyond Point-and-Shoot Morality: Six Rules for Modern Herders

In the beginning, there was primordial soup. Cooperative molecules formed larger molecules, some of which could make copies of themselves and surround themselves with protective films. Cooperative cells merged to form complex cells, and then cooperative clusters of cells. Life grew increasingly complex, finding again and again the magic corner in which individual sacrifice buys collective success, from bees to bonobos. But cooperative organisms are, by biological design, not universally cooperative. Cooperation evolved as a competitive weapon, as a strategy for outcompeting others. Thus, cooperation at the highest level is inevitably strained, opposed by forces favoring Us over Them.

Some animals evolved brains: computational control centers that absorb information and use it to guide behavior. Most brains are reflexive machines, automatically mapping inputs to outputs, with no ability to reflect on what they are doing or to imagine novel behaviors. But we humans evolved a fundamentally new kind of intelligence, a general-purpose reasoning capacity that can solve complex, novel problems, ones that can't be solved with reflexes. Intelligence fast and slow is a winning combination, but a dangerous one, too. Thanks to our big brains, we've defeated most of our natural enemies. We can make as much food as we need and build shelters to protect ourselves from the elements. We've outsmarted most of our predators, from lions to bacteria. Today our most formidable natural enemy is ourselves. Nearly all of our biggest problems are caused by, or at least preventable by, human choice.

Recently, we've made enormous progress in reducing human enmity, replacing warfare with gentle commerce, autocracy with democracy, and superstition with science. But there remains room for improvement. We have age-old global problems (poverty, disease, war, exploitation, personal violence), looming global problems (climate change, terrorism using weapons of mass destruction), and moral problems that are unique to modern life (bioethics, big government versus small government, the role of religion in public life). How can we do better?

Our brains, like our other organs, evolved to help us spread our genes. For familiar reasons, our brains endow us with selfish impulses, automatic programs that impel us to get what we need to survive and reproduce. For less obvious reasons, our brains impel us to care about others, and to care about whether others do the same. We have empathy, love, friendship, anger, social disgust, gratitude, vengefulness, honor, guilt, loyalty, humility, awe, judgmentalism, gossip, embarrassment, and righteous indignation. These universal features of human psychology allow Us to triumph over Me, putting us in the magic corner, averting the Tragedy of the Commons.

These cognitive gizmos are used by all healthy human brains, but we use them in different ways. Our respective tribes cooperate on different terms. We have different ideas and feelings about what people owe one another and about how honorable people respond to threats. We are devoted to different "proper nouns," local moral authorities. And we are, by design, tribalistic, favoring Us over Them. Even when we think we're being fair, we unconsciously favor the version of fairness most congenial to Us. Thus, we face the Tragedy of Commonsense Morality: moral tribes that can't agree on what's right or wrong.

———

S olving a problem is often a matter of framing it in the right way. In this book I've tried to provide a framework for thinking about our biggest moral problems. Once again, we face two fundamentally different kinds of moral problems: Me versus Us (Tragedy of the Commons) and Us versus Them (Tragedy of Commonsense Morality). We also have two fundamentally different kinds of moral thinking: fast (using emotional automatic settings) and slow (using manual-mode reasoning). And, once again, the key is to match the right kind of thinking to the right kind of problem: When it's Me versus Us, think fast. When it's Us versus Them, think slow.

Modern herders need to think slower and harder, but we need to do it in the right way. If we use our manual-mode reasoning to describe or rationalize our moral feelings, we'll get nowhere. Instead of organizing and justifying the products of our automatic settings, we need to transcend them. Thus stated, the solution to our problem seems obvious: We should put our divisive tribal feelings aside and do whatever produces the best overall results. But what is "best"?

Nearly everything that we value is valuable because of its impact on our experience. Thus, we might say that what's best is what makes our experience as good as possible, giving equal weight to each person's quality of life. Bentham and Mill turned this splendid idea into a systematic philosophy, and gave it an awful name. We've been misunderstanding and underappreciating their ideas ever since. The problem, however, runs deeper than bad marketing. Our gut reactions were not designed to form a coherent moral philosophy. Thus, any truly coherent philosophy is bound to offend us, sometimes in the real world but especially in the world of philosophical thought experiments, in which one can artificially pit our strongest feelings against the greater good. We've underestimated utilitarianism because we've overestimated our own minds. We've mistakenly assumed that our gut reactions are reliable guides to moral truth. As Chekhov said, to become better, we have to know what we're like.

At the start of this long and complicated book, I promised you greater clarity. I hope that you now see moral problems more clearly than you did on page 1. I hope you see the Tragedy of Commonsense Morality unfolding around you, the emotions that are its root cause, and the kind of reasoning that can move us forward. We've covered a lot of abstract ideas, a lot of "isms." As a social scientist and a pragmatist, I know all too well the gap between theory and practice. To be effective in the long term, our ideals must be embodied not just in our isms, but in our habits. With this is mind, I close with some simple, practical suggestions for life on the new pastures.

SIX RULES FOR MODERN HERDERS

Rule No. 1. In the face of moral controversy, consult, but do not trust, your moral instincts*

Your moral intuitions are fantastic cognitive gizmos, honed by millions of years of biological evolution, thousands of years of cultural evolution, and years of personal experience. In your personal life, you should trust your moral instincts and be wary of your manual mode, which is all too adept at figuring out how to put Me ahead of Us. But in the face of moral controversy, when it's Us versus Them, it's time to shift into manual mode. When our emotional moral compasses point in opposite directions, they can't both be right.

Rule No. 2. Rights are not for making arguments; they're for ending arguments

We have no non-question-begging way of figuring out who has which rights and which rights outweigh others. We love rights (and duties, rights' frumpy older sister), because they are handy rationalization devices, presenting our subjective feelings as perceptions of abstract moral objects. Whether or not such objects exist, there's little point in arguing about them. We can use "rights" as shields, protecting the moral progress we've made. And we can use "rights" as rhetorical weapons, when the time for rational argument has passed. But we should do this sparingly. And when we do, we should know what we're doing: When we appeal to rights, we're not making an argument; we're declaring that the argument is over.

Rule No. 3. Focus on the facts, and make others do the same

For deep pragmatists, one can't know whether a proposal is good or bad without knowing how it's supposed to work and what its effects are likely to be. Nevertheless, most of us readily pass judgment on policies—from environmental regulations to healthcare systems—that we barely understand. Public moral debate should be a lot wonkier. We should force ourselves, and one another, to know not only which policies we favor or oppose, but how these policies are supposed to work. We should provide—and demand—evidence about what works and what doesn't. And when we don't know how things work, in theory or in practice, we should emulate the wisdom of Socrates and acknowledge our ignorance.

Rule No. 4. Beware of biased fairness

There are different ways of being fair, and we tend to favor, often unconsciously, the version of fairness that suits us best. Because biased fairness is a kind of fairness, it's hard to see that it's biased, especially in ourselves. We do this as individuals, and we do this as loyal members of our respective tribes. Sometimes we make personal sacrifices to further the biased fairness of our tribes—a kind of biased selflessness.

Rule No. 5. Use common currency

We can argue about rights and justice forever, but we are bound together by two more basic things. First, we are bound together by the ups and downs of the human experience. We all want to be happy. None of us wants to suffer. Second, we all understand the Golden Rule and the ideal of impartiality behind it. Put these two ideas together and we have a common currency, a system for making principled compromises. We can agree, over the objections of our tribal instincts, to do whatever works best, whatever makes us happiest overall.

To figure out what works best, we need a common currency of value, but we also need a common currency of fact. There are many sources of knowledge, but the most widely trusted, by far, is science, and for good reason. Science is not infallible, and people readily reject scientific knowledge when it contradicts their tribal beliefs. Nevertheless, nearly everyone appeals to scientific evidence when it suits them. (Would creationists not jump for joy if, tomorrow, credible scientists were to announce that the earth is, in fact, just a few thousand years old?) No other source of knowledge has this distinction. In our tribal quarters, and in our hearts, we may believe whatever we like. But on the new pastures, truth should be determined using the common currency of observable evidence.

Rule No. 6. Give

As individuals, we don't get to make the rules by which we live. But each of us makes some important, life-and-death decisions. By making small sacrifices, we in the affluent world have the power to dramatically improve the lives of others. As creatures wired for tribal life, our sympathies for distant "statistical" strangers are weak. And yet few of us can honestly say that our most luxurious luxuries are more important than saving someone's life,* or giving someone without access to healthcare or education a brighter future. We can delude ourselves about the facts, denying that our donations really help. Or, if we're more philosophically ambitious, we can rationalize our self-serving choices. But the honest response, the enlightened response, is to acknowledge the harsh reality of our habits and do our best to change them, knowing that a partially successful honest effort is better than a fully successful denial.

———

Immanuel Kant marveled at "the starry heavens above" and the "moral law within." It's a lovely sentiment, but one that I cannot wholeheartedly share. We are marvelous in many ways, but the moral laws within us are a mixed blessing. More marvelous, to me, is our ability to question the laws written in our hearts and replace them with something better. The natural world is full of cooperation, from tiny cells to packs of wolves. But all of this teamwork, however impressive, evolved for the amoral purpose of successful competition. And yet somehow we, with our overgrown primate brains, can grasp the abstract principles behind nature's machines and make them our own. On these pastures, something new is growing under the sun: a global tribe that looks out for its members, not to gain advantage over others, but simply because it's good.

Author's Note

GiveWell recommends charities based on their track records, cost-effectiveness, and need for additional funds. For the latest information visit: www.givewell.org.

Oxfam International works to create long-term solutions to poverty and injustice. To find your nearest Oxfam affiliate visit: www.oxfam.org.

Acknowledgments

My first debt is to my parents, Laurie and Jonathan Greene, who encouraged me from the start to think for myself, and whose hard work and devotion granted me the privilege of devoting my working life to ideas. I am grateful to them, and to my brother, Dan, and sister, Liz, for a lifetime of love and support. Thanks, too, to my friend and sister-in-law Sara Sternberg Greene, and to the newest in-law, Aaron Falchook.

Part of this privilege has been learning from wonderful mentors and colleagues who inspired me, took my fledgling ideas seriously, and taught me how to think. I am forever grateful to my undergraduate mentors Jonathan Baron, Paul Rozin, Amartya Sen, Allison Simmons, and Derek Parfit for giving me my start. Many thanks to my graduate advisers, David Lewis and Gilbert Harman, and to Peter Singer, who served on my committee and who has since offered invaluable advice and encouragement. From those days, I also thank the community of philosophers at Princeton for good arguments and good times. I owe an enormous debt to my postdoctoral (and unofficial doctoral) adviser, Jonathan Cohen, for taking a chance on me, teaching me the art of science, and helping me understand how a brain can be a mind. Likewise, I am grateful to Leigh Nystrom, John Darley, Susan Fiske, and others in Princeton's Department of

Psychology for offering me a second home and helping me find my way. Since graduate school, my other adoptive home has been the Moral Psychology Research Group, a merry band of philosopher-scientists who continue to broaden my mind and buoy my spirits. In recent years especially, I've benefited from the mentorship and support of MPRG tribal elders Stephen Stich, John Doris, Shaun Nichols, and Walter Sinnott-Armstrong. Finally, I am grateful to the members of Harvard's Department of Psychology for taking (yet another) chance on a reconditioned philosopher, for their inspiration and guidance, and for encouraging me to write this book despite my juniority. I am especially grateful to Mahzarin Banaji, Josh Buckholtz, Randy Buckner, Susan Carey, Dan Gilbert, Christine Hooker, Stephen Kosslyn, Wendy Mendes, Jason Mitchell, Matt Nock, Steven Pinker, Jim Sidanius, Jesse Snedeker, Felix Warneken, and Dan Wegner for their support and advice. Many thanks to Max Bazerman as well. And thanks to our department's wonderful administrative staff for making everything run so astonishingly well.

I thank Robert Sapolsky for offering my name to Katinka Matson, of Brockman Inc., who subsequently became my agent and occasional guardian angel. I thank Katinka for taking (yet another) chance on me, for teaching me about the world of books, for her patience and smarts, and for believing that this book was possible.

Like the fabled Ship of Theseus, this book's planks have been replaced so many times that nothing remains of the original vessel. I am forever grateful to the many colleagues who helped me build and rebuild this book, graciously donating hundreds of hours of their precious time and attention. The following people read and commented on complete drafts: Jonathan Baron, Max Bazerman, Paul Bloom, Tommaso Bruni, Alek Chakroff, Moshe Cohen-Eliya, Fiery Cushman, John Doris, Dan Gilbert, Jonathan Haidt, Brett Halsey, Andrea Heberlein, Anna Jenkins, Alex Jordan, Richard Joyce, Simon Keller, Joshua Knobe, Victor Kumar, Shaun Nichols, Joe Paxton, Steven Pinker, Robert Sapolsky, Peter Singer, Walter Sinnott-Armstrong, Tamler Sommers, Dan Wegner, and Liane Young. In addition, I received helpful comments on one or more chapters from Dan Ames, Kurt Gray, Gilbert Harman, Dan Kelly, Matt Killingsworth, Katrina

Koslov, Lindsey Powell, Jesse Prinz, Tori McGeer, Edouard Machery, Ron Mallon, Maria Merritt, Alex Plakias, Erica Roedder, Adina Roskies, Tim Schroeder, Susanna Siegel, Chandra Sripada, Stephen Stich, and Valerie Tiberius. (Extra-special thanks to Dan Gilbert, Steven Pinker, and Peter Singer for reading *two* complete drafts.) And thanks to David Luberoff for seeing the chili pepper connection. My sincere apologies to any benefactors who have escaped this list. When I think of the collective brainpower supplied by these individuals, I am deeply humbled and can only hope that I've made good on their investments. These great minds challenged me at every turn and far too frequently saved me from (further) embarrassment.

Many thanks to the talented people at the Penguin Press, beginning with Nick Trautwein, whose editorial analysis of my first draft gave me direction and encouragement, and Will Palmer, whose judicious copyediting gave this work a much-needed buff and shine. I thank Eamon Dolan for his helpful comments during his brief tenure as my editor and for the invaluable gift of time. Following the departures of Trautwein and Dolan, this book landed on the desk of Benjamin Platt, whose fresh perspective made for a fruitful and rewarding collaboration. Ben saw in my book shortcomings and opportunities to which I was blind, and recognized, among other things, that my story begins with what I thought was the middle. For this insight, among many others, I am grateful. My greatest debt at Penguin, however, is to Scott Moyers, who took (yet another) chance on me, offering an unknown postdoc a book contract, and who graciously descended from his high office to co-edit this book during the past year. In many ways, Scott understood my book better than I did, which was both humbling and inspiring. I am grateful to Scott for lending me his prodigious talent, for his Mametesque wit and wisdom, and for believing in this work from start to finish.

Over the past seven years, a squad of bright and dedicated young scientists cast its lot with me as graduate students and postdocs in the Moral Cognition Lab. Neither I nor they could have anticipated the toll this book would take on them. And while I, at least, would emerge from this process as the author of a book, they got nothing but delays, absences, exhausted blank stares, and apologies. My labbies have been so patient and so supportive, I will never be able to repay their kindness. (I can hear them now: *"Ain't that the truth!"*) Thank you, Nobuhito Abe, Elinor Amit, Regan Bernhard, Donal Cahill, Alek Chakroff, Fiery Cushman, Steven Frankland, Shauna Gordon-Mckeon, Sara Gottlieb, Christine Ma-Kellams, Joe Paxton, David Rand, and Amitai Shenhav. You make me happy to go to work each day. Special thanks to Steven Frankland for brain imagery and to Sara Gottlieb for her heroic efforts preparing the figures, notes, and bibliography.

For what remains of my sanity, I thank dear friends who sustained me through this time with their warmth and good humor: Nicole Lamy, Michael Patti, Paula Fuchs, Josh and Ashley Buckholtz, Brett Halsey, and Valerie Levitt Halsey.

My children, Sam and Frida, have brought me joy like I'd never known before they arrived. There is nothing I love more than being their dad, and nothing I regret more than the thousands of hours of their childhoods—the pumpkin picking and beach days and bedtime stories—that I missed because of this book. From now on, Daddy will be home a lot more. And finally I thank, with everything I've got, my wife, my best friend, and the love of my life, Andrea Heberlein, to whom this book is dedicated a thousand times over. Andrea's brilliance and good sense made this a better book, but without her devotion there would be no book to improve. Andrea arranged our lives so that I could finish what I started, having no idea what I had gotten us into. For this, and so much more, I am forever grateful.

Notes

vii **"man will become":** Chekhov (1977) 27, quoted in Pinker (2002). Chekhov, A. (1977). *Portable Chekhov.* New York: Penguin. Pinker, S. (2002). *The Blank Slate: The Modern Denial of Human Nature.* New York: Viking.

INTRODUCTION

6 **awash in misinformation:** The most notorious false claim is that Obamacare establishes "death panels" that decide who gets to live and die (FactCheck.org, August 14, 2009). Democrats have made some false claims, too; for example, about details concerning whether people can keep their health insurance plans under Obamacare (FactCheck.org, August 18, 2009).

7 **exchange with Texas congressman Ron Paul:** Politisite (September 13, 2011).

8 **enormous bets on housing prices:** Financial Crisis Inquiry Commission (2011).

8 **government bailed out several of the investment banks:** This move had bipartisan support but was favored more strongly by Democrats (US House of Representatives, 2008; US Senate, 2008).

8 **tippy top . . . 400 percent:** Krugman (November 24, 2011).

9 **"Occupy a Desk!":** Kim (December 12, 2011).

9 **"class warfare":** http://www.huffingtonpost.com/2011/10/06/herman-cain-occupy-wall-street_n_998092.html.

9 **the "47 percent":** http://www.motherjones.com/politics/2012/09/watch-full-secret-video-private-romney-fundraiser.

9 **thanks to lower tax rates:** Buffett (August 14, 2011).

10 **"steal and rob people with a gun":** ABC News (December 5, 2011).

10 **"a parasite who hates her host":** *The Rush Limbaugh Show* (September 22, 2011).

10 **values may color our view of the facts:** Kahan, Wittlin, et al. (2011); Kahan, Hoffman, et al. (2012); Kahan, Jenkins-Smith, et al. (2012); Kahan, Peters, et al. (2012).

11 **"proper nouns":** Strictly speaking, these are the *referents* of proper nouns.

13 **better at getting along:** Pinker (2011).

13 **modern market economies . . . of human kindness:** Henrich, Boyd, et al. (2001); Henrich, Ensminger, et al. (2010); Herrmann et al. (2008).

13 **twentieth century . . . approximately 230 million people:** Leitenberg (2003).

13 **conflict in Darfur . . . 300,000 people:** Degomme and Guha-Sapir (2010).

13 **A billion people . . . live in extreme poverty:** World Bank (February 29, 2012) reporting data from 2008.

13 **More than twenty million people are forced into labor:** International Labour Organization (2012).

13 **more calls from employers:** Bertrand and Mullainathan (2003).

14 **What are we doing right?:** Pinker (2011).

15 *utilitarianism:* John Stuart Mill's utilitarianism and Charles Darwin's theory of natural selection emerged around the same time and have had highly overlapping fan bases from the start, beginning with Darwin's and Mill's mutual admiration. This is not an accident, I think. Both groundbreaking ideas favor manual mode over automatic settings. For a nice discussion, see Wright (1994), chapter 16.

PART I: MORAL PROBLEMS

CHAPTER 1: THE TRAGEDY OF THE COMMONS

19   **"The Tragedy of the Commons":** Hardin (1968).
20   **central problem of social existence:** Von Neumann and Morgenstern (1944); Wright (2000); Nowak (2006).
20   **This principle has guided the evolution:** Margulis (1970); Wilson (2003); Nowak and Sigmund (2005).
22   **the larger party will kill:** Mitani, Watts, et al. (2010).
22   **a phenomenon known as cancer:** Michor, Iwasa, et al. (2004).
22   **Darwin himself was absorbed:** Darwin (1871/1981).
23   **"red in tooth and claw":** A. L. Tennyson, *In Memoriam AHH,* in Tennyson and Edey (1938).
23   **"Morality is a set":** This view originated with Darwin (1871/1981) and has become the consensus view among behavioral scientists in recent decades. See Axelrod and Hamilton (1981); Frank (1988); Wright (1994); Sober and Wilson (1999); Wilson (2003); Gintis et al. (2005); Joyce (2006); de Waal (2009); Haidt (2012).
24   **not assuming . . . group selection:** Even if morality evolved simply through individual selection, favoring capacities for reciprocal altruism, the same argument applies.
25   **Wittgenstein's famous metaphor:** Wittgenstein (1922/1971).
25   **nature's "intentions":** Birth control might be used to enhance one's long-term genetic prospects through judicious family planning, but it certainly doesn't have to be used that way.

CHAPTER 2: MORAL MACHINERY

28   **Prisoner's Dilemma:** Puzzles of the Prisoner's Dilemma form were devised by M. Flood and M. Dresher of the Rand Corporation. See Poundstone (1992).
31   **"Golden Rule" . . . every major religion:** Blackburn (2001), 101.
31   **kin selection:** Fisher (1930); Haldane (1932); Hamilton (1964); Smith (1964). This mainstay of evolutionary biology has once again become controversial. See Nowak, Tarnita, et al. (2010).
32   *reciprocity,* **or reciprocal altruism:** See Trivers (1971) and Axelrod and Hamilton (1981). Encounters between potential cooperators may be chosen or forced. See Rand, Arbesman, et al. (2011). Here, as in the above Art and Bud story, I've made their encounters chosen, to be consistent with the Prisoner's Dilemma story, but in standard models of reciprocal altruism, the encounters are forced by circumstances. In either case, the same reciprocal logic applies.
33   **variations on the Tit for Tat theme:** See, for example, Nowak and Sigmund (1993).
33   *anger, disgust,* **or** *contempt:* See Rozin, Lowery, et al. (1999) and Chapman, Kim, et al. (2009). Note that negative feelings such as anger and disgust are not perfectly interchangeable. Anger is an "approach" emotion, motivating active aggression. Disgust, in contrast, is a "withdrawal" emotion that originally evolved to expel contaminating substances, such as feces and rotten meat, from the body. Which of these negative attitudes is most strategically appropriate will depend on the relative costs and benefits of active aggression versus selective disengagement.
33   *gratitude* **. . . willingness to cooperate:** Rand, Dreber, et al. (2009).
33   **food-sharing behavior in chimpanzees:** De Waal (1989). See also Packer (1977) and Seyfarth and Cheney (1984).
34   **emotional dispositions that we inherited:** Gintis, Bowles, et al. (2005).
34   **things don't always go as planned:** Nowak and Sigmund (1992); Rand, Ohtsuki, et al. (2009); Fudenberg, Rand, et al. (2010).
34   **De Waal and Roosmalen:** De Waal and Roosmalen (1979).
35   **program might be called** *friendship:* Seyfarth and Cheney (2012).
35   **world of our ancestors . . . more violent:** Daly and Wilson (1988); Pinker (2011).
36   **"long pig":** Stevenson (1891/2009).
36   **birth of modern military training:** Grossman (1995).
36   **laboratory study . . . human aversion to violence:** Cushman, Gray, et al. (2012).
36   **Figure 2.2:** Adapted with permission from Cushman, Gray, et al. (2012).
37   **"lost" letters:** Milgram, Mann, et al. (1965).
37   **tips at restaurants:** Pinker (2002), 259.
37   **some researchers have questioned:** Cialdini et al. (1987).
37   **we** *feel bad* **for them:** Batson et al. (1981); Batson (1991).
37   **more likely to cooperate . . . in a prisoner's dilemma:** Batson and Moran (1999).
37   **one experiences the feelings of others:** In some cases, empathizing does not entail having the same feeling as the person with whom one is empathizing. For example, when one empathizes with a child who is scared, one need not be scared oneself.

37   **neural circuits that are engaged:** Singer, Seymour, et al. (2004). Empathy for pain involves the affective but not sensory components of pain. *Science* 303(5661), 1157–1162.

38   **Oxytocin . . . in maternal care:** Pedersen, Ascher, et al. (1982).

38   **Genes . . . oxytocin:** Rodrigues, Saslow, et al. (2009).

38   **more likely to initiate cooperation:** Kosfeld, Heinrichs, et al. (2005). But see Singer, Snozzi, et al. (2008).

38   **capacity to care about others:** De Waal (1997, 2009); Keltner (2009).

38   **Ladygina-Kohts:** As described by de Waal (2009).

38   **Arnhem Zoo:** de Waal (2009).

38   **without expectation of a reward:** Warneken et al. (2006, 2007, 2009).

39   **capuchin monkeys:** Lakshminarayanan and Santos (2008).

39   **empathy in rats:** Bartal, Decety, et al. (2011).

40   **MAD:** The formal theory behind MAD comes from von Neumann and Morgenstern (1944).

41   **emotional machinery that performs the same function:** Frank (1988). See also Schelling (1968). My example here follows Pinker (1997), who compares angry passions to the "Doomsday Device" in Stanley Kubrick's film *Dr. Strangelove*. The device is designed to automatically launch a nuclear counterstrike in the event of a first strike.

41   **not the only ones with a taste for vengeance:** Jensen, Call, et al. (2007).

41   **chimps do much the same in the wild:** De Waal and Luttrel (1988).

41   **breaking a promise:** Baumgartner, Fischbacher, et al. (2009).

42   **familial love and friendship . . . irrational:** Pinker (2008).

42   **in-house bank-robbing expert:** As in the case of Frank Abagnale. See Abagnale and Redding (2000).

42   **As Steven Pinker observes:** Pinker (2008).

43   **sometimes regard their leaders:** Henrich and Gil-White (2001).

43   **things larger than ourselves:** Keltner and Haidt (2003); Haidt (2012).

44   **Reputations can also enhance cooperation:** Nowak and Sigmund (1998).

44   **Kevin Haley and Daniel Fessler:** Haley and Fessler (2005).

44   **"Dictator Game":** Forsythe, Horowitz, et al. (1994).

45   **Figure 2.3:** Reprinted with permission from Haley and Fessler (2005).

45   **"honor box" for buying drinks:** Bateson, Nettle, et al. (2006).

45   **65 percent of their conversation time:** Dunbar (2004); Dunbar, Marriott, et al. (1997).

45   *gossiping* **. . . for social control:** Feinberg et al. (2012b); Nowak and Sigmund (1998, 2005); Milinski, Semmann, et al. (2002).

46   **transgressor appears to be embarrassed:** Semin and Manstead (1982); Keltner (2009).

46   **we're judgmental as** *babies:* Hamlin, Wynn, et al. (2007, 2011). See also Sloane, Baillargeon, et al. (2012).

47   **Figure 2.4:** Reprinted with permission from Hamlin, Wynn, et al. (2007).

47   **the** *hindering* **square:** In some versions of the experiment, the colors and shapes were reversed, showing that it's not just a preference for certain shapes or colors. In another version of the experiment, they showed that the infants prefer the helper to a neutral shape and prefer a neutral shape to a hinderer.

49   **enthnocentrism as universal:** Brown (1991).

50   *parochial altruism:* Bernhard, Fischbacher, et al. (2006); Choi and Bowles (2007).

50   **infants . . . language to distinguish:** Kinzler, Dupoux, et al. (2007). See also Mahajan and Wynn (2012).

51   **separating Us from Them:** McElreath, Boyd, and Richerson (2003).

51   **Implicit Association Test:** Greenwald, McGhee, et al. (1998); Greenwald and Banaji (1995).

51   **Try it yourself:** https://implicit.harvard.edu/implicit.

51   **whites have an implicit preference for whites:** Greenwald, McGhee, et al. (1998).

52   **children . . . same kind of race-based biases:** Baron and Banaji (2006).

52   **IAT developed for monkeys:** Mahajan, Martinez, et al. (2011).

52   **(Emily, Greg) . . . (Lakisha, Jamal):** Bertrand and Mullainathan (2003).

52   **stereotypically black facial features:** Eberhardt, Davies, et al. (2006).

52   **Google searches:** Stephens-Davidowitz (2012).

53   **sensitivity to race . . . group membership:** Kurzban, Tooby, et al. (2001).

53–54   **Classic studies by Henri Tajfel:** Tajfel (1970, 1982); Tajfel and Turner (1979).

54   **oxytocin . . . out-group members:** De Dreu, Greer, et al. (2010, 2011).

55   **"nasty, brutish, and short":** Hobbes (1651/1994).

55   **religion may be a device:** Wilson (2003); Roes and Raymond (2003); Norenzayan and Shariff (2008).

56   **wary of people who are not "God-fearing":** Gervais, Shariff, et al. (2011).

56   **enforced cooperation:** Boyd and Richerson (1992).

56   **pre-agricultural societies are rather egalitarian:** Boehm (2001).

56 **odds of getting punished . . . very high:** This assumes that the punished cannot or will not retaliate in full force against the punishers. If they can and do, cooperation may break down. See Dreber, Rand, et al. (2008); Hermann, Thöni, et al. (2008).

56 *indirect reciprocity:* Gintis (2000); Bowles, Gintis. et al. (2003); Gintis, Bowles, et al. (2005).

57 **"altruistic punishment":** Fehr and Gächter (2002); Boyd, Gintis, et al. (2003).

57 **people are . . . pro-social punishers:** Fehr and Gächter (2002); Marlowe, Berbesque, et al. (2008). But see Kurzban, DeScioli, et al. (2007).

57 **"Public Goods Game":** Dawes, McTavish, et al. (1977).

58 **contributions typically go up:** Boyd and Richerson (1992); Fehr and Gächter (1999).

58 **pro-social punishment is just a by-product:** Kurzban, DeScioli, et al. (2007).

58 **pro-social punishment evolved through:** Boyd, Gintis, et al. (2003).

59 **life on Earth . . . story of increasingly complex cooperation:** Margulis (1970); Nowak (2006); Wright (2000).

62 **familiar features of human nature:** Many of them appear on Donald Brown's (1991) list of human universals (e.g., "empathy," "gossip," "shame," "revenge"), and all of them are closely related to, if not logically entailed by, items on his list.

62 **feelings that do this thinking for us . . . series of studies:** Rand, Greene, et al. (2012).

63 **Figure 2.5:** All data reported in ibid. Decision time (in log10 seconds) is on the x-axes. Contribution level is on the y-axes, expressed as a percentage of the maximum (top) or probability of cooperation given a yes/no choice (all others). Top: A one-shot Public Goods Game. Middle left: First decision from a series of one-shot Prisoner's Dilemmas. Middle right: A repeated Prisoner's Dilemma with execution errors. Bottom left: A repeated Prisoner's Dilemma with or without costly punishment. Bottom right: A repeated Public Goods Game with or without reward and/or punishment. Dot size is proportional to number of observations, which are indicated next to each dot. Error bars indicate standard error of the mean.

64 **prohibition against eating cows may increase the food supply:** Harris, Bose, et al. (1966).

64 **blocking alternative routes to sexual gratification:** Davidson and Ekelund (1997).

65 **divine will and chance:** Dawkins (1986).

CHAPTER 3: STRIFE ON THE NEW PASTURES

69 **not about whether . . . but about *why*:** Pinker (2002).

69 **altruism within groups could not have evolved:** Choi and Bowles (2007); Bowles (2009).

70 **anthropologists studying small-scale societies:** Henrich, Boyd, et al. (2001); Henrich, Gil-White, et al. (2001); Henrich, McElreath, et al. (2006); Henrich, Ensminger, et al. (2010).

70 **"Ultimatum Game":** Güth, Schmittberger, et al. (1982).

71 **"Wall Street Game" or the "Community Game":** Liberman, Samuels, et al. (2004).

71 **Americans tend to give nothing:** List (2007).

72 **most cooperative . . . most willing to punish:** Henrich, McElreath, et al. (2006).

72 **Dictator Game . . . involves no *co*-operation:** The Dictator Game is about altruism, giving something for nothing. But from a mathematical, game-theoretical perspective, cooperation and altruism are (or can be) equivalent. This is because cooperation (the interesting kind) requires paying a personal cost to benefit others. To be cooperative is to be altruistic, and successful cooperation is just mutually beneficial altruism. (One notable difference between the Dictator Game and true cooperation games such as the Prisoner's Dilemma and the Public Goods Game, however, is that the size of the pie is typically not fixed in true cooperation games. That is, true cooperation games are not "zero sum.") One might say that cooperation is not altruistic if one expects to get something of equal or greater value in return for one's contribution, and that's fair. But one can say the same for altruism. In one-shot interactions, cooperation is altruistic, and in repeated interactions neither cooperation nor altruism is necessarily altruistic in the strongest sense.

72 **A more recent study:** Henrich, Ensminger et al. (2010).

73 **cooperation and punishment in a set of large-scale societies:** Herrmann, Thöni, et al. (2008).

73 **Figure 3.1:** Adapted with permission from ibid. (2008). Here I show data from only nine of the sixteen cities studied to more clearly illustrate three prominent trends.

74 **punish cooperators on the first round:** Ellingsen, Herrmann, et al. (2012).

75 **Figure 3.2:** Adapted with permission from Herrmann, Thöni, et al. (2008).

75 **Greece . . . had become financially insolvent:** BBC News (November 27, 2012).

76 **airing their prejudices:** There is a long history of scholars presenting research that is suspiciously consistent with their own ideological and cultural commitments. Take, for example, the case of Ellsworth Huntington, an eminent early-twentieth-century geographer, Yale professor, and president of the board of directors of the American Eugenics Society. (Yes, they had a society.) Huntington believed that economic development is

determined primarily by climate. More specifically, he concluded that New Haven, Connecticut (home to Yale University), has a more or less ideal climate for intellectual innovation and economic development.

And here I am, a secular Jew and professor at Harvard University, in metropolitan Boston, telling you that predominantly Muslim cities such as Riyadh and Muscat have cultures that thwart cooperation, while other cities have cultures that are highly conducive to cooperation—cities such as, oh, let's say, Boston.

Amid such suspicion, let me begin with three straightforward points about this research: First, I didn't conduct this research. Second, these results are not cherry-picked. As far as I know, there are no similar studies showing substantially different results, and if you want to see for yourself, you can hop on Google Scholar and search for them. Third, the results presented above are mixed. Riyadh and Muscat come out near the bottom in this study of cooperative behavior, but so does Athens, the birthplace of democracy and the cradle of Western philosophy. And while it's true that my hometown comes out near the top, so does Bonn, not far from the former sites of Nazi concentration camps. But beyond this particular set of studies and my relation to them, there's a more general question about how we ought to respond to scientific research that makes some people look in some ways better than others, particularly when the research's proponents are among those who come out looking better.

Let's start with the two extreme positions on this issue. At one extreme, we have complete deference: If someone with scientific credentials says that something is well supported by scientific evidence, then it must be true. For obvious reasons, this is not a good policy. At the other extreme, we have complete suspicion and skepticism: Anytime a scientist presents supposedly scientific research that makes some people look in some way better than others, we should be highly skeptical, and if the research makes its proponents look better than others, we should assume that it's just a bunch of politically motivated, self-serving, phony baloney.

As you might expect, I think that our attitude should be somewhere in the middle. To be open-minded, we must allow for the possibility that cross-cultural research will reveal cultural differences that make some cultures look better than others—not better in some ultimate sense, but better in some ways. Moreover, we must allow for the possibility that researchers who discover such differences may be among those who end up looking better. (For all I know, Huntington might be right.) At the same time, we should bear in mind that scientists are people, and that like all people they are subject to bias, including unconscious bias. (See later in this chapter.)

In reacting to cross-cultural social science, we should be clear about what follows from the science and what depends on our own moral assumptions. For example, the studies described above tell us that the people tested in Boston and Copenhagen made more money by cooperating in Public Goods Games than did the people tested in Riyadh and Athens. However, these studies don't tell us whether the people in Boston and Copenhagen played *better* or whether they have cultural attributes that are generally better to have. Those are value judgments that go beyond the data.

We should be aware that scientific studies may have serious flaws. (What if the people in Boston understood the directions better than the people in Athens?) At the same time, we should have some respect for the scientific process. Scientific journals select papers for publication based on anonymous peer review, and—believe me—scientists are plenty critical of one another, especially when they're anonymous. Papers published in respectable journals may have serious flaws, but they are unlikely to have *obvious* flaws. (As it happens, in all of the studies described above, participants were tested to ensure that they understood the directions, which is standard practice in this kind of research.)

In sum, bias in cross-cultural social scientific research is a legitimate concern, but the solution is not to dismiss all such research as biased political posturing. This is no better than blindly believing that everything well-credentialed scientists have ever said is true.

76   **studies examining cultural differences among Americans:** Cohen and Nisbett (1994); Nisbett and Cohen (1996).

78   **In surveys, southerners:** Cohen and Nisbett (1994).

78–79 **"Southern ideas of honor":** Fischer (1991).

79   **southern support for their anti-Soviet foreign policies:** Lind (1999).

79   **based on cooperative agriculture:** Nisbett, Peng, et al. (2001).

79   **American and Chinese . . . Magistrates and the Mob:** Described in Doris and Plakias (2007).

80   **"corrupt mind":** Anscombe (1958), cited in ibid. Note that Anscombe says this about killing an innocent person to quell the mob. It's not clear what she would say about imprisoning an innocent person.

80   **sensitive nature of these topics:** See "airing their prejudices" above in this chapter.

81   **reading of the Koran to non-Muslims:** Denmark TV2 (October 9, 2004) (translated with Google Translate).

82   **reward . . . to anyone who would behead "the Danish cartoonist":** *Indian Express* (February 18, 2006).

82   **boycotts of Danish goods cost . . . $170 million:** BBC News (September 9, 2006).

82   **YouTube video:** *International Business Times* (September 21, 2012).

83　**"If someone sues you":** *U.S. News & World Report* (January 30, 1995), cited by Bazerman and Moore (2006), 74.

83　**both versions of the question:** Hsee, Loewenstein, et al. (1999).

84　**experiments have documented this tendency:** Walster, Berscheid, et al. (1973); Messick and Sentis (1979).

84　**"Performance-based pay is fair":** Van Yperen, van den Bos, et al. (2005).

84　**series of negotiation experiments:** Babcock, Loewenstein, et al. (1995); Babcock, Wang, et al. (1996); Babcock and Loewenstein (1997).

85　**didn't know which side they would be on:** Note the parallel with Rawls's (1971) "veil of ignorance."

85　**better able to remember . . . material that supported their side:** Thompson and Loewenstein (1992).

86　**biased fairness . . . environmental commons problem:** Wade-Benzoni, Tenbrunsel, et al. (1996).

86　**negotiate over penalties for . . . criminals:** Harinck, De Dreu, et al. (2000).

89　**tribalistic brand of biased fairness:** Cohen (2003).

89　**It's all unconscious:** The results of Cohen's study are fascinating and sobering, but I would be amazed if they generalized to issues in which partisan disagreements are qualitative rather than quantitative. For example, I doubt that partisan repackaging could reverse people's views on abortion or gay marriage, at least not in the short run.

89　**believed that Saddam Hussein had been personally involved:** Millbank and Deanne (September 6, 2003).

90　**2008 World Public Opinion poll:** Reuters (September 10, 2008).

90　**students . . . watched footage from a football game:** Hastorf and Cantril (1954).

90　**more confident in their views . . . considering the mixed evidence:** Lord, Ross, et al. (1979).

90　**"hostile media effect":** Vallone, Ross, et al. (1985).

90　**people watched footage of protesters:** Kahan and Hoffman (2012).

91　**consensus among experts:** Intergovernmental Panel on Climate Change (2007); Powell (November 15, 2012).

91　**Republicans believe:** Jones (March 11, 2010), reporting on Gallup poll.

91　**commons problem of its own:** Kahan, Wittlin, et al. (2011).

94　**Figure 3.3:** Adapted with permission from Kahan, Peters, et al. (2012).

94–95　**In 1998, Republicans and Democrats . . . in 2010:** Dunlap (May 29, 2008) and Jones (March 11, 2010), reporting on Gallup polls.

95　**deep geologic isolation:** Kahan, Jenkins-Smith, et al. (2011).

95　**biased perception in the escalation of conflict:** Shergill, Bays, et al. (2003).

96　**Figure 3.4:** Adapted with permission from ibid. (2003).

96　**brain automatically anticipates:** Blakemore, Wolpert, et al. (1998).

97　**only partially aware of the contributions of others:** Forsyth and Schlenker (1977); Brawley (1984); Caruso, Epley, et al. (2006).

97　**As Steven Pinker explains:** Pinker (2011).

98　**some of our biggest problems:** Copenhagen Consensus Center (2012).

99　**"perfect moral storm":** Gardiner (2011).

100　**any .number of arrangements:** Singer (2004).

100　**the world's second-largest carbon emitter:** Union of Concerned Scientists (2008), citing data originally compiled by the US Energy Information Agency (2008).

100　**"cleaning up the world's air":** Second presidential debate in 2000, quoted in Singer (2004), 26.

101　**"An attempt to point out":** Fisher (1971), 113.

101　**"Officials think of themselves":** Ibid., 112.

101　**"Laying down the moral law":** Schlesinger (1971), 73.

PART II: MORALITY FAST AND SLOW

CHAPTER 4: TROLLEYOLOGY

106　**Jeremy Bentham and John Stuart Mill:** Mill (1861/1998); Bentham (1781/1996). The third of the great founding utilitarians is Henry Sidgwick (1907), whose exposition of utilitarianism is more thorough and precise than those of his more famous predecessors. Thanks to Katarzyna de Lazari-Radek and Peter Singer for highlighting many of the points on which Sidgwick anticipates points made here.

109　**without a girlfriend:** Or boyfriend, as the case may be.

111　**theory of parental investment:** Trivers (1972).

112　**in some birds and fish:** Eens, M., & Pinxten, R. (2000). Sex-role reversal in vertebrates: Behavioural and endocrinological accounts. *Behavioural processes*, 51(1), 135–147.

112　**Peter Singer first posed it:** Singer (1972).

113 **Two rivers, twenty rivers: It all sounds the same:** This attitude would make sense if the question were about making a small contribution to a larger (two-river or twenty-river) effort. But the question here is about how much you would pay to have the whole job completed if somehow your payment alone would do it.

113 **my first scientific publication:** Baron and Greene (1996).

113 **heuristic thinking:** Gilovich, Griffin, et al. (2002); Kahneman (2011).

113 **"The Trolley Problem":** Thomson (1985). The first papers on the Trolley Problem were by Philippa Foot (1967) and Thomson (1976). This gave rise to a large literature in ethics. See Fischer and Ravizza (1992); Unger (1996); and Kamm (1998, 2001, 2006).

114 **wrong to push the man off the footbridge:** Thomson (1985); Petrinovich, O'Neill, et al. (1993); Mikhail (2000, 2011); Greene, Somerville, et al. (2001).

115 **"Act so that you treat humanity":** Kant (1785/2002). This is the second of four formulations of the categorical imperative described in the *Groundwork*.

116 **people all over the world agree:** O'Neill and Petrinovich (1998); Hauser, Cushman, et al. (2007).

117 **the case of Phineas Gage:** See Damasio (1994) and Macmillan (2002).

118 **due to emotional deficits:** Saver and Damasio (1991); Bechara, Damasio, et al. (1994).

118 **"to know, but not to feel":** Damasio (1994), 45.

119 **Cognitive control:** Miller and Cohen (2001).

119 **color-naming Stroop task:** Stroop (1935).

120 **enabled by . . . DLPFC:** Miller and Cohen (2001).

122 **fMRI:** fMRI uses an MRI scanner of the sort routinely used in modern hospitals. For most clinical purposes, an MRI takes a still, three-dimensional picture of the body, a "structural scan." fMRI takes "movies" of the brain in action. The movies have a modest spatial resolution, composed of "voxels" (volumetric pixels) of about 2 to 5 millimeters. The temporal resolution is very low, with an image (a "frame" in the movie) acquired about once every 1 to 3 seconds. The images produced by fMRI look like pixilated blobs, which are typically overlaid on top of a higher-resolution structural scan, allowing one to see where the blobs are in the brain. The blobs are not the direct result of "looking" at the brain. They are the products of statistical processing. What a blob in a brain region typically means is that there is, on average, more "activity" in that region when someone is performing one task (e.g., looking at human faces) as compared with another task (e.g., looking at animal faces). The "activity" in question is the electrical activity of neurons in the brain, but this activity is not measured directly. Instead, it's measured indirectly, by tracking changes in the flow of oxygenated blood. For more information, see Huettel, Song, et al. (2004).

122 **including parts of the VMPFC:** Greene, Somerville, et al. (2001). Many other brain regions exhibited effects in this contrast, including most of what is now called the "default network" (Gusnard, Raichle, et al., 2001). Many of these regions appear to be involved not in emotional response per se, but in the representation of nonpresent realities (Buckner, Andrews-Hanna, et al., 2008).

122 **Our second experiment:** Greene, Nystrom, et al. (2004).

123 **We addressed this question in later work:** Greene, Cushman, et al. (2009).

123 **ice cream doesn't cause drowning:** I don't know who first used this example.

124 **follow-up studies of our own:** Some philosophers have raised doubts about the evidence for the dual-process theory (McGuire, Langdon, et al., 2009; Kahane and Shackel, 2010; Kahane, Wiech, et al., 2012; Berker, 2009; Kamm, 2009). For replies, see Paxton, Bruni, and Greene (under review); Greene (2009); and Greene (under review). For further details on Berker, see a set of notes (Greene, 2010) assembled for a conference at Arizona State University, available on my webpage or by request.

124 **patients with . . . (FTD):** Mendez, Anderson, et al. (2005).

125 **dilemmas . . . patients with VMPFC damage:** Koenigs, Young, et al. (2007); Ciaramelli, Muciolli, et al. (2007).

125 **sweaty palms:** Moretto, Ladavas, et al. (2010).

125 **same conclusion:** See also Schaich Borg, Hynes, et al. (2006); Conway and Gawronski (2012); Trémolière, Neys, et al. (2012).

125 **turning the trolley onto family members:** Thomas, Croft, et al. (2011).

125 **Low-anxiety psychopaths:** Koenigs, Kruepke, et al. (2012). See also Glenn, Raine, et al. (2009).

125 **alexithymia:** Koven (2011).

125 **physiological arousal . . . fewer utilitarian judgments:** Cushman, Gray, et al. (2012). See also Navarrete, McDonald, et al. (2012).

126 **gut feelings:** Bartels (2008).

126 **Inducing people to feel mirth:** Valdesolo and DeSteno (2006); Strohminger, Lewis, et al. (2011).

126 **the amygdala:** Adolphs (2003).

126 **psychopathic tendencies:** Glenn, Raine et al. (2009).

126 **amygdala . . . correlates negatively with utilitarian judgments:** Shenhav and Greene (in prep.).

126 **citalopram . . . fewer utilitarian judgments:** Crockett, Clark, et al. (2010).

126 **lorazepam has the opposite effect:** Perkins, Leonard, et al. (2012).

126 **role of visual imagery:** Amit and Greene (2012).

126 **"Do whatever will produce the most good":** By this I don't mean that people who make utilitarian judgments are card-carrying utilitarians, subscribing to the full philosophy. I mean only that they are applying an impartial "cost-benefit" decision rule.

127 **DLPFC . . . success in the Stroop task:** MacDonald, Cohen, et al. (2000).

127 **brain imaging studies . . . similar results:** Shenhav and Greene (2010); Sarlo, Lotto, et al. (2012); Shenhav and Greene (in prep.).

127 **simultaneous secondary task:** Greene, Morelli, et al. (2008). See also Trémolière, Neys, et al. (2012).

127 **removing time pressure and encouraging deliberation:** Suter and Hertwig (2011).

127 **experience of being led astray:** Method follows Pinillos, Smith, et al. (2011).

127 **tricky math problems:** Frederick (2005).

127 **people who solved these tricky math problems:** Paxton, Ungar, and Greene (2011). In the case of the *footbridge* dilemma, the tricky math problems didn't change people's judgments. Instead, we found that people who were generally better at solving the tricky math problems gave more utilitarian judgments in response to the *footbridge* case. See also Hardman (2008). Paxton and I also used the CRT method with a "white lie dilemma" devised by Kahane, Wiech, et al. (2012), a case in which the non-utilitarian response was alleged to be counterintuitive. Our results indicate the opposite, consistent with the original dual-process theory. See Paxton, Bruni, and Greene (under review).

127 **people who generally favor effortful thinking:** Bartels (2008); Moore, Clark, et al. (2008).

128 **moral reasons of which people are conscious:** Cushman, Young, et al. (2006).

128 **justifying that judgment in a consistent way:** Ibid.; Hauser, Cushman, et al. (2007).

128 **on a molecular level:** Crockett, Clark, et al. (2010); Perkins, Leonard, et al. (2010); Marsh, Crowe, et al. (2011); De Dreu, Greer, et al. (2011).

129 **moral judgments of medical doctors:** Manuscript in preparation, based on Ransohoff (2011). For a review of bioethical issues from a neuroscientific perspective, see Gazzaniga (2006).

129 **minimize the risk of actively harming:** It's often said that the Hippocratic Oath, taken up by doctors upon entering the profession, commands them to "First, do no harm." However, these words don't actually appear in the oath. See: http://www.nlm.nih.gov/hmd/greek/greek_oath.html.

CHAPTER 5: EFFICIENCY, FLEXIBILITY, AND THE DUAL-PROCESS BRAIN

132 *Everything Bug:* Winner (2004).

132 **one of the most important ideas to emerge:** Posner and Snyder (1975); Shiffrin and Schneider (1977); Sloman (1996); Loewenstein (1996); Chaiken and Trope (1999); Metcalfe and Mischel (1999); Lieberman, Gaunt, et al. (2002); Stanovich and West (2000); Kahneman (2003, 2011).

134 *Thinking, Fast and Slow:* Kahneman (2011).

134 **get rid of the concept of "emotion":** Griffiths (1997).

135 **but it's not emotional:** Such processing may trigger emotional responses, but the visual processing itself is not emotional.

135 *action tendencies:* Darwin (1872/2002); Frijda (1987); Plutchik (1980).

135 **Fear . . . enhancing the sense of smell:** Susskind, Lee, et al. (2008).

135 **influenced . . . decisions by influencing their moods:** Lerner, Small, et al. (2004).

137 **"slave of the passions":** Hume (1739/1978).

137 **cannot produce good decisions without . . . emotional input:** VMPFC patients like Phineas Gage are, in general, very bad decision makers. They can give reasons for choosing one thing over another, and the reasons they give often sound good. But these reasons are fragmented. Instead of adding up to a good decision, they float free in a jumble, resulting in foolish behavior. (See Damasio, 1994.) In a telling experiment, Lesley Fellows and Martha Farah (2007) showed that VMPFC patients are more likely than others to exhibit "intransitive" preferences—that is, to say that they prefer A to B, B to C, and C to A. With respect to decision making, this is the hallmark of irrationality. What's more, the DLPFC, the seat of abstract reasoning, is deeply interconnected with the dopamine system, which is responsible for placing values on objects and actions (Rangel, Camerer, et al., 2008; Padoa-Schioppa, 2011). From a neural and evolutionary perspective, our reasoning systems are not independent logic machines. They are outgrowths of more primitive mammalian systems for selecting rewarding behaviors—cognitive prostheses for enterprising mammals. In other words, Hume seems to have gotten it right.

137 **fruit salad or chocolate cake:** Shiv and Fedorikhin (1999).

138 **two different kinds of decisions:** McClure, Laibson, et al. (2004).

139   **yielding immediate rewards:** Here the immediate rewards are not so immediate. In a later study, using food rewards (McClure, Ericson, et al., 2007), they were more immediate.

139   **Figure 5.1:** Images adapted with permission from Ochsner, Bunge, et al. (2002); McClure, Laibson, et al. (2004); and Cunningham, Johnson, et al. (2004).

140   *intra*personal . . . *inter*personal: Nagel (1979).

140   **we see the same pattern:** Cohen (2005).

140   **reinterpret the pictures in a more positive way:** Ochsner, Bunge, et al. (2002).

140   **presented white people with pictures:** Cunningham, Johnson, et al. (2004).

141   **interacting with a black person . . . cognitive load:** Richeson and Shelton (2003).

141   **automatic settings that tell us how to proceed:** Bargh and Chartrand (1999).

141   **amygdala . . . 1.7 hundredths of a second:** Whalen, Kagan, et al. (2004).

141   **VMPFC . . . decisions involving risk:** Bechara, Damasio, et al. (1994); Bechara, Damasio, et al. (1997); Damasio (1994).

141   **as revealed in their sweaty palms:** Small differences in palm sweat can be detected by passing a small current through the skin, which conducts current more effectively when moist. This technique is known as the measurement of "skin conductance response" (SCR) or "galvanic skin response" (GSR).

142   **Figure 5.2:** Adapted with permission from Whalen, Kagan, et al. and Rathmann (1994).

142   **we need our emotional automatic settings:** Woodward and Allman (2007).

143   **shaped by cultural learning:** Olsson and Phelps (2004, 2007).

PART III: COMMON CURRENCY

CHAPTER 6: A SPLENDID IDEA

148   **within-group cooperation:** And also within-group competition.

149   **"moral relativist":** I here refer to *relativism* in the colloquial sense. In philosophy, *relativism* can be rather different. See Harman (1975).

153   **"pragmatism" often has a different meaning:** Pragmatist theories of truth are, roughly, ones according to which claims are true or false depending on the practical effects of believing in them.

154   **make things go as well as possible:** Strictly speaking, I am talking about act consequentialism.

155   **earliest opponents of slavery:** Driver (2009).

156   **two centuries' worth of philosophical debate:** Smart and Williams (1973).

161   **the value that gives other values their value:** Mill (1861/1987), chap. 4, 307–314; Bentham (1781/1996), chap. 1.

161   **"It is better to be a human being dissatisfied":** Mill (1861/1998), 281.

162   **an argument that Mill dashes off in passing:** Ibid., 282–283.

162   **"broaden and build":** Fredrickson (2001).

164   **Measuring happiness:** Easterlin (1974); Diener, Suh, et al. (1999); Diener (2000); Seligman (2002); Kahneman, Diener, et al. (2003); Gilbert (2006); Layard (2006); Stevenson and Wolfers (2008); Easterlin, McVey, et al. (2010).

165   **science of happiness excels:** See previous note.

165   **unemployment is often emotionally devastating:** Clark and Oswald (1994); Winkelmann and Winkelmann (1995); Clark, Georgellis, et al. (2003).

165   **making a bit less money:** Here the debate is between those who say that additional money for the well-off buys no additional happiness (Easterlin, 1974; Easterlin, McVey, et al., 2010) and those who say that it buys some but not much (Stevenson and Wolfers, 2008). At best, happiness seems to increase as a logarithmic function of income, meaning that gaining another unit of happiness requires ten times more income than it took to gain the last unit.

166   **we may soon have such measures:** At least for happiness in the moment. Neural measures of life satisfaction pose a far greater challenge.

166   **This stereotype . . . is undeserved:** Mill (1861/1998), 294.

167   **attempt to outcalculate . . . at our peril:** Ibid., 294–298; Hare (1981); Bazerman and Greene (2010).

169   **down to first principles:** There is a sense in which utilitarians are more closely aligned with ideological collectivists than with ideological individualists. Both utilitarians and ideological collectivists aim for the greater good. The difference is that ideological collectivists are committed to a collectivist way of life as a matter of first principles. In contrast to both utilitarians and ideological collectivists, ideological individualists do not aim for the greater good per se. If some people are foolish and lazy and they get less, that's perfectly fine with individualists, even if their getting less reduces aggregate happiness. For ideological individualists,

the goal is not to maximize happiness but to give people the happiness or unhappiness they *deserve*. The utilitarian take on communism follows the old quip "Great in theory, terrible in practice." Ideological individualists won't even say "Great in theory."

170 **values . . . derive their value from their effects on our experience:** See Sidgwick (1907), 401.

170 **no impact on our experience . . . would not be valuable:** This statement and the one preceding it are not, in fact, equivalent. It could be that all values must have an impact on our experience in order to be valuable, but from this it doesn't follow that the value of a value is derived solely from its impact on our experience. In other words, impact on experience may be necessary for having value, but not sufficient for determining the value of a value.

170 **second utilitarian ingredient is impartiality:** Sidgwick (1907) calls this the axiom of justice.

172 **connection between manual-mode thinking and *utilitarian* thinking:** This association has been challenged by Kahane, Wiech, et al. (2012). They argue that manual-mode thinking favors utilitarian judgments in some cases—like the *footbridge* dilemma—but not in general. To make this point, they conducted a neuroimaging study using a new set of dilemmas in which, according to them, the deontological judgment (the nonutilitarian judgment favoring rights or duties over the greater good) is less intuitive than the utilitarian judgment. However, Joe Paxton, Tommaso Bruni, and I have since conducted an experiment that casts serious doubt on their conclusions, which were not well supported by the neuroimaging data to begin with (Paxton, Bruni, and Greene, under review). We used something called the Cognitive Reflection Test (Frederick, 2005), which can both measure and induce (Pinillos, Smith, et al., 2005) reflective thinking, to test one of their new dilemmas. This is a "white lie" case in which the greater good is served by telling a lie. As a control, we tested one of our standard *footbridge*-like dilemmas. We showed that in *both* cases, being more reflective is associated with more utilitarian judgment. This is a striking victory for the dual-process theory presented here, because it employs a dilemma that both I (Greene, 2007) and these critics thought would work as a counterexample.

173 **the amygdala and the VMPFC:** Actually, the situation is a bit more complicated. Current research suggests that the amygdala functions more like an alarm bell, while the VMPFC is actually more of an integrator of emotional signals, translating motivational information into a common affective currency (Chib, Rangel, et al., 2009). Thus, VMPFC damage may block the influence of automatic settings by blocking the route by which they influence decisions. Decision rules applied by the DLPFC can influence the affective integration in the VMPFC (Hare, Camerer, et al., 2009), but such rules can also be applied without the VMPFC. See Shenhav and Greene (in prep.).

173 **ambiguous Golden Rule:** The Golden Rule is ambiguous because people's situations are always different, and the Golden Rule doesn't tell us which situational differences justify differences in treatment. For almost any disparity in treatment, one can find a formally impartial rule that justifies it: "Yes, and if *you* were king, and *I* were a peasant, then *you* would have the right to do whatever you want to *me*!" The Golden Rule works only when there is agreement about which features of our situations matter morally. In other words, the Golden Rule doesn't set the terms of cooperation. It just says that pure selfishness, as in "I get more just because I'm me," is not allowed. When it comes to resolving conflicts, that's not very helpful, because, as explained in chapter 3, no tribe's values are purely selfish.

CHAPTER 7: IN SEARCH OF COMMON CURRENCY

175 **aware of this problem:** Obama (2006) continued: "Now this is going to be difficult for some who believe in the inerrancy of the Bible, as many evangelicals do. But in a pluralistic democracy, we have no choice. Politics depends on our ability to persuade each other of common aims based on a common reality. It involves the compromise, the art of what's possible. At some fundamental level, religion does not allow for compromise. It's the art of the impossible. If God has spoken, then followers are expected to live up to God's edicts, regardless of the consequences. To base one's life on such uncompromising commitments may be sublime, but to base our policy making on such commitments would be a dangerous thing."

176 **"only people of non-faith can . . . make their case":** See Greenberg (February 27, 2012). Santorum was responding directly to President Kennedy's views, which were similar to Obama's.

176 **Rights . . . *trump* consequences:** Dworkin (1978).

177 **For many modern moral thinkers:** Kant (1785/2002); Hare (1952); Gewirth (1980); Smith (1994); Korsgaard (1996). See also a forthcoming book (still untitled) by Katarzyna de Lazari-Radek and Peter Singer in which they, inspired by Henry Sidgwick, defend utilitarianism as an axiomatizable system.

177 **Other tribes say that earthquakes:** Espresso Education (n.d.).

178 **whether or not it's the moral truth:** What do we mean by "works?" How can we say whether a metamorality "works" without applying some kind of evaluative standard? And how can we apply such a standard without assuming some kind of moral truth or, at least, a metamorality? We'll discuss this problem in more detail later,

but for now the short answer is this: A metamorality "works" if we're generally satisfied with it. And one metamorality works better than another if, in general, we're more satisfied with it. An analogy here is with law. You don't have to believe "Thou shalt not drink under twenty-one" is the moral truth to be satisfied with a law setting the legal drinking age at twenty-one. And different people can be satisfied with such a law for different reasons. General satisfaction with a moral system does not presuppose agreement on moral first principles.

178   **Plato:** Plato's *Euthyphro,* in Allen and Platon (1970).

179   **How can we know God's will?:** Craig and Sinnott-Armstrong (2004).

180   **open letter to Dr. Laura:** The letter is available on many websites in many forms. For one reprinting and a discussion of its origins, see Snopes.com (November 7, 2012).

182   **"Abraham is ordered by God":** Obama (2006).

182   **reflection discourages belief:** Shenhav, Rand, and Greene (2012); Gervais and Norenzayan (2012).

183   **most of them, at least:** Of course, some religions are less tribalistic than others. A decidedly untribal religion is the Unitarian Universalist church.

183   **"pure practical reasoning":** Kant (1785/2002).

183   **views that can't be rationally defended:** Of course, some people *are* committing rational errors, maintaining sets of moral beliefs that are internally inconsistent. But the hard-line rationalist believes that some specific moral conclusions (as opposed to combinations of conclusions) could never be rationally defended. This requires that morality be like math, with substantive conclusions derivable from self-evident first principles. More on this shortly.

184   **manageable set of self-evident moral truths:** I say "manageable" because morality is not like math if the axioms are an enormous set of statements too expansive to be written down.

184   **no one has found . . . axioms:** Why has no one found such axioms? What kind of principles would make good axioms? Given that the axioms need to be self-evident, we might hope for axioms that are "analytic," that is, true by virtue of the meanings of the words used to express them. (The validity of the analytic/synthetic distinction was famously questioned by Quine [1951], but it seems very hard to get by without it [Grice and Strawson, 1956].)

For example, the statement "All bachelors are unmarried" is analytic. You might say that analytic statements are "true by definition," with the caveat that the truth of some analytic statements may be nonobvious, especially if they are long and complicated. It's also possible to have self-evident truths that are not analytic—for example, Euclid's axiom stating that it's possible to connect any two points with a straight line. This is obviously true, but there's nothing in the definitions of "point" and "two" from which one can derive this truth. Put another way, the concept POINT does not *contain* the concept STRAIGHT (or the concept LINE) in the way that the concept BACHELOR *contains* the concept UNMARRIED. Thus, using Euclid's as our model, we might hope to find moral axioms that are self-evidently true but not true by definition. Or we might hope to find axioms that are true by definition. If we're looking for moral axioms, those are our options.

The early-twentieth-century philosopher G. E. Moore (1903/1993) put forth an argument, known as the Open Question Argument, that provides a test for aspiring moral axioms. We can start by thinking of the Open Question Argument as a test for self-evidence, although that's not how Moore thought of it. (Moore thought that propositions that failed the test couldn't be true, but he overlooked the possibility that they could be true but not self-evidently or obviously true.) A test for self-evidence is what we need if we're looking to model morality on math, because, once again, the moral axioms need to be self-evident.

Moore's test works as follows. Take a moral principle that purports to tell you what sorts of things are right, wrong, good, bad, etc. For example, a utilitarian principle like this one:

What's right is what maximizes overall happiness.

If you have utilitarian inclinations, you might think that this principle is not only true but also self-evident. To you, Moore poses the following challenge: Suppose we know that a certain action will maximize overall happiness. Isn't it still an open question whether the action is right? If the answer is yes, then it can't be self-evident that what maximizes happiness is what's right. In this case, you can feel the pull of Moore's Open Question Argument by considering counterexamples. Take the case of pushing the man off the footbridge to save the five. Let's grant that this action will maximize overall happiness. Have we then granted that it's right? Clearly not. The moral question remains open for now, regardless of what we may conclude in the end.

Let's try an even more abstract principle. This one is borrowed, with liberties taken, from Michael Smith (1994):

What's right is what we would want if we were fully informed and fully rational.

This principle may seem to be self-evidently true, but is it really? Suppose we have an action that is favored by someone who is fully informed and fully rational. Is it not an "open question" whether this action is right?

Think, for example, of your classic evil masterminds, such as Hannibal Lecter. Maybe it's true that a baddie like Lecter, who kills and eats innocent people, must be making some kind of logical error, or must be ignorant of some nonmoral facts. Maybe, but maybe not. The point is that it's not self-evident that this is so. Shaun Nichols has shown that many ordinary people believe that psychopaths know right from wrong but simply don't care, consistent with the idea that one can be morally deficient without being irrational or non-morally ignorant (Nichols, 2002). Even if we become fully informed and fully rational, it's still an "open question" whether we have therefore become morally perfect. And that means that the principle above can't be self-evident. It may appear to be self-evident (to some of us, at any rate), because we think that being more rational and more informed can only help. But that's very different from saying that full rationality and full information are, self-evidently, all you need for perfect moral judgment and motivation. We could be fully informed and fully rational and still make some moral mistakes. In any case, it's not self-evident that what I just said is false. And that means that the above principle can't be a moral axiom because, even if it's true, it's not self-evidently true.

What's more, even if we were to accept this kind of very abstract principle as an axiom, it wouldn't give us a common currency. It wouldn't tell us how to make trade-offs among competing values. It would simply tell us that the right trade-offs are the trade-offs we would make if we were fully informed and fully rational, which is not much help.

Moore thought that, for any moral principle, the question it purports to answer would always remain open. To see why, we need to think about what a moral principle is. For Moore, a moral principle is one that connects a "natural property" to a "moral property." Take, for example, the principle "Lying is wrong." An action's being an instance of telling a lie is a "natural property" of that action. An action's being wrong is a "moral property" of that action. And what the principle "Lying is wrong" means, in property-speak, is this: If an action has the natural property of being an instance of lying, then it has the moral property of being wrong. What Moore's Open Question Argument suggests is that ascribing natural properties to things will never get us all the way to moral properties, at least not in a way that is self-evident. (Moore's terminology implies that moral properties must be "unnatural," but this is not an essential part of his argument. Instead of speaking about "natural" properties, we can instead speak about apparently factual properties that can be ascribed without controversy, the "facts of the case," as lawyers say.)

Suppose that Joe lied to the police in order to protect his friend. We disagree about whether this is wrong. It is, however, a noncontroversial "fact of the case" that Joe lied. From this fact, are we forced to conclude that what Joe did is wrong? No, says Moore. It's an "open question." Moore argues that all substantive moral principles—not just "Lying is wrong"—have the same limitation. The reason is that all substantive moral principles must span the is-ought gap, with the "natural properties" (the noncontroversial facts of the case) on one side and the "moral properties" on the other. The facts of the case are always about what "is"—facts like the fact that Joe lied. Moral conclusions, in contrast, are always about what "ought" to be, such as the fact that Joe ought not to have lied. Now, it may be true that lying is wrong, but the key point here is that this moral principle can't be self-evidently true. Why? Because even if we agree that Joe lied (a fact about what "is"), it's still an open question whether this act of lying was wrong (a fact about what "ought" to be).

"Lying is wrong" is not self-evidently true, but that's just one candidate. Maybe there are other familiar moral principles that are self-evidently true. We might start with a principle concerning something that seems obviously and unconditionally wrong. How about this: "Torturing kittens is wrong." Isn't it self-evident that it's wrong to torture kittens? Well . . . What if the only way to save a million people is to torture one kitten? Would that be wrong? And is it *self-evident* that it's wrong? Perhaps we need a little tweak: "Torturing kittens is wrong, unless you have a really good reason." But what's a "really good reason"? One that's good enough to justify torturing a kitten? If so, then we have, instead of a substantive moral principle, an empty tautology: "Torturing kittens is wrong unless there is a reason sufficient to justify kitten torturing." We could attempt to be more specific. But how much good does your kitten torturing have to do? And how does the amount of good vary with the amount of torture? And are there answers to these questions that are self-evidently correct? Perhaps, with enough tweaking, we can get a moral statement about kitten torturing (or whatever) that passes the open-question test. But what we're going to end up with is not a moral axiom, a foundational principle from which more specific moral truths can be derived. Rather, it's going to be a very specific and highly quali-fied statement about the ethics of torturing kittens (or whatever).

We've been talking about the Open Question Argument as a test for self-evidence, and self-evidence is what we need from our axioms. But strictly speaking, a statement can be self-evidently true even if it's being true remains an "open question." How? Consider this statement: All bachelors are not not not not not not not not not not married. Unless you counted the *not*s, the truth of the previous sentence is, for you, an open question. It turns out that this sentence is true, and it's self-evidently true, meaning that you don't need any evidence beyond what's in the sentence to verify that it's true. Thus, strictly speaking, a statement can be self-evident even if its truth is an open question. And, thus, the Open Question Argument is not, strictly speaking, a test for self-evidence. Rather, it's a test for something more like "obviousness." This leaves open the possibility that there are useful moral axioms that are self-evidently true but not obviously true. What would such a principle be like? Unlike the "not not" sentence above, such a principle would not be reducible to something simpler. Otherwise the simpler version could be used as the axiom. Thus, such a principle would have to be an irreducibly complex moral statement that can be seen as true without any further evidence or argument, but whose truth is not obvious, due to its complexity. And from this statement (along with others like it, perhaps) and nonmoral facts, we will be able to derive answers to controversial moral questions. I cannot prove that moral axioms of this kind do not exist, but it's fair to say that we should not count on their arriving anytime soon.

In sum, the prospects for modeling morality on math don't look good. To pull this off, we'll need moral axioms that are both self-evident and useful, but there don't seem to be any such axioms around. To be useful, moral principles must enable us to connect the "is" to the "ought," taking us from the "facts of the case" to specific moral answers. And principles that are powerful enough to do that don't seem to be self-evident, however plausible they may be. I cannot prove that morality will never be axiomatized, and therefore made like math. But we best not hold our collective breath.

P.S. Katarzyna de Lazari-Radek and Peter Singer have argued (in an unpublished and untitled book manuscript) that utilitarianism rests on a set of self-evident axioms. Their argument, which is inspired by Henry Sidgwick (1907), is, in my opinion, about as compelling as such arguments get. But in the end I can't agree, for reasons suggested by the foregoing discussion.

186 **competition between groups:** Once again, I'm not assuming a group-selectionist account of moral evolution. Here, a "group" may consist of two people, a tribe of thousands, or anything in between.

186 **favor people with genocidal tendencies:** Joyce (2011).

186 **argument, which is controversial, also applies to cultural evolution:** Casebeer (2003). See also Ruse and Wilson (1986) and a rebuttal by Kitcher (1994). Suppose that what evolved biologically was not morality, but a general capacity to acquire cultural practices. And suppose that morality evolved purely culturally, meaning that it is a set of "memes" (cultural variations) that spread because morality outcompeted other memes in the struggle for existence in human brains. In that case, the same argument still applies. The ultimate function of morality would then be to make more copies of itself in the brains of other humans, rather than to make more copies of its associated genes. Moral tendencies might spread simply because they are good at "infecting" brains, rather like catchy tunes and conspiracy theories. Or they might survive because they help their hosts survive. More specifically, moral memes might survive because they help their hosts outcompete the competition. But whatever the case, and whether or not morality does some good along the way, the bottom line of cultural evolution is the spreading of cultural memes, just as the bottom line of biological evolution is the spreading of genes.

186 **the "is-ought" problem:** Hume (1739/1978).

186 **"naturalistic fallacy":** According to G. E. Moore (1903/1993), who coined the term, the naturalistic fallacy is to infer that an entity is good from facts about its "natural" properties—for example, inferring that chocolate is good because it is tasty.

186 **social Darwinists:** The term "social Darwinism" is used primarily as a pejorative. It is often attributed, rather unfairly, to Herbert Spencer. Insofar as Social Darwinism existed as an ideology, it was in the minds of elite capitalists who saw in Darwin confirmation of their preexisting moral and political beliefs. See Wright (1994).

188 **I thought this was *the* question:** Greene (2002).

188 **What really matters . . . path through the morass:** Thanks to Walter Sinnott-Armstrong, Peter Singer, and Simon Keller for pressing me on this point.

188 **Do we call what's left "the moral truth"?:** Here's the dilemma: On the one hand, it seems that some moral views are clearly *better* than others. If a moral position is internally inconsistent, or if its appeal depends on false assumptions, then such a view is in some sense objectively inferior to others that lack such problems. But if we believe in "objectively better" and "objectively worse," then why not believe in "objectively right"? Why not say that the moral truth is just whatever our moral beliefs become after we've objectively improved them as much as possible?

On the other hand, you might think that moral truth requires more than postimprovement agreement. (See Mackie, 1977; Horgan and Timmons, 1992; and Joyce, 2001, 2006.) If what you believe is truly *true*, then it should be impossible for someone to disagree with you without making some kind of objective mistake. Suppose that a well-informed, perfectly rational psychopath—one whose thinking meets our standards for "fully improved"—disagrees with us about, say, the wrongness of torturing kittens. There are two possibilities. First, we might deny that such a person could exist. If his thinking is fully objectively improved, we say, then he must agree that torturing kittens is wrong. But on what grounds can we say that? This is the kind of thing that we can say only if we have direct access to moral truth, which we apparently don't have. Our other option is to accept that a psychopath with fully improved moral thinking (knows all the facts, no internal inconsistencies, etc.) could exist. But this means that someone can reject the moral truth without making

374   NOTES

(in any non-question-begging sense) a mistake. But that doesn't sound like *truth*. It's like saying this: "It's true that the sun is larger than the earth, but if you think that the sun is not larger than the earth, you're not necessarily wrong." Huh?

So is there moral truth? In my dissertation (Greene, 2002), I argued that there isn't any moral truth, for the reasons given "on the other hand" above. But now I think that what matters most for practical purposes is the possibility of objective improvement, not the possibility of objective correctness. And this inclines me to say that, for practical purposes, there can be something very much like moral truth (Blackburn, 1993), which might more or less be the moral truth, if not the Moral Truth. But, really, I think it's the wrong question on which to focus, which is why I've paid it relatively little attention in this book. What matters is what we do with the morass, not whether we call the final product the "moral truth."

189   **at the highest level:** "At the highest level" is key. Almost no one thinks that we must be completely impartial in a day-to-day sense, caring for strangers no less than we care for ourselves and our loved ones. But at the same time, we all recognize that, from a moral perspective, we must all be subject to the same rules, even as we occupy different positions within the system established by those rules. If I'm allowed to favor my loved ones over strangers, then so are you, so long as our positions are symmetrical. And if I'm allowed to favor mine more than you're allowed to favor yours, it's because we occupy objectively different positions (e.g., a president of a private club vs. a federal judge). In short, we're allowed to be partial at a low level, but the rules that define when and where partiality is acceptable must be applied impartially.

189   **"moral truth" . . . an open question:** See note "Do we call what's left 'the moral truth'?".

CHAPTER 8: COMMON CURRENCY FOUND

190   **episode of *The Twilight Zone*:** The episode (Medak, 1986), written by Richard Matheson, was based on his earlier short story and was the basis for the movie *The Box*.

191   ***"If all else is equal":*** We're assuming that there is no hidden upside here. Breaking your kneecap won't improve your character. You won't meet the love of your life in the hospital. Here, breaking your kneecap is just an unmitigated reduction in your happiness that you can avoid by pushing the button.

193   **go out and check:** At the very least, I expect at least some members of all tribes to follow the utilitarian logic *within the tribe*. There may be some tribes that are so tribalistic that they are essentially psychopathic at the intertribal level. But as noted above, if that's how they are, they are simply not part of the "we" of this conversation. At least not yet. As noted in chapter 3, kindness toward strangers seems to be supported, if not created, by modern market societies (Henrich, Ensminger et al., 2010). In testing this conjecture, methods will have to be tailored to the population(s) being tested.

193   **substantial moral common ground:** You might disagree. You might observe that "if all else is equal," commitments are a dime a dozen. We're all opposed to lying, if all else is equal. We're all in favor of letting people spend their money however they please, if all else is equal. And so on. In other words, many values, perhaps most values, are shared to some extent, and something that we all value to some extent is something to which we're all committed "if all else is equal." Conflict arises primarily because people prioritize different values in different ways. Thus, there's nothing special about our "if all else is equal" commitment to maximizing happiness. We have "if all else is equal" commitments to many, many values. The problem is that all else is never equal.

It is indeed true that "if all else is equal" moral commitments are easy to come by. Nevertheless, our "if all else is equal" commitment to maximizing happiness is special. "No lying" is not a moral system. Nor is "Spend your money however you want," and so on. By contrast, "Maximize happiness" is a moral system. Why is it a system? Because a commitment to maximizing happiness tells us how to prioritize different values—in other words, how to make *trade-offs*. It gives us answers to questions such as "When is it permissible to lie?" and "When does economic freedom go too far?" and so on. Thus, our "if all else is equal" commitment to maximizing happiness is not just a default commitment to one among many moral values. It's a default commitment to what is, or could be made into, a complete system of moral values. That's profoundly important.

194   **objections . . . ultimately driven by *automatic settings*:** The key word here is *ultimately*. I understand that there are some abstract, theoretical arguments against utilitarianism. My claim, however, is that these theories are ultimately motivated by gut feelings. See chapter 11.

196   **make the world that way:** Russell and Norvig (2010).

196   **simple kind of problem solver is a thermostat:** Dennett (1987).

198   **what the human PFC does:** Miller and Cohen (2001).

199   **recognize these errors as errors:** Kahneman (2011).

200   **one way to get . . . *impartiality*:** For developments of this idea, see Gauthier (1987) and Boehm and Boehm (2001) on egalitarianism in hunter-gatherer societies.

200 *The Expanding Circle:* Singer (1981).

201 **does not entail abandoning one's *subjective* reasons:** Here, there may be no objective reason to favor one's self, but, for all we've said, there's no objective reason *not* to favor one's self. One might conclude that, objectively speaking, all concerned are equally entitled to be completely selfish.

201 ***empathy,* the ability to feel what others feel:** Batson, Duncan, et al. (1981); Hoffman (2000); Decety and Jackson (2004); de Waal (2010).

201 **to choose one's finger:** Smith (1759/1976), section III.3.4; Pinker (2011), 669–670; and Bloom (in press) make the same point, also citing Smith.

205 **According to John Rawls:** Rawls (1971).

206 **save a life for about $2,500:** Givewell.org (n.d.)

206 **Against Malaria Foundation:** Ibid.

207 **spend that money helping desperately needy people:** Singer (1972, 2009); Unger (1996).

207 **not the only un-splendid implications:** Utilitarianism has other famously counterintuitive implications. First, it fails to distinguish between natural and artificially generated experience (Nozick, 1974). Second, it allows sufficiently large numbers of individuals (e.g., rabbits) with minimally positive experiences to take precedence over many people living good lives, a "repugnant conclusion" (Parfit, 1984). Third, it also allows one individual (a "utility monster") with extremely high-quality experience to take precedence over many people living good lives (Nozick, 1974). I addressed these issues in my undergraduate thesis (Greene, 1997), and Felipe De Brigard (2010) makes a nice, empirically based argument along similar lines concerning the issue of artificial versus real experience. I won't say much about these issues here, because they are, in my estimation, less closely related to real-world moral issues, as they invoke premises that take us deep into the realm of science fiction, pushing the imagination past its emotional, if not conceptual, limits. For further discussion of the utility monster and the repugnant conclusion see note "the 'in principle' version of this objection."

207–08 **an abstract idea . . . specific problems:** For a general discussion of the tension between abstract and concrete thinking, see Sinnott-Armstrong (2008).

208 **hypothetical questions . . . widely underappreciated:** For an excellent discussion of the allergy to hypothetical questions, see Kinsley (2003).

208 **utilitarianism defended:** Two recent popular books—*The Moral Landscape,* by Sam Harris (2010), and *The Righteous Mind,* by Jonathan Haidt (2012)—discuss recent advances in moral psychology/neuroscience and end up favoring a version of utilitarianism, as I do in this book. Haidt does not attempt to defend utilitarianism against the objections listed above. Harris explains why utilitarianism's foundational principles are reasonable, as I do here in part 3, and as Bentham and Mill did long ago. Harris, however, pays little attention to the many compelling objections to utilitarianism listed above. Harris aims to show that science can "determine human values," but I don't think he does this, at least not in the sense that has been controversial among moral philosophers. He shows that, given an assumption of utilitarian values (which is neither supported nor undermined by science), science can determine further values. In other words, science can help us figure out what makes people happy. I'm sympathetic to Harris's practical conclusions, but in my opinion he has ignored, rather than solved, the problem he seems to want to address. Many have offered similar assessments of his book, and he has responded. See Harris (January 29, 2011). In part 4, I attempt to give utilitarianism a more thorough (though inevitably incomplete) defense, drawing on the new science of moral cognition.

PART IV: MORAL CONVICTIONS

CHAPTER 9: ALARMING ACTS

211 *accommodation* and *reform:* My use of these terms comes from Brink (2011). See also Bazerman and Greene (2010) on utilitarian accommodation.

211 **greater good . . . in the long run:** This still leaves us with the problem of endorsing such actions in principle, a problem that I take seriously. More on this later.

212 **apparent intellectual inferiority of women:** Mill (1895).

212 **too inflexible to serve as the ultimate arbiters:** These arguments resemble and build on ones made previously by Jonathan Baron (1994), Cass Sunstein (2005), Peter Singer (2005), Walter Sinnott-Armstrong (2004), and Stephen Stich (2006), among others. See also Greene (1997, 2002, 2007, under review).

212 **race of the defendant:** Baldus, Woodworth, et al. (1998); Eberhardt, Davies, et al. (2006). See also US General Accounting Office (1990).

214 **controlled for people's real-world expectations:** In these studies (Greene, Cushman, et al., 2009) we asked people versions of these three questions: In the real world, what are the odds that this attempt to save

five lives would go as planned? What are the odds that it would go worse than planned? What are the odds that it would go better? We then used people's answers to these questions to statistically control for people's real-world expectations. In other words, we asked whether we can predict people's judgments simply by knowing their real-world expectations. What we found was that we could a little bit, but not very much. It seems that when people say no to trading one life for five in these cases, it's not because of their real-world expectations. It's primarily because of the features of the dilemmas described below.

215 **"personalness" of the harmful action:** Greene, Cushman, et al. (2009). The meaning of "personal" that comes out of these more recent studies is different from the one tentatively proposed earlier (Greene, Sommerville, et al., 2001).

216 **what seems to matter is touching:** See also Cushman, Young, et al. (2006); Moore, Clark, et al. (2008); and Royzman and Baron (2002).

216 **without touching:** Of course, there's a sense in which this involves touching, namely touching with a pole.

216 **a big drop:** Greene, Cushman, et al. (2009).

217 *accommodation:* That is, utilitarianism can accommodate the fact—I assume it's a fact—that a willingness to engage in personally harmful utilitarian actions likely indicates a more general antisocial willingness to harm people. See Bartels and Pizarro (2011) on Machiavellian utilitarianism. Conway and Gawronski (2012), however, show that the Machiavellians are not really utilitarian but rather undeontological.

218 **forbidden by international law:** McMahan (2009).

218 **American Medical Association:** American Medical Association (1991)

220 **no to the *footbridge* case:** Greene, Cushman, et al. (2009).

220 *loop* case: Thomson (1985).

221 **no point in hitting the switch:** If you're thinking that this buys the five workmen more time, we can bulge out the main track in the other direction.

221 **81 percent . . . approved:** This result is consistent with Thomson's (1985) intuition, and with Waldmann and Dieterich (2007), but not with Hauser, Cushman, et al. (2007). See Greene, Cushman, et al. (2009) for an explanation.

221 **"Doctrine of Triple Effect":** Kamm (2000).

222 **87 percent . . . approved:** As of this writing, these data have not been published. This experiment was conducted concurrently with those reported in Greene, Cushman, et al. (2009), using identical methods. Testing materials and data available by request.

222 **magic combination:** If you've been paying really close attention, you'll have noticed a gap in this pattern. The *collision alarm* case gets 86 percent approval, and the *remote footbridge* case gets 63 percent approval. And yet, these are both means cases without personal force. Why the difference? It seems that there is another factor that interacts with the means/side-effect distinction: whether or not the victim gets dropped from a footbridge. If you do a dropping version of the *collision alarm* case, the approval ratings drop to roughly the level of *remote footbridge*. But if you do a dropping version of the *switch* case, the drop has little to no effect. More generally, it seems that multiple "forcey" factors (force of muscles, force of gravity) interact with the means/side-effect factor. Even more generally, it seems that the effect of the means/side-effect factor does not depend entirely on the presence of personal force. But it does depend on the presence of some other factors, as demonstrated by the *collision alarm* and *loop* cases.

223 **bound up with . . . personal force:** But not completely bound up. See previous note.

223 **no knowledge of the doctrine:** Cushman, Young, et al. (2006); Hauser, Cushman, et al. (2007).

223 **intuitions that justify the principle:** Cushman and Greene (2011).

224 **when we contemplate harming:** This idea is similar to Blair's earlier idea of a violence inhibition mechanism (Blair, 1995). Cushman (in press) has a model according to which the emotional response to intentional personal harm is triggered not by a dedicated alarm system but by a learned negative emotional response encoded within a more general emotional learning system (more specifically, a "model free" learning system). Cushman's model preserves the critical features of what I'm here calling the myopic module. First, the emotional response is blind to side effects (myopia) and, for the reasons given here, related to the analysis of action plans. Second, this system's internal operations are not accessible to introspection. That is, they are "informationally encapsulated" (modularity). But if Cushman is right—and I suspect that he is—this system is not specifically dedicated to serving the function that it serves here. That said, one can think of this learned association as a kind of acquired module.

225 **as Hobbes observed:** Hobbes (1651/1994).

225 **smash his head in with a rock:** It's not clear whether other species, such as chimps, can engage in this kind of premeditated violence. Chimps go on raiding parties, killing members of neighboring groups, but whether these killings are performed with a conscious goal in mind is unclear. They may be more like animal migrations—functional, complicated, and socially coordinated but not consciously carried out with a purpose.

226 **dangerous for . . . the attacker:** DeScioli and Kurzban (2009).

226   **how you treat others:** Dreber, Rand, et al. (2008).

226   **contemplating an act of violence:** Blair (1995).

227   **at least somewhat "modular":** Fodor (1983).

228   **Mikhail . . . Goldman . . . Bratman:** Mikhail (2000, 2011); Goldman (1970); and Bratman (1987).

230   **approving more of . . . harmful side effects:** Schaich Borg, Hynes, et al. (2006); Cushman, Young, et al. (2006).

233   **sounds the alarm:** That is, the *between-dilemma* variability is determined by the strength of the automatic responses. But in fact, the evidence suggests that much of the *within-dilemma* variability is determined by individual differences in reliance on manual mode. See Paxton, Ungar, and Greene (2011) and Bartels (2008).

234   **adding some pushing to the loop case:** At the same time, adding a push to *loop* does appear to have *some* effect, which complicates things for the modular myopia hypothesis. Adding a push to *switch* has little or no effect, and, ideally, the same would be true for adding a push to *loop*. We do see an effectfully consistent with the modular myopia hypothesis when it comes to adding *drops* (dropping onto the tracks from a footbridge through a trapdoor). That is, adding a drop to *switch* has little or no effect, and adding a drop to *loop* has little or no effect, but adding a drop to *collision alarm* significantly lowers approval ratings. This is all a work in progress, and it is not entirely clear to me what is going on. The critical point, for now, is that personal force and dropping do not seem to supply a fully adequate account of the gap between *footbridge* and *loop*. For now, I'm putting these unresolved ambiguities aside, because my purpose in this section is to lay out the modular myopia hypothesis as a *hypothesis* and not as a theory that currently explains everything.

236   **why . . . chain with the harm . . . *secondary* one?:** The secondary causal chain is secondary because it is parasitic on the primary causal chain. The turning of the trolley away from the five makes sense as a goal-directed action all by itself, without reference to the secondary causal chain, that is, to what happens after the trolley is turned. But the secondary causal chain cannot stand alone. This is because the secondary causal chain, to make sense as a complete action, must extend all the way back to the body movement, which is the hitting of the switch. But the hitting of the switch makes sense only with reference to the primary causal chain, that is, to the fact that the unturned trolley will proceed down the main track and kill the five if nothing is done.

236   **a means that is *structured like a side effect*:** Kamm (2000) refers to this kind of structure as a case of "triple effect," one in which there is a foreseen event that is recognized as causally necessary for the achievement of the goal and yet is, in some morally relevant sense, not intended.

238   **cost-benefit . . . sufficiently compelling:** Nichols and Mallon (2006); Paxton, Ungar, and Greene (2011).

239   **nested multitasking:** Koechlin, Ody, et al. (2003).

240   **"Doctrine of Doing and Allowing":** Howard-Snyder (May 14, 2002).

242   **choose between pairs of objects:** Feiman et al. (in prep.).

242   **infants' brains represented . . . experimenter wanted:** This effect was first demonstrated by Woodward and Somerville (2000).

244   **people evaluated both active and passive harmful actions:** Cushman, Murray, et al. (2011).

244   **ignoring . . . requires more manual-mode DLPFC activity:** An earlier study provided more ambiguous evidence. Cushman, Young, et al. (2006) had people evaluate harmful actions and harmful omissions and then justify their ratings. About 80 percent of the time, people who distinguished between actions and omissions in their ratings were able to justify their judgments with an explicit appeal to the action/omission distinction. However, this means that about 20 percent of these people did what they did without knowing what they were doing. Clearly, these people were not consciously applying the action/omission principle in manual mode. It's not clear whether some or all of the 80 percent were doing so, or if they became conscious of the action/omission principle only after drawing the distinction intuitively. The brain-imaging data suggest the latter.

244   **tongue, fingers, and feet:** Hauk, Johnsrude, et al. (2004).

245   **jacking up the price . . . feels less bad if done indirectly:** Paheria, Kasam, et al. (2009).

246   ***specifically intended:*** There is a technical problem with calling any of these harms "specifically intended." For example, in the *footbridge* case, one might say that what is specifically intended is using the man's body to block the trolley, which does not *logically* entail any harm to the man. (What if it's Superman?) By this interpretation, the death of the man and the pain he experiences are just contingent *side effects*, unfortunate by-products of using the man's body to stop the trolley. While this interpretation is possible in principle, this is clearly not how our brains represent these events. Thus, there is an interesting psychological problem here: namely, to understand the mechanism that parses events in these contexts.

247   **explanation of our sensitivity to the action/omission distinction:** Cushman, Young, et al. (2006) have found effects of means versus side effect for passive harms, but these are likely cases in which the omission is unusually purposeful, very specifically failing to do what one would ordinarily do in order to save more lives. Thus, it may be possible, but unusual, to have omissions as part of an action plan, as in a recipe ("Do not remove the fritters from the oven until they are golden brown"). Also, it's worth noting that the means/side-effect effect is much weaker for the omission cases.

247   **represent causes in terms of *forces*:** Talmy (1988); Wolff (2007); Pinker (2007).
247   **hitting, slapping, punching:** This doesn't mean that it can't learn to respond to other kinds of violence, such as gun violence. It's possible that guns are sufficiently familiar that we incorporate the explosive force of a gun into the body schema, conceptualizing it as a force that we personally control. The same may also happen with gravity. These are interesting empirical questions. For a fascinating and, I predict, very important theory of how we learn to recoil at certain kinds of harms, I recommend Cushman ("Action, Outcome").
247–48 **hard . . . to think of actions that don't feel violent:** The best candidate that I can think of is surgery, but surgery *does* feel violent. It's just that surgeons learn to get over that feeling (if they're not psychopathic), and we don't blame them for what they do, because we know that their actions are for the good of the patient.
248   **millions of people can be saved by pushing:** Paxton, Ungar, and Greene (2011). See also Nichols and Mallon (2006). The 70 percent figure comes from unpublished data using the same methods as Greene, Cushman, et al. (2009).
248   **would all be more psychopathic:** Bartels and Pizarro (2011); Glenn, Raine, et al. (2009); Koenigs, Kruepke, et al. (2012).
249   **inferences about moral character:** Pizarro, D.A. and Tannenbaum, D. (2011). Bringing character back: How the motivation to evaluate character influences judgments of moral blame. In M. Mikulincer & Shaver, P. (Eds) The Social psychology of morality: Exploring the causes of good and evil. APA Press.
253   **abortion and the Doctrine of Double Effect:** Foot (1967).
253   **If harming the environment felt like pushing:** See also Gilbert (July 2, 2006).

CHAPTER 10: JUSTICE AND FAIRNESS

255   **a dollar spent in the right way:** Givewell.org (n.d.); Sachs (2006); Singer (2009).
255   **more accountable than ever:** www.givewell.org.
256   **"decency to admit that I'm a hypocrite":** Exchange described by Simon Keller.
258   **neighbors are already doing it:** Cialdini (2003).
258   **moral problem as . . . originally posed:** Singer (1972). I've made some minor modifications to Singer's thought experiment.
259   **hard to justify treating the . . . faraway starving child differently:** Jamieson (1999).
260   **"trolleyology" of Peter Singer's problem:** Manuscript in preparation, based on Musen (2010). Much of the work done in these experiments was based on thought experiments carried out by Peter Singer (1972) and Peter Unger (1996).
261   **heavily influenced by mere physical distance:** Nagel and Waldmann (2012) claim that mere physical distance does not matter and that the relevant factor is informational directness. However, my experiments with Musen show effects of mere distance independent of informational directness. In any case, it would be hard to argue that informational directness per se is a normatively significant factor. Therefore, the main conclusions reached here would not change if Nagel and Waldmann's conclusions are correct.
261   **faraway starving children don't:** Note that in Trolleyland, spatial distance didn't seem to matter much, whereas here it does. This is likely because we're dealing with a different automatic setting, one that responds to preventable harm rather than actions that cause harm. It could also be because the distances in these two types of cases differ by at least two orders of magnitude.
262   **from cultural evolution:** Pinker (2011); Henrich, Ensminger, et al. (2010).
262   **unidentified "statistical" people:** Some aid organizations deliberately pair individual donors with individual recipients in order to make the experience more personal.
262   **economist Thomas Schelling:** Schelling (1968).
262   **"Baby Jessica":** Small and Loewenstein (2003); *Variety* (1989).
262   **"statistical death":** Schelling (1968); Small and Loewenstein (2003).
263   **identifiable . . . "statistical," victims:** Ibid.
264   **one sick child . . . or *eight*:** Kogut and Ritov (2005).
264   **numbers as small as *two*:** Slovic (2007). Note: My recommendations mirror Slovic's specific suggestions about how we might change our approach to the world's needy.
265   **give us legitimate moral obligations and options:** Smart and Williams (1973).
265   **lives defined by relationships . . . must take this . . . into account:** See Sidgwick (1907), 434.
266   **counterproductive to pooh-pooh philanthropists:** Note: See Sidgwick (1907), 221, 428, and 493.
268   **if only they knew:** A. Marsh, personal communication, January 31, 2013.
268   **Wesley Autrey:** Buckley (January 3, 2007).
268   **only human:** Note: See Parfit (1984) on blameless wrongdoing, 32.
268   **utilitarian rationale for punishing people:** Bentham (1830).

268   **favored by many:** Kant (1785/2002).

268   **a little extra justice before pushing off:** Cited in Falk (1990), 137.

270   **Prisoners are frequently sexually abused:** Mariner (2001); Gaes and Goldberg (2004).

271   **is barbaric, we say:** You might object that this is not a fair comparison, because prison rape as it occurs now is a chancy affair, while state-sanctioned rape would be a sure thing. Fair enough. We can use a roulette wheel to introduce an element of chance into our official state raping policy. Now do you like it?

271   **our current criminal justice system, which is highly retributive:** Tonry (2004).

272   **natural function of punishment is quasi-utilitarian:** I say "quasi-utilitarian," however, because our sense of justice is not necessarily designed to make us happier. The Us who benefits from punishment does not necessarily include everyone, and our sense of justice does not necessarily weight everyone's well-being equally. Still, overall, the existence of punishment is a good thing, by utilitarian standards or any reasonable standard.

272   **obvious utilitarian answer:** Carlsmith, Darley, et al. (2002).

272   **punishing based solely on how they *feel*:** Baron and Ritov (1993); Carlsmith, Darley, et al. (2002); Kahneman, Schkade, et al. (1998).

273   **Crimes with lower detection rates:** Carlsmith, Darley, et al. (2002).

273   **people punished "determined" transgressors about twice as much:** Small and Loewenstein (2005).

273   **presented . . . with . . . "deterministic" universe:** Nichols and Knobe (2007).

274   **abstract judgment goes out the window:** See also Sinnott-Armstrong (2008).

275   **maximizing happiness can lead to gross injustice:** Rawls (1971), 158–161.

276   **one-to-one ratio is a *conservative* assumption:** The more typical situation, historically, is that one slaveholder owns many slaves. This only makes it harder for slavery to maximize happiness. One could imagine, however, going in the opposite direction, with a "time share" arrangement in which, say, five people own a slave collectively. But this doesn't change the fundamental math described below. This would mean, let's say, that each slave-share owner gets the equivalent of $10,000 additional income per year. Would you be willing to spend a fifth of your life as a slave in order to get a $10,000 raise?

277   **additional income . . . adds relatively little to one's happiness:** Easterlin (1974); Layard (2006); Stevenson and Wolfers (2008); Easterlin, McVey, et al. (2010). As noted earlier (see chapter 6, "making a bit less money"), the debate is about whether additional income for those already well-off adds nothing, or relatively little, to one's happiness.

277   **not a tentative finding:** Rawls (1971, 158–161) suggests that it is. But he was writing before research on happiness took off.

278   **"utility monsters":** Nozick (1974).

278   **no goods to be extracted . . . outweigh the horrors:** Still not convinced? Let's try a bit harder to think of realistic examples of utility-maximizing oppression. What about the old transplant case (Thomson, 1985)? What if a single healthy body could provide lifesaving organs to twenty people? Would a utilitarian allow us to randomly kidnap people and kill them for their organs, assuming this maximizes happiness? No, because there are clearly better alternatives. Before resorting to kidnapping, which would cause widespread panic and grief, we could establish a legal market for organs. You may or may not think this is a good idea, but it's not grossly unjust, à la slavery. Reasonable people can disagree about whether there ought to be a well-regulated market for human organs.

         What about the kind of oppression in which oppressing one person can benefit thousands of people? What about the crowds who cheer as lions tear into the entrails of hapless gladiators? Or people who enjoy child pornography? If there are enough gleeful onlookers, can such suffering be justified? Only if you think it's a net gain when people take joy in the exploitation and suffering of others. We can imagine a hypothetical world in which taking joy from the suffering of innocent people has no detrimental effects, but that's not the real world.

280   **easy to mistake utilitarianism for "wealthitarianism":** Greene and Baron (2001).

281   **counting dollars, rather than happiness:** Rawls (1971, 158–161, 167–168) is aware of the argument that people exhibit diminishing marginal returns of utility from goods and that therefore utilitarian policies will tend to favor egalitarian outcomes. But he dismisses this argument as providing insufficient moral assurance. In doing this, he makes two assumptions. First, even if utilitarianism is generally egalitarian, it will sometimes favor social inequality. Second, sometimes the inequalities that utilitarianism favors will be morally repugnant, à la slavery. Rawls is right about the first assumption (see above), but he takes this first assumption as license to make the second assumption, which is highly doubtful, at least in the real world. Rawls thinks the second assumption is also reasonable, but that, I claim, is because he's making the same error as everyone else: confusing utility with wealth.

283   **pattern . . . predicted . . . based on reading Rawls:** Rawls's error is committed most starkly when he argues (Rawls, 1999, 144) that people *should* be risk averse with respect to utility, implying that some utility is worth more (i.e., has more utility) than other utilities.

284 **the "in principle" version of this objection:** Suppose that it really were possible to maximize happiness by oppressing some people. Wouldn't that still be wrong? And doesn't that show that there's something rotten at the core of utilitarianism? Here the classic example is Nozick's (1974) utility monster, which I alluded to above. Nozick's utility monster gains enormous quantities of happiness by eating people. But it seems that it would be wrong to feed innocent people to the utility monster, even if doing so would, by hypothesis, maximize happiness. Another famous case comes from Derek Parfit (1984), who imagines a choice between two types of worlds: one in which very many people are very happy and one in which many, many more people lead lives that are "barely worth living." The "repugnant conclusion" that follows from utilitarianism is this: No matter how good the world is, there is always a better world, consisting of many, many more individuals whose lives are only minimally good. To drive the point home further, one can even substitute animals for people, so long as we agree that animal experience counts for something. One can imagine an immense warehouse full of trillions of rabbits whose brains are hooked up to stimulators that intermittently produce mild levels of rabbit gratification. What each rabbit gets is not that great, but there are *so many rabbits*. Thus, utilitarian revolutionaries could, in principle, justify destroying our world in order to realize their dream of building an enormous rabbit gratification factory. This prospect strikes most people as unjust. (Of course, no one has bothered to ask the rabbits!)

I have two responses to these "in principle" objections. First, once again, I'm not claiming that utilitarianism is the moral truth. Nor do I claim that it perfectly captures and balances all of human values. My claim is simply that it provides a good common currency for resolving real-world moral disagreements. If the utility monsters and the rabbits ever arrive, demanding their utilitarian due, we may have to amend our principles. Or maybe they would have a good point, albeit one that we have a hard time appreciating.

Which brings me to my second response to these "in principle" objections. We should be very wary of trusting our intuitions about things that defy intuitive comprehension. The utility monster and the rabbits both push our intuitive thinking beyond its limits. More specifically, they push along orthogonal dimensions: *quality* and *quantity*. The utility monster is a single individual (low quantity) with an incomprehensibly high *quality of life*. He gets more out of a single meal than you get out of your entire existence. The rabbits, by contrast, have a rather low quality of life, but the *quantity* of rabbits defies intuitive comprehension. Of course, there is a sense in which we can understand these things. After all, I just described them to you, and you understood my description. But it's your *manual mode* that is doing the understanding. You cannot understand *intuitively* what it's like to eat a meal that produces more happiness than an entire happy human lifetime. Likewise, you can't *intuitively* distinguish between a million rabbits and a trillion rabbits. We can think about these things in an abstract way, but asking us to have gut feelings about such things is like asking a bird to imagine a worm that's a mile long.

284 **Rawls's argument from the "original position":** If you know your Rawls, you'll have noticed that I didn't actually address his official argument against utilitarianism. Rawls argues that the most just organizing principles for a society are those that people would choose from behind a "veil of ignorance," not knowing which positions in society they will occupy. And he argues that people in this "original position" would not choose a utilitarian society, because the possible downside of living in a utilitarian society is too great. In other words, Rawls's official argument depends on the same mistaken assumption described above, which is that a utilitarian society could be oppressive in the real world, with human nature as it actually exists. Rawls's argument also involves some serious fudging related to risk aversion and the structure of the original position. For more on this, see chapter 11, "central argument in A Theory of Justice. . . ."

PART V: MORAL SOLUTIONS

CHAPTER 11: DEEP PRAGMATISM

289 **Ten percent . . . control 70 percent:** Davies, Shorrocks, et al. (2007). See also Norton and Ariely (2011).
289 **Alex Kozinski:** Alex Kozinski and Sean Gallagher, "For an Honest Death Penalty." *New York Times*, March 8, 1995.
292 **second moral compass:** Thanks to Scott Moyers for suggesting the "two compasses" metaphor.
295 **how the brain gets out of this pickle:** Botvinick, Braver, et al. (2001). This theory is somewhat controversial, but that need not concern us here. Our interest is in the cognitive strategy, regardless of whether the brain actually uses it. That said, I know of no alternative solution to the regress problem described above.
296 **engage the ACC and DLPFC:** Greene, Nystrom, et al. (2004); Greene and Paxton (2009); Cushman, Murray, et al. (2011).
297 **wiser when we acknowledge our ignorance:** Plato (1987).
297 **"illusion of explanatory depth":** Rozenblit and Keil (2002); Keil (2003).
297 **applied this idea to politics:** Fernbach, Rogers, et al. (in press).

297   **left their strong opinions intact:** The demand for reasons did moderate some people's views, but these tended to be people who couldn't produce any reasons at all when asked.

297   **may even do the opposite:** Tesser, Martin, et al. (1995).

297   **an alternative approach to public debate:** Sloman and Fernbach (2012); Fernbach (May 27, 2012).

298   **men crossing two different bridges:** Dutton and Aron (1974).

298   **make up a plausible-sounding story and go with it:** For a classic demonstration of this kind of interpretive effect, see Schachter and Singer (1962).

298   **not an isolated phenomenon:** Bargh and Williams (2006); Wilson (2002).

298   **choose . . . panty hose:** Nisbett and Wilson (1977).

299   **"change into my work clothes":** Stuss, Alexander, et al. (1978).

299   **"split-brain" patients:** Gazzaniga and Le Doux (1978).

300   **plausible narrative:** Bem (1967); Wilson (2002).

300   **consummate moral rationalizers:** Haidt (2001, 2012).

300   **"Concerning Wanton Self-Abuse":** Kant's "Concerning Wanton Self-Abuse" is a section in the *Metaphysics of Morals* originally published in 1797. See Kant (1994).

301   **using someone as a means:** Ibid.

301   **"Kant's joke":** Nietzsche (1882/1974).

301   **"born slaves":** Bernasconi (2002).

301   **Rationalization . . . enemy of moral progress, and thus of deep pragmatism:** The argument made in this section, along with other parts of this chapter, was originally made in Greene (2007).

302   **"outweighed" by the rights of the five:** Thomson (1985, 1990). Note that Thomson has changed her mind and now thinks that it's wrong to turn the trolley (Thomson, 2008). What this essentially does is put the rights theorist's explanatory burden on the act/omission distinction—that is, unless one thinks that we're obliged to actively turn the trolley away from the one and onto the five.

303   **we have no duty to save them:** Jamieson (1999).

303   **The rights and the duties follow the emotions:** Kahane and colleagues (2010, 2012) have argued that there is no special relationship between automatic emotional responses and characteristically deontological moral judgments, and that the appearance of such a relationship is the product of a biased selection of stimuli. For evidence to the contrary, see chapter 6, "connection between manual-mode thinking," on the "white lie" case, and Paxton, Bruni, and Greene (under revision).

303   **sexiness is in the mind of the beholder:** This doesn't mean that perceptions of sexiness are *arbitrary*. As evolutionary psychologists have pointed out (Miller and Todd, 1998), what we find sexually attractive is typically indicative of high reproductive potential. But the fact that sexual attraction is non-arbitrary and biologically functional does not imply that it's *objectively correct*. There's no meaningful sense in which we're objectively (absolutely, nonrelatively) correct about who's sexy while baboons are objectively incorrect—or vice versa.

304   **cognitive apparatus . . . concrete objects and events:** Lakoff and Johnson (1980).

304   **nonnegotiable facts:** Of course, there are facts about which rights and duties are granted by law, but in the midst of a moral controversy, such legal facts almost never settle the question. Public moral controversies are about what the law *ought to be*, not about what it is.

305   **not arguments, but weapons:** In a provocative paper, Mercier and Sperber (2011) claim that reasoning is just one big weapon for persuading others to do what we want. This strikes me as highly implausible. What makes their argument go is that they exclude from the category of "reasoning" all of the boring, everyday things for which we use our manual modes, such as figuring out the best order in which to run one's errands. ("I'd better go food shopping last or else the ice cream will melt in the car.") This argumentative theory of reasoning also makes little evolutionary sense. Reasoning did not emerge de novo in humans. Indeed, the neural structures that we use for reasoning are the same ones that our primate relatives use to solve their own (fairly) complex problems. However, chimps and macaques very clearly do not engage in persuasive verbal jousting.

306   **Dershowitz once told a handful:** This was told to me and other undergraduates at a "meet the professor" lunch in 1994. The details are as close as I can recall.

306   **costs of lavishing time and attention:** To spell out this point more explicitly: Dershowitz's response was clever because it distinguished the benefits from the costs. He essentially said: I'm not refusing to debate you because I'm afraid. I'm refusing to debate you because there are costs associated with taking cranks like you seriously. But if you're willing to debate me in a way that denies you the credibility that you seek (the cost), then I'm happy to have a free exchange of ideas (the benefit).

307   **good . . . to reject some ideas out of hand:** See also Dennett (1995) on "good nonsense."

309   **truly moral considerations on both sides:** Here, by "truly moral" I mean not just tribalistically moral.

309   **those of Peter Singer:** Singer (1979), chap. 6; Singer (1994).

310   **manual-mode scrutiny:** For arguments along the lines of those made here, see Singer (1979), chap. 6; Singer (1994).

310   **late-term abortions are not:** If you're okay with late-term abortion, what about infanticide? Most of the arguments below apply equally well.

310   **Both early- and late-term abortions prevent a human life:** The odds of living may differ, but surely this difference in the odds of successful birth (say, 60 percent vs. 95 percent) can't be the difference between having a right to life or not. And what if a late-term fetus, for some reason, had the same odds as a typical early-term fetus? Now would it be okay to abort it?

310   **Viability is as much a function of technology:** You might say that what matters is the ability to survive without special technology. If that's right, then what about a nine-month-old fetus that, due to an atypical medical condition, can survive outside the womb, but only with the temporary help of readily available technology? Is it okay to abort that late-term fetus simply because it's not viable without technology?

310   **born as early as twenty-two weeks can survive:** Stoll, Hansen, et al. (2010).

310   **thanks to new technology . . . first-trimester abortions have become immoral?:** Perhaps you're inclined to say yes. After all, you might think that the possibility of being kept alive from that stage of development is morally significant. But note that it's not that technology enables the fetus to survive from that stage of development. Rather, technology enables the fetus to survive from that stage *outside the womb*. The fetus is already able to survive from that stage of development without the fancy technology. It just has to stay inside the womb! We already have the "technology" to keep early-term fetuses alive.

311   **more than being a moral vegetarian:** At least it requires giving up certain kinds of meat. One could perhaps make room for certain other kinds of meat, such as that coming from animals that lack the relevant features.

311   **pro-choicers are unwilling to go that far:** And even then, it's not clear that the argument works. Animal rights activists typically focus on the suffering that animals experience while being raised for food. If that's the reason why it's wrong to eat meat, then the same argument would not apply to late-term abortion, as long as the abortion process did not involve suffering, or did not involve a lot of suffering.

312   **Deanna Troi . . . not human:** Okay, okay: not *fully* human. She's only half Betazoid. But the point applies to her maternal relatives. *Jeez.*

312   **can move their bodies:** Dongen and Goudie (1980).

314   **which human shall be, if anyone is to be:** Of course, with only one sperm on deck, the odds of fertilization are lower, but so what? Pro-lifers would not allow us to abort a zygote simply because it has, for whatever reason, low odds of surviving.

314   **robbed an innocent person of his life?:** Moreover, the idea that conception determines identity seems to have more to do with our limited knowledge than with facts about what has or has not been determined. When a couple sets out to conceive a child the old-fashioned way, they may not know which sperm and egg are going to join, and we may have no way of knowing. But whichever child is going to result from a given act of sexual intercourse, *that's* the child that's going to result. And if the couple decides not to go through with it, it's *that child* who will not exist as a result of their backing out. When this happens, no one knows, or will ever know, who "that child" is, but so what? If their having sex would have led to some specific child's existing, then their refraining from having sex led to the nonexistence of that specific child. (I'll refrain from getting into the problem of determinism at this point.)

314   **full biology lesson:** Gilbert (2010), 6, 14, 123–158, 301.

316   **"I think even if life begins in that horrible situation of rape":** Madison (2012).

317   **campaign went up in flames:** Haberkorn (2012).

317   **"problem of evil":** Tooley (2008).

318   **silent drama:** Heider and Simmel (1944).

318   **attributions happen so automatically:** Heberlein and Adolphs (2004).

319   **animals . . . move and have eyes:** Many of us would have a hard time killing the animals we eat, but that's probably just because we're not used to it. Our ancestors did this for millions of years.

320   **Most tribes believe in souls:** Bloom (2004).

327   **what's right and wrong when it comes to life and death:** Beauchamp, Walters, et al. (1989); Baron (2006); Kuhse and Singer (2006).

327   **less optimistic . . . about . . . sophisticated moral theory:** See also Greene (under review).

328   **"reflective equilibrium":** Daniels (2008).

328   **"considered judgments":** Gut reactions are not the same as "considered judgments," but they play a dominant role in determining them.

330   **Aristotle . . . great champion of common sense:** Aristotle (1941).

330   **Aristotle is essentially a tribal philosopher:** MacIntyre (1981).

330   **Aristotelian virtue theory . . . revival:** As part of the Aristotelian revival, I include not just virtue ethics proper (Crisp and Slote, 1997; Hursthouse, 2000), but also "sensibility" theories (Wiggins, 1987),

particularism (Dancy, 2001), and the like—all of the approaches to normative ethics that have given up on discovering or constructing explicit moral principles that tell us what to do. The revival is due in large part to Alasdair MacIntyre (1981), who has a similar diagnosis of modern moral problems but thinks that a revamped form of virtue theory is the best we can do, following on the failures of Enlightenment moral theory.

332 **cannot be "universalized":** Kant (1785/2002).

332 **Kant's argument requires impossibility:** Kant's universalization argument is not simply a version of the familiar "What if everyone did that?" argument. He's not merely saying that it would be *bad* if everyone were to lie, break promises, et cetera. That's a *utilitarian* argument against lying—rule-utilitarian or act-utilitarian, depending on how you interpret it. This isn't good enough for Kant, because he wants an absolute prohibition against lying, one that doesn't depend on how things happen to work out in the real world. (Things tend to go badly if everyone lies, et cetera.) He wants morality to be like math: necessarily true and knowable with certainty. See Korsgaard (1996), chapter 3.

333 **well aware of the flaws . . . they have their replies:** See, for example, Korsgaard (1996).

333 **central argument in *A Theory of Justice* . . . essentially a rationalization:** While there is much to admire about Rawls's work, and the man himself, I believe that his central argument is essentially a rationalization, an attempt to derive from first principles the kind of practical moral conclusions that he intuitively favors, which he mistakenly believes to be at odds with utilitarianism. (See pp. 279–84.) Rawls's main argument is laid out in chapters 1, 2, and 3 of *A Theory of Justice* (1971).

I mentioned earlier that utilitarianism begins with two very general moral ideas. First, happiness is what ultimately matters and is worth maximizing. Second, morality must be impartial. Essentially, Rawls keeps the impartiality assumption but drops the assumption that happiness is what ultimately matters. He replaces the assumption that happiness is inherently valuable with the assumption that *choice* is inherently valuable. Thus, for Rawls, the best organizing principles for a society are the ones that people would choose if they were to choose impartially. This is a great idea, with roots in the philosophies of both Kant and John Locke. (Rawls, like Locke, is a "contractarian.")

So how do we figure out what people would choose if they were to choose impartially? To answer this question, Rawls constructs a thought experiment. He imagines a situation, called the original position, in which it's *impossible* to choose in a directly self-serving way and then asks what people would choose. Choosing in a straightforwardly selfish manner is impossible in the original position, because one chooses from behind a *veil of ignorance*. That is, the parties in the original position must negotiate an agreement about how their society will be organized without knowing their own races, genders, ethnic backgrounds, social positions, economic statuses, or the nature and extent of their natural talents. The idea, then, is that the negotiators have been denied all of the information they could use to *bias* the agreement in their respective favors. The decision makers are expected to choose rationally and selfishly, but because they are choosing from behind a veil of ignorance, the kind of social structure they choose is, according to Rawls, necessarily fair and just. Agreeing on a social structure from behind a veil of ignorance is rather like using the "I cut, you choose" method for dividing a piece of cake. The fairness emanates from the decision procedure rather than from the goodwill of the decision makers.

This core idea (modeling social choice as bias-free selfish choice) was developed independently, and slightly earlier, by the Hungarian economist John Harsanyi (1953, 1955), who would later win the Nobel Prize in economics for his contributions to game theory. Harsanyi, unlike Rawls, saw his version of the original position as providing a rational grounding for utilitarianism. Harsanyi imagined people choosing their society's organizing principles while not knowing which positions in society they would occupy (rich or poor, etc.) but knowing that they would have an *equal probability* of occupying any position in society. Given this assumption, if people are utility maximizers (each seeking to maximize his/her own happiness), the decision makers will choose a society organized so as to maximize utility, one that is as happy as possible overall. (This maximizes both the *total* and the *average* amount of happiness, assuming that the population size is fixed.)

Rawls, however, argued for a very different conclusion about the kind of society that selfish people in the original position would choose. Rawls says that the original positioners would choose a society organized by a "maximin" principle rather than a utilitarian principle. The maximin principle ranks societies based solely on the status of the society's least well-off person. According to this principle, one's preferences for one societal arrangement over another will be based entirely on the "worst-case scenario" within each arrangement. Rawls acknowledges that this is not, in general, a good decision rule, as illustrated by the following example.

Suppose that you are buying a car, but in the following unusual way. You must buy a lottery ticket that will give you one car randomly chosen from a lot of a thousand cars. Ticket A takes you to a lot that has a thousand mediocre cars. On a scale of 1 to 10, each of these cars is a 4. If you buy ticket A, you get one of those. Ticket B takes you to a lot that also has a thousand cars. This lot has 999 cars that score a perfect 10 on your scale; however, it also has one car that scores a 3. So if you buy ticket B, you have a 99.9 percent chance of getting your dream car, but you have a 0.1 percent chance of getting a car that's okay but slightly worse than the one you're guaranteed to get with ticket A. Which do you choose? Obviously, you would choose ticket B. However, according to the "maximin" rule, you would choose ticket A, because the worst-case scenario in buying ticket A is better than the worst-case scenario in buying ticket B. Not so smart.

The problem with the maximin rule is that it's maximally *risk averse*. Rawls agrees that such risk aversion is not appropriate in general (e.g., when buying cars by lottery), but he argues that it *is* appropriate for people who are choosing their society's organizing principles without knowing which positions in society they will occupy. Rawls thinks that life in a utilitarian society might be "intolerable" (pp. 156, 175). If you're randomly plopped into a utilitarian society, Rawls warns, you could end up as a slave. Such outcomes are so bad that no one would choose to take that risk. Instead, people choosing from behind the veil of ignorance would use the maximin rule, favoring the society with the best worst-case scenario. Rawls makes this argument with respect to what he calls "basic liberties." Rather than leave the allocation of liberties up to utilitarian calculations, the people in the original position would choose principles that directly secured "basic liberties." He makes the same kind of argument about educational/economic opportunities and about economic outcomes. Here, too, says Rawls, the worst-case scenario in a utilitarian society could be so bad that it's not worth taking the risk.

First, let's note that Rawls's formal argument depends on the error described in the last chapter, confusing wealth and utility. More specifically, Rawls assumes that the people in the original position would make the same mistake that he makes. Once again, the reason for favoring the maximin rule is that the worst-case scenario in a utilitarian society might be "intolerable." It's not hard to see how the worst-case scenario in a *wealthitarian* society would be intolerable. Maximizing GDP might require some oppression, but, as explained earlier, it's simply not plausible that making the world as happy as possible could, in the real world, require oppression. Human psychology would have to be completely rewired such that the suffering caused by being a slave is smaller than the benefits one derives from owning a slave, and so on. (Once again, would you give up half your life to slavery in order to have a slave for the other half? Could you imagine a situation in the real world in which this is a tough decision?)

That's Rawls's first mistake. (I don't think this is a *rationalization*. I think it's just a mistake.) But now let's suppose that Rawls is right and that life in a maximally happy society could be, for some, "intolerable." Even if you make this implausible assumption, Rawls's argument still doesn't work. Once again, Rawls's maximin rule evaluates each societal arrangement based solely on its worst-case scenario—the quality of life of that society's least well-off person. In other words, Rawls assumes that people will be maximally risk averse so long as there are intolerable outcomes in the mix. But as Harsanyi (1953, 1955) and others have pointed out, that is simply not a reasonable assumption. Every time you get in a car, you increase your risk of being horribly maimed in a car accident, an outcome that most of us would regard as "intolerable" in Rawls's sense. And yet we accept such risks for things as trivial as a late-night pint of ice cream. (You might point out that one can be horribly maimed by staying home. For example, the roof could collapse. Thus, the worst-case scenario is actually the same whether you make your ice cream run or not. That's fine, but then you have to apply the same logic to Rawls's argument. Life could be "intolerable" even in a society governed by maximin. Your roof could collapse.)

To prime the risk aversion pump, Rawls adds an unnecessary twist to his version of the original position. In Harsanyi's version, if you recall, the decision makers choose while knowing that they will have an *equal probability* of occupying each position in society. Rawls, however, does something different. He assumes that people in the original position have no information at all about the range of possible outcomes and their associated odds, leaving them in a state of complete actuarial ignorance. In this state of maximal ignorance, Rawls argues, the people in the original position would, and should, be highly risk averse. ("Anything could happen!") In technical terms, Rawls makes the original position an *ambiguous* situation rather than merely an *uncertain* one.

Why does Rawls make the decision in the original position maximally ambiguous? Why not simply assume, as Harsanyi does, that the people in the original position know the range of possible social positions and know that they have equal odds of landing in any one of them? Rawls addresses this issue, and as far as I can tell, his argument is completely circular. He *defines* the original position as one in which people have no information about the attendant probabilities, and then argues, on that assumption, that they should not rely on probability estimates, because, really, they have no way of knowing what the probabilities are (pp. 155, 168–169). As Harsanyi (1975) points out, even under this assumption of complete ignorance, an assumption of equal probability for all outcomes would be far more rationally defensible than the assumption that the worst outcome has an effective probability of 100 percent—the assumption built into the maximin rule. But we can put that aside. Why, in the first place, should Rawls define the original position as one in which the outcome probabilities are unknown? The whole point of the original position is to constrain the choice so that the decision makers are effectively impartial. To be impartial is to give equal weight to each person's interests. Thus, it makes perfect sense to *define* the original position as one in which each chooser knows that she has an equal probability of occupying each position in society. This would in no way bias people's choices. On the contrary, it embodies the ideal of impartiality in the clearest possible way.

As far as I can tell, Rawls makes the probabilistic structure of the original position maximally ambiguous for reasons that have nothing to do with justice or fairness or impartiality. As far as I can tell, this is just a fudge, an ad hoc attempt to make his intuitively correct answer more plausible. Coming into the world of political philosophy, Rawls has no particular reason to adopt an extreme theory of risk aversion. But once he's committed to the original position as a device for working out a theory of justice—which is a very nice idea— he suddenly finds himself in an awkward position. He wants a society in which priority is given, as a matter of first principles, to people with the worst outcomes. But this desire, filtered through the logic of his thought experiment, requires Rawls's hypothetical selfish decision makers to be inordinately preoccupied with the worst outcomes as they make their self-interested choices. That is, he needs them to be inordinately risk averse. And thus, to get his desired result, Rawls adds a layer of gratuitous ambiguity to his thought experiment to make extreme risk aversion seem more plausible.

As in Kant's case, this kind of finagling makes it clear what Rawls is really up to. He's not starting with first principles and then following them to their logical conclusions. He knows where he wants the argument to go, and he's doing everything he can to get it there.

Thus, Rawls's well-intentioned rationalizing illustrates two points. First, it's another nice example of what happens when very smart people are determined to vindicate their moral emotions through reasoning. Second, it suggests that Harsanyi may be right. If you run the original-position thought experiment properly by (a) not confusing wealth and utility, (b) not assuming that people are inordinately risk averse, and (c) not making the hypothetical decision gratuitously ambiguous, you just might end up with a utilitarian conclusion. In other words, if you replace the happiness assumption with an assumption favoring choice, you end up with utilitarianism, because impartial people, with no ideological commitments, will naturally choose a society that maximizes their prospects for happiness.

334   **wasn't always a liberal:** In my youth I was something of a libertarian conservative. My libertarian claim to fame: In my senior year of high school I won third prize in an Ayn Rand essay contest. However, by the time the prize came through, I was already starting to change my mind. I shared my doubts with the woman who called to congratulate me. This did not go over well.

334   **Jonathan Haidt:** Haidt (2001, 2007, 2012).

336   **"Emotional Dog":** Haidt (2001).

336   **this characterization of his view:** According to Haidt, moral reasoning plays an important role in his landmark theory of moral psychology, the Social Intuitionist Model (SIM) (Haidt, 2001). Whether this is true depends on what counts as moral "reasoning" (Paxton and Greene, 2010).

According to the SIM, moral judgment works as follows: Moral judgments are, in general, caused by moral intuitions, and when we engage in moral reasoning, our reasoning is typically deployed post hoc to justify the moral judgments that we have already made on an intuitive basis. (See discussion of moral rationalization earlier in this chapter.) Haidt says that people do sometimes engage in private moral reasoning, but that this is "rare, occurring primarily in cases in which the intuition is weak and processing capacity is high" (p. 819). This is why I say that, according to Haidt, moral reasoning plays a minor role in moral life.

However, there are two additional psychological processes to consider, the ones that put the "Social" in the SIM. First, according to the SIM, Person A's overtly making a moral judgment can influence Person B's moral intuitions, which can in turn influence Person B's moral judgment. Haidt calls this "social persuasion." This is clearly not moral reasoning, as there is no argument, just an intuitive response to observing another's judgment or behavior. Second—and this is the key part—Haidt says that people engage in "reasoned persuasion." Here, Person A provides a verbal justification for her judgment; Person B hears this justification; and this modifies his moral intuitions, which in turn influences his moral judgment. Haidt calls this "reasoned persuasion," but that label, I think, is misleading. Here, Person A influences Person B's judgment by modifying Person B's *feelings* (automatic settings) and not by engaging Person B's capacity for explicit reasoning (manual mode). Here the "reasons" that Person A produces function like a song that succeeds in moving Person B.

Haidt believes that this process is widespread and highly influential (which it may be). This is why he says that moral reasoning plays an important role in moral life. But, as I've said, I don't think that this qualifies as "moral reasoning." And that is why I say, over Haidt's protests, that his view is not one according to which moral reasoning plays a major role. According to the SIM, I cannot change your mind on a moral issue (such as gay marriage, abortion, or eating animals) without first changing your feelings. I cannot appeal directly to your capacity for reasoning and thus cause you to override your feelings. I think that this picture of moral psychology is incorrect. This point is illustrated by an experiment in which my collaborators and I used a rather abstract argument to persuade people (at least temporarily) to accept a counterintuitive moral conclusion (Paxton, Ungar, and Greene, 2011).

336   **Some liberals say . . . some social conservatives believe:** Here are two examples: http://www.libchrist.com/other/abortion/choice.html; http://k2globalcommunicationsllc.wordpress.com/2012/08/28/abortion-nihilist -argument-eliminate-poverty-kill-the-poor. See also John Paul II (1995).

336   **liberals have impoverished moral sensibilities:** Haidt and Graham (2007); Graham, Haidt, et al. (2009); Haidt (2012).

337   **doubts about this six-part theory . . . important aspect . . . well-supported:** The survey data (Graham et al., 2009, 2011) that Haidt uses to support his theory (the original version with five foundations) show an enormous division between two clusters: the care-fairness cluster and the loyalty-authority-sanctity cluster. There is, by contrast, relatively little evidence for a two-way division within the first cluster or a three-way division within the second cluster, and what evidence there is can be accounted for by the fact that the surveys used to collect these data were designed with five clusters in mind. To provide strong evidence for a five-factor (or six- or n-factor) theory of morality, one would have to use a "bottom-up" approach, testing the theory using testing materials that were not designed with any particular theory in mind. Haidt says that, to a first approximation, the moral world has five (or six) "continents." In Haidt's data, I see evidence for two continents, which may or may not have two or three interesting bulges.

337   **questions like these:** Graham, Haidt, et al. (2009) posed these questions in a different way ("How much money would it take for you to . . . ?").

338   **Bentham and Kant:** I think that Haidt's (2012) psychological portrait of Kant (p. 120) is off the mark. Kant might have had some autistic tendencies, and he was certainly a "systematizer," but he was very authoritarian and no stranger to moral disgust. He was also, not incidentally, very religious.

338   **WEIRD:** See Henrich, Heine, et al. (2010).

338   **predicting what liberals will say . . . vice versa:** Graham, Nosek, et al. (2012).

339   **dismisses . . . science:** Mooney (2012).

339   **"real" Americans:** Devos and Banaji (2005).

339   **45 percent of Republicans believe:** Condon (April 21, 2011).

339   **little respect for . . . United Nations:** Wike (September 21, 2009).

339   **Muslim American . . . should not be trusted:** Arab American Institute (August 22, 2012).

340   **If Iranians . . . want to protest:** Swami (June 15, 2009).

340   **minority . . . report believing in God:** European Commission (2005).

340   **lowest crime rates . . . happiness:** Economic Intelligence Unit (2005); United Nations Office of Drugs and Crime (2011); United Nations (2011); Ingelhart, Foa, et al. (2008). Murder rates, educational attainment, and test scores: World Values Survey on happiness.

340   **"yang" . . . "yin":** Haidt (2012), 294.

340   **no Republicans hold elected office:** Based on my own online searches and confirmed by Henry Irving, Republican city committee chair, Cambridge, MA (personal communication, March 24, 2013).

340   **bonds are rated AAA:** http://www.cambridgema.gov/citynewsandpublications/news/2012/02/cambridge maintainsraredistinctionofearningthreetriplearatings.aspx.

340   **social conservatives are very good at:** Putnam (2000); Putnam and Campbell (2010).

342   **"Spread my work ethic":** Haidt (2012), 137.

342   **Refusing to buy . . . more harm than good:** Nicholas D. Kristof, "Where Sweatshops Are a Dream." *New York Times*, January, 14, 2009.

343   **"welfare queens":** The term was coined by Ronald Reagan during his 1976 presidential campaign: " 'Welfare Queen' Becomes Issue in Reagan Campaign." *New York Times*, February 15, 1976.

343   **former slave states:** Lind (2012).

343   **government hands off my Medicare:** Krugman (July 28, 2009).

343   **billionaires . . . lower rate than their secretaries:** Tienabeso (January 25, 2012); Buffett (August 14, 2011).

344   **reliable returns:** Yes, such donors might benefit enormously from a Romney administration's policies, but the odds that any single donor's donation would sway the election are extremely small.

344   **endorses . . . utilitarianism:** Haidt (2012) endorses, for the purposes of making policy, what he calls "Durkheimian utilitarianism" (p. 272), which is utilitarianism that accounts for the value of conservative social institutions such as religion. Durkheimian utilitarianism is actually just utilitarianism wisely applied. Nevertheless, the point is worth making because not all self-styled utilitarians appreciate the value of conservative social institutions. Mill (1885), however, certainly did, as explained, for example, in his essay "The Utility of Religion."

345   **"I don't know":** Haidt (2012), 272.

345   **"90 percent chimp and 10 percent bee":** Ibid., xv.

346   **moral reasoning . . . ineffective, though not completely:** See Paxton, Ungar, and Greene (2011).

346   **a good argument can change the shape of things:** An alternative analogy: A good argument is like a piece of technology. Few of us will ever invent a new piece of technology, and on any given day it's unlikely that we'll adopt one. Nevertheless, the world we inhabit is defined by technological change. Likewise, I believe that the world we inhabit is a product of good moral arguments. It's hard to catch someone in the midst of reasoned moral persuasion, and harder still to observe the genesis of a good argument. But I believe that without our capacity for moral reasoning, the world would be a very different place. See also Pinker (2011), chaps. 9–10; Pizarro and Bloom (2003); Finnemore and Sikkink (1998).

346   **"I have been tormenting":** Bentham (1978).

346   **"But it would be a mistake":** Mill (1895), 1–2.

<small>CHAPTER 12: BEYOND POINT-AND-SHOOT MORALITY: SIX RULES FOR MODERN HERDERS</small>

349   **Chekhov:** See introduction, "man will become."

350   **consult, but do not trust, your moral instincts:** This rule is available in bumper sticker form: "Don't believe everything you think." Available at www.northernsun.com.

350   **manual mode . . . Me ahead of Us:** Valdesolo and DeSteno (2007).

352   **more important than saving someone's life:** Likewise, few of us can honestly say that animals should suffer enormous pain because pork is tastier than tofu or because an extra dollar is too much to spend on a cruelty-free cheeseburger (not widely available, but only for lack of demand).

Bibliography

Abagnale, F. W., and S. Redding (2000). *Catch me if you can.* New York: Broadway.

ABC News (2011, December 5). Ron Paul: Why Elizabeth Warren is wrong. Retrieved Feburary 3, 2013, from http://www.youtube.com/watch?v=glvkLEUC 6Q&list=UUoIpecKvJiBIAOhaFXw-bAg&index=34.

Abrams, D., and M. A. Hogg (2012). *Social identifications: A social psychology of intergroup relations and group processes.* London: Routledge.

Adolphs, R. (2003). Cognitive neuroscience of human social behaviour. *Nature Reviews Neuroscience* 4(3): 165–178.

Allen, R. E., and N. Platon (1970). *Plato's "Euthyphro" and the earlier theory of forms.* New York: Humanities Press.

American Medical Association (1991). Decisions near the end of life. http://www.ama-assn.org/resources/doc/code-medical-ethics/221a.pdf

Amit, E., and J. D. Greene (2012). You see, the ends don't justify the means visual imagery and moral judgment. *Psychological Science* 23(8): 861–868.

Anscombe, G. E. M. (1958). Modern moral philosophy. *Philosophy* 33(124): 1–19.

Arab American Institute (2012, August 22). The American divide: How we view Arabs and Muslims. http://aai.3cdn.net/82424c9036660402e5_a7m6b1i7z.pdf

Aristotle (1941). Nichomachean ethics. In R. McKeon, ed., *The basic works of Aristotle* (pp. 927–1112). New York: Random House.

Axelrod, R., and W. Hamilton (1981). The evolution of cooperation. *Science* 211(4489): 1390–1396.

Babcock, L., and G. Loewenstein (1997). Explaining bargaining impasse: The role of self-serving biases. *The Journal of Economic Perspectives* 11(1): 109–126.

Babcock, L., G. Loewenstein, et al. (1995). Biased judgments of fairness in bargaining. *The American Economic Review*: 85 (5):1337–1343.

Babcock, L., X. Wang, et al. (1996). Choosing the wrong pond: Social comparisons in negotiations that reflect a self-serving bias. *The Quarterly Journal of Economics* 111(1): 1–19.

Baldus, D. C., G. Woodworth, et al. (1998). Racial discrimination and the death penalty in the post-Furman era: An empirical and legal overview, with recent findings from Philadelphia. *Cornell Law Review* 83: 1638–1821.

Bargh, J. A., and T. L. Chartrand (1999). The unbearable automaticity of being. *American Psychologist* 54(7): 462.

Bargh, J. A., and E. L. Williams (2006). The automaticity of social life. *Current Directions in Psychological Science* 15(1): 1–4.

Baron, A. S., and M. R. Banaji (2006). The development of implicit attitudes: Evidence of race evaluations from ages 6 and 10 and adulthood. *Psychological Science* 17(1): 53–58.

Baron, J. (1994). Nonconsequentialist decisions. *Behavioral and Brain Sciences* 17: 1–42.

Baron, J. (2006). *Against bioethics.* Cambridge, MA: MIT Press.

Baron, J., and J. Greene (1996). Determinants of insensitivity to quantity in valuation of public goods: Contribution, warm glow, budget constraints, availability, and prominence. *Journal of Experimental Psychology: Applied* 2(2): 107.

Baron, J., and I. Ritov (1993). Intuitions about penalties and compensation in the context of tort law. *Journal of Risk and Uncertainty* 7: 17–33.

Bartal, I. B. A., J. Decety, et al. (2011). Empathy and pro-social behavior in rats. *Science* 334(6061): 1427–1430.

Bartels, D. M. (2008). Principled moral sentiment and the flexibility of moral judgment and decision making. *Cognition* 108: 381–417.

Bartels, D. M., and D. A. Pizarro (2011). The mismeasure of morals: Antisocial personality traits predict utilitarian responses to moral dilemmas. *Cognition* 121(1): 154–161.

Bateson, M., D. Nettle, et al. (2006). Cues of being watched enhance cooperation in a real-world setting. *Biology Letters* 2(3): 412–414.

Batson, C. D. (1991). *The altruism question: Toward a social-psychological answer.* Hillsdale, NJ: Lawrence Erlbaum Associates, Inc.

Batson, C. D., B. D. Duncan, et al. (1981). Is empathic emotion a source of altruistic motivation? *Journal of Personality and Social Psychology* 40(2): 290.

Batson, C. D., and T. Moran (1999). Empathy-induced altruism in a prisoner's dilemma. *European Journal of Social Psychology* 29(7): 909–924.

Baumgartner, T., U. Fischbacher, et al. (2009). The neural circuitry of a broken promise. *Neuron* 64(5): 756–770.

Bazerman, M. H., and D. A. Moore (2006). *Judgment in managerial decision making.* Hoboken, NJ: Wiley.

Bazerman, M. H., and J. D. Greene (2010). In favor of clear thinking: Incorporating moral rules into a wise cost-benefit analysis—Commentary on Bennis, Medin, & Bartels (2010). *Perspectives on Psychological Science* 5(2): 209–212.

BBC News (2006, September 9). Cartoons row hits Danish exports. Retrieved February 3, 2013, from http://news.bbc.co.uk/2/hi/europe/5329642.stm.

BBC News (2012, November 27). Eurozone crisis explained. Retrieved on February 3, 2013, from http://www.bbc.co.uk/news/business-13798000.

Beauchamp, T. L., and L. R. Walters (1989). *Contemporary issues in bioethics.* Belmont, CA: Wadsworth Pub. Co.

Bechara, A., A. R. Damasio, et al. (1994). Insensitivity to future consequences following damage to human prefrontal cortex. *Cognition* 50(1): 7–15.

Bechara, A., H. Damasio, et al. (1997). Deciding advantageously before knowing the advantageous strategy. *Science* 275(5304): 1293–1295.

Bem, D. J. (1967). Self-perception: An alternative interpretation of cognitive dissonance phenomena. *Psychological Review* 74(3): 183.

Bentham, J. (1781/1996). *An introduction to the principles of morals and legislation (Collected works of Jeremy Bentham).* Oxford, UK: Clarendon Press.

Bentham, J. (1830). *The rationale of punishment.* London: Robert Heward.

Bentham, J. (1978). Offences against one's self. *Journal of Homosexuality* 3(4): 389–406.

Berker, S. (2009). The normative insignificance of neuroscience. *Philosophy & Public Affairs,* 37(4): 293–329.

Bernasconi, R. (2002). Kant as an unfamiliar source of racism. In J. Ward and T. Lott, eds., *Philosophers on Race: Critical Essays* (pp. 145–166). Oxford, UK: Blackwell.

Bernhard, H., U. Fischbacher, et al. (2006). Parochial altruism in humans. *Nature* 442(7105): 912–915.

Bertrand, M., and S. Mullainathan (2003). Are Emily and Greg more employable than Lakisha and Jamal? A field experiment on labor market discrimination. National Bureau of Economic Research.

Blackburn, S. (1993). *Essays in quasi-realism.* New York: Oxford University Press.

Blackburn, S. (2001). *Ethics: A very short introduction.* Oxford, UK: Oxford University Press.

Blair, R. J. R. (1995). A cognitive developmental approach to morality: Investigating the psychopath. *Cognition* 57(1): 1–29.

Blakemore, S. J., D. M. Wolpert, et al. (1998). Central cancellation of self-produced tickle sensation. *Nature Neuroscience* 1(7): 635–640.

Bloom, P. (2004). *Descartes' baby: How the science of child development explains what makes us human.* New York: Basic Books.

Bloom, P. (in press). *Just babies.* New York: Crown.

Boehm, C. (2001). *Hierarchy in the forest: The evolution of egalitarian behavior.* Cambridge, MA: Harvard University Press.

Botvinick, M. M., T. S. Braver, et al. (2001). Conflict monitoring and cognitive control. *Psychological Review* 108(3): 624–652.

Bowles, S. (2009). Did warfare among ancestral hunter-gatherers affect the evolution of human social behaviors? *Science* 324(5932): 1293–1298.

Boyd, R., H. Gintis, et al. (2003). The evolution of altruistic punishment. *Proceedings of the National Academy of Sciences* 100(6): 3531–3535.

Boyd, R., and P. J. Richerson (1992). Punishment allows the evolution of cooperation (or anything else) in sizable groups. *Ethology and Sociobiology* 13(3): 171–195.

Bratman, M. (1987). *Intention, plans, and practical reason.* Cambridge, MA: Harvard University Press.

Brawley, L. (1984). Unintentional egocentric biases in attributions. *Journal of Sport and Exercise Psychology* 6 (3): 264–278.

Brink, D. O. (2011). *Principles and intuition in ethics.* (unpublished manuscript)

Brown, D. E. (1991). *Human universals.* Philadelphia: Temple University Press.

Buckley, C. (2007, January 3). Man is rescued by stranger on subway tracks. *New York Times.*

Buckner, R. L., J. R. Andrews-Hanna, et al. (2008). The brain's default network. *Annals of the New York Academy of Sciences* 1124(1): 1–38.

Buffett, W. E. (2011, August 14). Stop coddling the super-rich. *New York Times* 14.

Carlsmith, K. M., J. M. Darley, et al. (2002). Why do we punish? Deterrence and just deserts as motives for punishment. *Journal of Personality and Social Psychology* 83(2): 284–299.

Caruso, E., N. Epley, et al. (2006). The costs and benefits of undoing egocentric responsibility assessments in groups. *Journal of Personality and Social Psychology* 91(5): 857.

Casebeer, W. D. (2003). *Natural ethical facts: Evolution, connectionism, and moral cognition.* Cambridge, MA: MIT Press.

Chaiken, S., and Y. Trope (1999). *Dual-process theories in social psychology.* New York: Guilford Press.

Chapman, H. A., D. A. Kim, et al. (2009). In bad taste: Evidence for the oral origins of moral disgust. *Science* 323(5918): 1222–1226.

Chekhov, A. (1977). *Portable Chekhov.* New York: Penguin.

Chib, V. S., A. Rangel, et al. (2009). Evidence for a common representation of decision values for dissimilar goods in human ventromedial prefrontal cortex. *Journal of Neuroscience* 29(39): 12315–12320.

Choi, J. K., and S. Bowles (2007). The coevolution of parochial altruism and war. *Science* 318(5850): 636–640.

Cialdini, R. B. (2003). Crafting normative messages to protect the environment. *Current Directions in Psychological Science* 12(4): 105–109.

Cialdini, R. B., M. Schaller, et al. (1987). Empathy-based helping: Is it selflessly or selfishly motivated? *Journal of Personality and Social Psychology* 52(4): 749.

Ciaramelli, E., M. Muccioli, et al. (2007). Selective deficit in personal moral judgment following damage to ventromedial prefrontal cortex. *Social Cognitive and Affective Neuroscience* 2(2): 84–92.

Clark, A., Y. Georgellis, et al. (2003). Scarring: The psychological impact of past unemployment. *Economica* 68(270): 221–241.

Clark, A. E., and A. J. Oswald (1994). Unhappiness and unemployment. *The Economic Journal* 104 (May): 648–659.

Cohen, D., and R. E. Nisbett (1994). Self-protection and the culture of honor: Explaining southern violence. *Personality and Social Psychology Bulletin* 20(5): 551–567.

Cohen, G. L. (2003). Party over policy: The dominating impact of group influence on political beliefs. *Journal of Personality and Social Psychology* 85(5): 808.

Cohen, J. D. (2005). The vulcanization of the human brain: A neural perspective on interactions between cognition and emotion. *The Journal of Economic Perspectives* 19(4): 3–24.

Condon, S. (2011, April 21). Poll: One in four Americans think Obama was not born in U.S. CBS News.

Conway, P., and B. Gawronski (2012). Deontological and utilitarian inclinations in moral decision making: A process dissociation approach. *Journal of Personality and Social Psychology* doi:10.1037/a0031021.

Copenhagen Consensus Center (2012). Copenhagen Census 2012 Report. Retrieved February 3, 2013, from http://www.copenhagenconsensus.com/Admin/Public/DWSDownload.aspx?File=%2fFiles%2fFiler%2fCC12+papers%2fOutcome Document Updated 1105.pdf.

Craig, W. L., and W. Sinnott-Armstrong (2004). *God? A debate between a Christian and an atheist.* Oxford, UK: Oxford University Press.

Crisp, R., and M. A. Slote (1997). *Virtue ethics.* Oxford, UK: Oxford University Press.

Crockett, M. J., L. Clark, et al. (2010). Serotonin selectively influences moral judgment and behavior through effects on harm aversion. *Proceedings of the National Academy of Sciences* 107(40): 17433–17438.

Cunningham, W., M. K. Johnson, et al. (2004). Separable neural components in the processing of black and white faces. *Psychological Science* 15(12): 806–813.

Cushman, F. (in press). Action, outcome and value: A dual-system framework for morality. *Personality and Social Psychology Review.*

Cushman, F., K. Gray, et al. (2012). Simulating murder: The aversion to harmful action. *Emotion* 12(1): 2.

Cushman, F., and J. D. Greene (2012). Finding faults: How moral dilemmas illuminate cognitive structure. *Social Neuroscience* 7(3): 269–279.

Cushman, F., D. Murray, et al. (2011). Judgment before principle: engagement of the frontoparietal control network in condemning harms of omission. *Social Cognitive and Affective Neuroscience.* doi:10.1093/scan/nsr072.

Cushman, F., L. Young, et al. (2006). The role of conscious reasoning and intuition in moral judgment testing three principles of harm. *Psychological Science* 17(12): 1082–1089.

Cushman, F. A., and J. D. Greene (2011). The philosopher in the theater. In M. Mikulincer and P. R. Shaver, eds., *The Social Psychology of Morality.* Washington, DC: APA Press.

Daly, M., and M. Wilson (1988). *Homicide.* New Brunswick, NJ: Aldine.

Damasio, A. R. (1994). *Descartes' error: Emotion, reason, and the human brain.* New York: G.P. Putnam.

Dancy, J. (2009). Moral Particularism. In Edward N. Zalta (ed.), *The Stanford Encyclopedia of Philosophy* http://plato.stanford.edu/archives/spr2009/entries/moral-particularism/.

Daniels, N. (2008). Reflective equilibrium. *Stanford Encyclopedia of Philosophy.*

Daniels, N. (2011). Reflective Equilibrium. In Edward N. Zalta (ed.), *The Stanford Encyclopedia of Philosophy* http://plato.stanford.edu/archives/spr2011/entries/reflective-equilibrium/.

Darwin, C. (1871/1981). *The descent of man, and selection in relation to sex.* Princeton, NJ: Princeton Univesity Press.

Darwin, C. (1872/2002). *The expression of the emotions in man and animals.* New York: Oxford University Press USA.

Davidson, A. B., and R. B. Ekelund (1997). The medieval church and rents from marriage market regulations. *Journal of Economic Behavior & Organization* 32(2): 215–245.

Davies, J. B., A. Shorrocks, et al. (2007). The World Distribution of Household Wealth. UC Santa Cruz: Center for Global, International and Regional Studies. Retrieved February 3, 2013, from http://escholarship.org/uc/item/3jv048hx

Dawes, R. M., J. McTavish, et al. (1977). Behavior, communication, and assumptions about other people's behavior in a commons dilemma situation. *Journal of Personality and Social Psychology* 35(1): 1.

Dawkins, R. (1986). *The blind watchmaker: Why the evidence of evolution reveals a universe without design.* New York: WW Norton & Company.

De Brigard, F. (2010). If you like it, does it matter if it's real? *Philosophical Psychology* 23(1): 43–57.

De Dreu, C. K. W., L. L. Greer, et al. (2010). The neuropeptide oxytocin regulates parochial altruism in intergroup conflict among humans. *Science* 328(5984): 1408–1411.

De Dreu, C. K. W., L. L. Greer, et al. (2011). Oxytocin promotes human ethnocentrism. *Proceedings of the National Academy of Sciences* 108(4): 1262–1266.

de Waal, F. (1989). Food sharing and reciprocal obligations among chimpanzees. *Journal of Human Evolution* 18(5): 433–459.

de Waal, F. (1997). *Good natured: The origins of right and wrong in humans and other animals.* Cambridge, MA: Harvard University Press.

de Waal, F. (2009). *Primates and philosophers: How morality evolved.* Princeton, NJ: Princeton University Press.

de Waal, F. (2010). *The age of empathy: Nature's lessons for a kinder society.* New York: Three Rivers Press.

de Waal, F., and A. Roosmalen (1979). Reconciliation and consolation among chimpanzees. *Behavioral Ecology and Sociobiology* 5(1): 55–66.

de Waal, F., and L. M. Luttrell (1988). Mechanisms of social reciprocity in three primate species: Symmetrical relationship characteristics or cognition? *Ethology and Sociobiology* 9(2): 101–118.

Decety, J. (2011). Dissecting the neural mechanisms mediating empathy. *Emotion Review* 3(1): 92–108.

Decety, J., and P. L. Jackson (2004). The functional architecture of human empathy. *Behavioral and Cognitive Neuroscience Reviews* 3(2): 71–100.

Degomme, O., and D. Guha-Sapir (2010). Patterns of mortality rates in Darfur conflict. *The Lancet* 375(9711): 294–300.

Denmark TV2 (2004, October 9). Overfaldet efter Koran-læsning. Retrieved February 3, 2013, from http://nyhederne.tv2.dk/article.php/id-1424089:overfaldet-efter-koranl%C3%A6sning.html.

Dennett, D. C. (1987). *The intentional stance.* Cambridge, MA: MIT Press.

Dennett, D. C. (1995). *Darwin's dangerous idea: Evolution and the meanings of life.* New York: Simon & Schuster.

DeScioli, P., and R. Kurzban (2009). Mysteries of morality. *Cognition* 112(2): 281–299.

Devos, T., and M. R. Banaji (2005). American=white? *Journal of Personality and Social Psychology* 88(3): 447.

Diener, E. (2000). Subjective well-being: The science of happiness and a proposal for a national index. *American Psychologist* 55(1): 34.

Diener, E., E. M. Suh, et al. (1999). Subjective well-being: Three decades of progress. *Psychological Bulletin* 125(2): 276.

Dongen, L. G. R., and E. G. Goudie (1980). Fetal movement patterns in the first trimester of pregnancy. *BJOG: An International Journal of Obstetrics & Gynaecology* 87(3): 191–193.

Doris, J., and A. Plakias (2007). How to argue about disagreement: Evaluative diversity and moral realism. In W. Sinnott-Armstrong, ed., *Moral Psychology, vol. 2: The Cognitive Science of Morality.* Cambridge, MA: MIT Press.

Dreber, A., D. G. Rand, et al. (2008). Winners don't punish. *Nature* 452(7185): 348–351.

Driver, J. (2009). The History of Utilitarianism. In Edward N. Zalta (ed.), *The Stanford Encyclopedia of Philosophy.* http://plato.stanford.edu/archives/sum2009/entries/utilitarianism-history/

Dunbar, R. I. M. (2004). Gossip in evolutionary perspective. *Review of General Psychology* 8(2): 100.

Dunbar, R. I. M., A. Marriott, et al. (1997). Human conversational behavior. *Human Nature* 8(3): 231–246.

Dunlap, R. (2008, May 29). Climate-change views: Republican-Democratic gaps expand. Retrieved February 3, 2013, from http://www.gallup.com/poll/107569/ClimateChange-Views-RepublicanDemocratic-Gaps -Expand.aspx.

Dutton, D. G., and A. P. Aron (1974). Some evidence for heightened sexual attraction under conditions of high anxiety. *Journal of Personality and Social Psychology* 30(4): 510.

Dworkin, R. (1978). *Taking rights seriously.* Cambridge, MA: Harvard University Press.

Dworkin, R. "Rights as Trumps" (1984). In J. Waldron (ed.), *Theories of Rights,* 153–167. Oxford: Oxford University Press.

Easterlin, R. A. (1974). Does economic growth improve the human lot? In P. David and M. Reder, eds., *Nations and Households in Economic Growth: Essays in Honour of Moses Abramovitz.* New York: Academic Press.

Easterlin, R. A., L. A. McVey, et al. (2010). The happiness-income paradox revisited. *Proceedings of the National Academy of Sciences* 107(52): 22463–22468.

Eberhardt, J. L., P. G. Davies, et al. (2006). Looking deathworthy: Perceived stereotypicality of black defendants predicts capital-sentencing outcomes. *Psychological Science* 17(5): 383–386.

The Economist (2005). The Economist Intelligence Unit's Quality-of-Life Index. http://www.economist .com/media/pdf/QUALITY_OF_LIFE.pdf

Ellingsen, T., B. Herrmann, et al. (2012). Civic Capital in Two Cultures: The Nature of Cooperation in Romania and USA. Available at SSRN.

Espresso Education (n.d.). Earthquake legends. Retrieved February 3, 2013, from http://content .espressoeducation.com/espresso/modules/t2 special reports/natural disasters/eqlegnd.html.

European Commission, (2005). Special Eurobarometer 225: Social values, science and technology. Brussels: Directorate General Press.

FactCheck.org (2009, August 14). Palin vs. Obama: Death panels. Retrieved February 3, 2013, from http://www .factcheck.org/2009/08/palin-vs-obama-death-panels.

FactCheck.org (2009, August 18). Keep your insutance? Not everyone. Retrieved February 3, 2013, from http:// www.factcheck.org/2009/08/keep-your-insurance-not-everyone.

Falk, G. (1990). Murder, an analysis of its forms, conditions, and causes. Jefferson, NC: McFarland & Company Incorporated Pub.

Fehr, E., and S. Gächter (2002). Altruistic punishment in humans. *Nature* 415(6868): 137–140.

Fehr, E., and S. Gächter (1999). Cooperation and punishment in public goods experiments. Institute for Empirical Research in Economics Working Paper (10).

Feiman, R., Cushman, F., Carey, S. (in prep): Infants fail to represent a negative goal, but not a negative event.

Feinberg, M., R. Willer, et al. (2012a). Flustered and faithful: Embarrassment as a signal of prosociality. *Journal of Personality and Social Psychology* 102(1): 81.

Feinberg, M., R. Willer, et al. (2012b). The virtues of gossip: Reputational information sharing as prosocial behavior. *Journal of Personality and Social Psychology* 102(5): 1015.

Fellows, L. K., and M. J. Farah (2007). The role of ventromedial prefrontal cortex in decision making: judgment under uncertainty or judgment per se? *Cerebral Cortex* 17(11): 2669–2674.

Fernbach, P. (2012, May 27) Weak evidence. WAMC Northeast Public Radio.

Fernbach, P. M., T. Rogers, C. R. Fox, and S. A. Sloman (in press). Political extremism is supported by an illusion of understanding. *Psychological Science.*

Financial Crisis Inquiry Commission (2011). Financial crisis inquiry report. Retrieved Feburary 3, 2013, from http://fcic-static.law.stanford.edu/cdn media/fcic-reports/fcic final report full.pdf.

Finnemore, M., and K. Sikkink (1998). International norm dynamics and political change. *International Organization* 52(4): 887–917.

Fischer, D. H. (1991). *Albion's seed: Four British folkways in America.* New York: Oxford University Press USA.

Fischer, J. M., and M. Ravizza, eds. (1992). *Ethics: Problems and principles.* Fort Worth, TX: Harcourt Brace Jovanovich College Publishers.

Fisher, R. (1930). *The genetical theory of natural selection.* Oxford, UK: Clarendon Press.

Fisher, R. (1971). *Basic negotiation strategy: International conflict for beginners.* London: Allen Lane.

Fodor, J. A. (1983). *Modularity of mind: An essay on faculty psychology.* Cambridge, MA: MIT Press.

Foot, P. (1967). The problem of abortion and the doctrine of double effect. *Oxford Review* 5: 5–15.

Forsyth, D. R., and B. R. Schlenker (1977). Attributional egocentrism following performance of a competitive task. *The Journal of Social Psychology* 102(2): 215–222.

Forsythe, R., J. L. Horowitz, et al. (1994). Fairness in simple bargaining experiments. *Games and Economic Behavior* 6(3): 347–369.

Frank, R. H. (1988). *Passions within reason: The strategic role of the emotions.* New York: WW Norton & Company.

Frederick, S. (2005). Cognitive reflection and decision making. *The Journal of Economic Perspectives* 19(4): 25–42.

Fredrickson, B. L. (2001). The role of positive emotions in positive psychology: The broaden-and-build theory of positive emotions. *American Psychologist* 56(3): 218.

Frijda, N. H. (1987). *The emotions.* Cambridge, UK: Cambridge University Press.

Fudenberg, D., D. Rand, et al. (2010). Slow to anger and fast to forgive: cooperation in an uncertain world. *American Economic Review* 102(2): 720–749.

Gaes, G. G., and A. L. Goldberg (2004). Prison rape: A critical review of the literature. Washington, DC: National Institute of Justice.

Gardiner, S. M. (2011). *A perfect moral storm: The ethical tragedy of climate change.* New York: Oxford University Press USA.

Gauthier, D. (1987). *Morals by agreement.* Oxford, UK: Clarendon Press.

Gazzaniga, M. S. (2006). *The ethical brain: The science of our moral dilemmas.* New York: Harper Perennial.

Gazzaniga, M. S., and J. E. Le Doux (1978). *The integrated mind.* New York: Plenum.

Gervais, W. M., and A. Norenzayan (2012). Analytic thinking promotes religious disbelief. *Science* 336(6080): 493–496.

Gervais, W. M., A. F. Shariff, et al. (2011). Do you believe in atheists? Distrust is central to anti-atheist prejudice. *Journal of Personality and Social Psychology* 101(6): 1189.

Gewirth, A. (1980). *Reason and morality.* Chicago: University of Chicago Press.

Gilbert, D. (2006, July 2). If only gay sex caused global warming. *Los Angeles Times* 2.

Gilbert, D. (2006). *Stumbling on happiness.* New York: Knopf.

Gilbert, S. (2010). *Developmental biology,* 9th ed. Sunderland, MA: Sunderland, Sinauer Associates.

Gilovich, T., D. Griffin, et al. (2002). *Heuristics and biases: The psychology of intuitive judgment.* Cambridge, UK: Cambridge University Press.

Gintis, H. (2000). Strong reciprocity and human sociality. *Journal of Theoretical Biology* 206(2): 169–179.

Gintis, H., S. Bowles, et al. (2005). Moral sentiments and material interests: The foundations of cooperation in economic life, MIT press.

Givewell.org (2012). Top charities. Retrieved February 3, 2013, from http://www.givewell.org/charities/top-charities.

Givewell.org (n.d.). Against Malaria Foundation. Retrieved February 3, 2013, from http://www.givewell.org/international/top-charities/AMF.

Glenn, A. L., A. Raine, et al. (2009). The neural correlates of moral decision-making in psychopathy. *Molecular Psychiatry* 14(1): 5–6.

Goldman, A. I. (1970). *A theory of human action,* Prentice-Hall Englewood Cliffs, NJ.

Graham, J., J. Haidt, et al. (2009). Liberals and conservatives rely on different sets of moral foundations. *Journal of Personality and Social Psychology* 96(5): 1029.

Graham, J., B. A. Nosek, et al. (2012). The moral stereotypes of liberals and conservatives: Exaggeration of differences across the political spectrum. *PLOS ONE* 7(12): e50092.

Graham, J., B. A. Nosek, et al. (2011). Mapping the moral domain. *Journal of Personality and Social Psychology* 101(2): 366.

Greenberg, D. (2012, February 27). Sick to his stomach. *Slate.* Retrieved February 3, 2013, from http://www.slate.com/articles/news_and_politics/history_lesson/2012/02/how_santorum_misunderstands_kennedy_s_speech_on_religious_freedom_.html.

Greene, J. (1997). Moral psychology and moral progress. Undergraduate thesis, Department of Philosophy, Harvard University.

Greene, J. (2002). The terrible, horrible, no good, very bad truth about morality and what to do about it. Doctoral Thesis, Department of Philosophy, Princeton University.

Greene, J. (2007). The secret joke of Kant's soul. In W. Sinnott-Armstrong, ed., *Moral Psychology, vol. 3: The Neuroscience of Morality: Emotion, Disease, and Development.* Cambridge, MA: MIT Press.

Greene, J. D. (2009). Dual-process morality and the personal/impersonal distinction: A reply to McGuire, Langdon, Coltheart, and Mackenzie. *Journal of Experimental Social Psychology* 45(3): 581–584.

Greene, J. D. (2010). Notes on "The Normative Insignificance of Neuroscience" by Selim Berker. Retrieved from http://www.wjh.harvard.edu/~jgreene/GreeneWJH/Greene-Notes-on-Berker-Nov10.pdf.

Greene, J. D. (under review). Beyond point-and-shoot morality: Why cognitive (neuro)science matters for ethics.

Greene, J. D., and J. Baron (2001). Intuitions about declining marginal utility. *Journal of Behavioral Decision Making* 14: 243–255.

Greene, J. D., and J. Paxton (2009). Patterns of neural activity associated with honest and dishonest moral decisions. *Proceedings of the National Academy of Sciences* 106(30): 12506–12511.

Greene, J. D., F. A. Cushman, et al. (2009). Pushing moral buttons: The interaction between personal force and intention in moral judgment. *Cognition* 111(3): 364–371.

Greene, J. D., S. A. Morelli, et al. (2008). Cognitive load selectively interferes with utilitarian moral judgment. *Cognition* 107: 1144–1154.

Greene, J. D., L. E. Nystrom, et al. (2004). The neural bases of cognitive conflict and control in moral judgment. *Neuron* 44(2): 389–400.

Greene, J. D., R. B. Sommerville, et al. (2001). An fMRI investigation of emotional engagement in moral judgment. *Science* 293(5537): 2105–2108.

Greenwald, A. G., and M. R. Banaji (1995). Implicit social cognition: Attitudes, self-esteem, and stereotypes. *Psychological Review* 102(1): 4–27.

Greenwald, A. G., D. E. McGhee, et al. (1998). Measuring individual differences in implicit cognition: The implicit association test. *Journal of Personality and Social Psychology* 74(6): 1464.

Grice, H. P., and P. F. Strawson (1956). In defense of a dogma. *The Philosophical Review* 65(2): 141–158.

Griffiths, P. E. (1997). *What emotions really are: The problem of psychological categories.* Chicago: University of Chicago Press.

Grossman, D. (1995). On killing. E-reads/E-rights. New York: Little, Brown.

Gusnard, D. A., M. E. Raichle, et al. (2001). Searching for a baseline: Functional imaging and the resting human brain. *Nature Reviews Neuroscience* 2(10): 685–694.

Güth, W., R. Schmittberger, et al. (1982). An experimental analysis of ultimatum bargaining. *Journal of Economic Behavior & Organization* 3(4): 367–388.

Haberkorn, J. (2012, November 6). Abortion, rape controversy shaped key races. *Politico.*

Haidt, J. (2001). The emotional dog and its rational tail: A social intuitionist approach to moral judgment. *Psychological Review* 108: 814–834.

Haidt, J. (2006). *The happiness hypothesis.* New York: Basic Books.

Haidt, J. (2007). The new synthesis in moral psychology. *Science* 316: 998–1002.

Haidt, J. (2012). *The righteous mind: Why good people are divided by politics and religion.* New York: Pantheon.

Haidt, J., and J. Graham (2007). When morality opposes justice: Conservatives have moral intuitions that liberals may not recognize. *Social Justice Research* 20(1): 98–116.

Haldane, J. (1932). *The causes of evolution.* London: Longmans, Green & Co.

Haley, K. J., and D. M. T. Fessler (2005). Nobody's watching? Subtle cues affect generosity in an anonymous economic game. *Evolution and Human Behavior* 26(3): 245–256.

Hamilton, W. (1964). The genetical evoution of social behavior. *Journal of Theoretical Biology* 7(1): 1–16.

Hamlin, J. K., K. Wynn, et al. (2007). Social evaluation by preverbal infants. *Nature* 450(7169): 557–559.

Hamlin, J. K., K. Wynn, et al. (2011). How infants and toddlers react to antisocial others. *Proceedings of the National Academy of Sciences* 108(50): 19931–19936.

Hardin, G. (1968). The tragedy of the commons. *Science* 162: 1243–1248.

Hardman, D. (2008). Moral dilemmas: Who makes utilitarian choices? (unpublished manuscript).

Hare, R. M. (1952). *The language of morals.* Oxford, UK: Clarendon Press.

Hare, R. M. (1981). *Moral thinking: Its levels, method, and point.* Oxford, UK: Oxford University Press.

Hare, T. A., C. F. Camerer, et al. (2009). Self-control in decision-making involves modulation of the vmPFC valuation system. *Science* 324(5927): 646–648.

Harinck, F., C. K. W. De Dreu, et al. (2000). The impact of conflict issues on fixed-pie perceptions, problem solving, and integrative outcomes in negotiation. *Organizational Behavior and Human Decision Processes* 81(2): 329–358.

Harman, G. (1975). Moral relativism defended. *The Philosophical Review* 84(1): 3–22.

Harris, J. R., and C. D. Sutton (1995). Unravelling the ethical decision-making process: Clues from an empirical study comparing Fortune 1000 executives and MBA students. *Journal of Business Ethics* 14(10): 805–817.

Harris, M., N. K. Bose, et al. (1966). The cultural ecology of India's sacred cattle. *Current Anthropology:* 51–66.

Harris, S. (2010). *The moral landscape: How science can determine human values.* New York: Free Press.

Harris, S. (2011, January 29). A response to critics. *Huffington Post.* Retrieved February 3, 2013, from http://www.huffingtonpost.com/sam-harris/a-response-to-critics b 815742.html.

Harsanyi, J. (1953). Cardinal utility in welfare economics and in the theory of risk-taking. *Journal of Political Economy* 61: 434–435.

Harsanyi, J. (1955). Cardinal welfare, individualistic ethics, and interpersonal comparisons of utility. *Journal of Political Economy* 63: 309–321.

Harsanyi, J. C. (1975). Can the maximin principle serve as a basis for morality? A critique of John Rawls's theory. *The American Political Science Review* 69(2), 594–606.

Hastorf, A. H., and H. Cantril (1954). They saw a game; a case study. *The Journal of Abnormal and Social Psychology* 49(1): 129.

Hauk, O., I. Johnsrude, et al. (2004). Somatotopic representation of action words in human motor and premotor cortex. *Neuron* 41(2): 301–307.

Hauser, M., F. Cushman, et al. (2007). A dissociation between moral judgments and justifications. *Mind & Language* 22(1): 1–21.

Heberlein, A. S., and R. Adolphs (2004). Impaired spontaneous anthropomorphizing despite intact perception and social knowledge. *Proceedings of the National Academy of Sciences* 101(19): 7487–7491.

Heider, F., and M. Simmel (1944). An experimental study of apparent behavior. *The American Journal of Psychology* 57(2): 243–259.

Henrich, J., R. Boyd, et al. (2001). In search of homo economicus: Behavioral experiments in 15 small–scale societies. *American Economic Review* 91(2): 73–78.

Henrich, J., J. Ensminger, et al. (2010). Markets, religion, community size, and the evolution of fairness and punishment. *Science* 327(5972): 1480–1484.

Henrich, J., and F. J. Gil-White (2001). The evolution of prestige: Freely conferred deference as a mechanism for enhancing the benefits of cultural transmission. *Evolution and Human Behavior* 22(3): 165–196.

Henrich, J., S. J. Heine, et al. (2010). The weirdest people in the world. *Behavioral and Brain Sciences* 33(2–3): 61–83.

Henrich, J., R. McElreath, et al. (2006). Costly punishment across human societies. *Science* 312(5781): 1767–1770.

Herrmann, B., C. Thöni, et al. (2008). Antisocial punishment across societies. *Science* 319(5868): 1362–1367.

Hobbes (1651/1994). *Leviathan*. Indianapolis: Hackett.

Hoffman, M. L. (2000). *Empathy and moral development: Implications for caring and justice*. New York: Cambridge University Press.

Horgan, T., and M. Timmons (1992). Troubles on Moral Twin Earth: Moral queerness revived. *Synthese* 92(2): 221–260.

Howard-Snyder, F. (2002, May 14). Doing vs. allowing harm. *Stanford Encyclopedia of Philosophy*.

Hsee, C. K., G. F. Loewenstein, et al. (1999). Preference reversals between joint and separate evaluations of options: A review and theoretical analysis. *Psychological Bulletin* 125(5): 576.

Huettel, S. A., A. W. Song, et al. (2004). *Functional magnetic resonance imaging*. Sunderland, MA: Sinauer Associates, Inc.

Hume, D. (1739/1978). *A treatise of human nature*, ed. L. A. Selby-Bigge and P. H. Nidditch. Oxford, UK: Oxford University Press.

Hursthouse, R. (2000). *On virtue ethics*. New York: Oxford University Press, USA.

Indian Express (2006, February 18). Rs 51-crore reward for Danish cartoonist's head, says UP Minister. Retrieved February 3, 2013, from http://www.indianexpress.com/storyOld.php?storyId=88158.

Inglehart, R., R. Foa, et al. (2008). Development, freedom, and rising happiness: A global perspective (1981–2007). *Perspectives on Psychological Science* 3(4): 264–285.

Intergovernmental Panel on Climate Change (2007). Synthesis report. Retrieved February 3, 2013, from http://www.ipcc.ch/pdf/assessment-report/ar4/syr/ar4 syr.pdf.

International Business Times (2012, September 21). "Innocence of Muslims" protests: Death toll rising in Pakistan. Retrieved February 3, 2013, from http://www.ibtimes.com/%E2%80%98innocence-muslims%E2%80%99-protests-death-toll-rising-pakistan-794296.

International Labour Organization (2012). Summary of the ILO 2012 Global Estimate of Forced Labour. Retrieved February 3, 2013, from http://www.ilo.org/sapfl/Informationresources/ILOPublications/WCMS_181953/lang--en/index.htm.

Jamieson, D. (1999). *Singer and his critics*. Oxford, UK: Wiley-Blackwell.

Jensen, K., J. Call, et al. (2007). Chimpanzees are vengeful but not spiteful. *Proceedings of the National Academy of Sciences* 104(32): 13046–13050.

John Paul II (1995). The gospel of life. *Evangelium vitae* 73.

Jones, J. (2010, March 11). Conservatives' doubts about global warming grow. Gallup. Retrieved October 29, 2011, from http://www.gallup.com/poll/126563/conservatives-doubts-global-warming-grow.aspx.

Joyce, R. (2001). *The myth of morality*. Cambridge, UK: Cambridge University Press.

Joyce, R. (2006). *The evolution of morality*. Cambridge, MA: MIT Press.

Joyce, R. (2011). The accidental error theorist. *Oxford Studies in Metaethics* 6: 153.

Kahan, D. M., M. Wittlin, et al. (2011). The tragedy of the risk-perception commons: Culture conflict, rationality conflict, and climate change. Temple University Legal Studies Research Paper (2011–26).

Kahan, D. M., D. A. Hoffman, et al. (2012). They saw a protest: Cognitive illiberalism and the speech-conduct distinction. *Stanford Law Review* 64: 851.

Kahan, D. M., H. Jenkins-Smith, et al. (2011). Cultural cognition of scientific consensus. *Journal of Risk Research* 14(2): 147–174.

Kahan, D. M., E. Peters, et al. (2012). The polarizing impact of science literacy and numeracy on perceived climate change risks. *Nature Climate Change* 2(10): 732–735.

Kahane, G., and N. Shackel (2010). Methodological issues in the neuroscience of moral judgment. *Mind and Language* 25(5): 561–582.

Kahane, G., K. Wiech, et al. (2012). The neural basis of intuitive and counterintuitive moral judgment. *Social Cognitive and Affective Neuroscience* 7(4): 393–402.

Kahneman, D. (2003). A perspective on judgment and choice: Mapping bounded rationality. *American Psychologist* 58(9): 697–720.

Kahneman, D. (2011). *Thinking, fast and slow.* New York: Farrar, Straus & Giroux.

Kahneman, D., E. Diener, et al. (2003). *Well-being: The foundations of hedonic psychology.* New York: Russell Sage Foundation Publications.

Kahneman, D., D. Schkade, et al. (1998). Shared outrage and erratic rewards: The psychology of punitive damages. *Journal of Risk and Uncertainty* 16: 49–86.

Kahneman, D., and A. Tversky. (2000). *Choices, Values, and Frames.* New York: Cambridge University Press.

Kamm, F. M. (1998). *Morality, mortality, vol. I: Death and whom to save from it.* New York: Oxford University Press USA.

Kamm, F. M. (2000). The doctrine of triple effect and why a rational agent need not intend the means to his end. *Proceedings of the Aristotelian Society* 74(Suppl. S.): 21–39.

Kamm, F. M. (2001). *Morality, mortality, vol. II: Rights, duties, and status.* New York: Oxford University Press USA.

Kamm, F. M. (2006). *Intricate ethics: Rights, responsibilities, and permissible harm.* New York: Oxford University Press USA.

Kamm, F. M. (2009). Neuroscience and moral reasoning: a note on recent research. *Philosophy & Public Affairs* 37(4): 330–345.

Kant, I. (1785/2002). *Groundwork for the metaphysics of morals.* New Haven, CT: Yale University Press.

Kant, I. (1994). *The metaphysics of morals: Ethical philosophy.* Indianapolis: Hackett.

Keil, F. C. (2003). Folkscience: Coarse interpretations of a complex reality. *Trends in Cognitive Sciences* 7(8): 368–373.

Keltner, D. (2009). *Born to be good: The science of a meaningful life.* New York: WW Norton & Company.

Keltner, D., and J. Haidt (2003). Approaching awe, a moral, spiritual, and aesthetic emotion. *Cognition & Emotion* 17(2): 297–314.

Kim, S. (2011, December 12). "Occupy a desk" job fair coms to Zuccotti Park. ABC News. Retrieved Feburary 3, 2013, from http://abcnews.go.com/Business/york-anti-occupy-wall-street-campaign-hosts-job/story?id=15121278.

Kinsley, M. (2003, October 2). Just supposin': In defense of hypothetical questions. *Slate.* Retrieved from http://www.slate.com/articles/news_and_politics/readme/2003/10/just_supposin.html

Kinzler, K. D., E. Dupoux, et al. (2007). The native language of social cognition. *Proceedings of the National Academy of Sciences* 104(30): 12577–12580.

Kitcher, P. (1994). Four ways of "biologicizing" ethics. *Conceptual Issues in Evolutionary Biology* 439–450.

Koechlin, E., C. Ody, et al. (2003). The architecture of cognitive control in the human prefrontal cortex. *Science* 302(5648): 1181–1185.

Koenigs, M., M. Kruepke, et al. (2012). Utilitarian moral judgment in psychopathy. *Social Cognitive and Affective Neuroscience* 7(6): 708–714.

Koenigs, M., L. Young, et al. (2007). Damage to the prefrontal cortex increases utilitarian moral judgements. *Nature* 446(7138): 908–911.

Kogut, T., and I. Ritov (2005). The singularity effect of identified victims in separate and joint evaluations. *Organizational Behavior and Human Decision Processes* 97(2): 106–116.

Korsgaard, C. M. (1996). *Creating the kingdom of ends.* New York: Cambridge University Press.

Kosfeld, M., M. Heinrichs, et al. (2005). Oxytocin increases trust in humans. *Nature* 435(7042): 673–676.

Koven, N. S. (2011). Specificity of meta-emotion effects on moral decision-making. *Emotion* 11(5): 1255.

Krugman, P. (2009, July 28). Why Americans hate single-payer insurance. *New York Times.*

Krugman, P. (2011, November 24). We are the 99.9%. *New York Times.*

Kuhse. H, P. Singer (2006). *Bioethics. An Anthology,* 2nd edition. Oxford: Blackwell Publishing.

Kurzban, R., P. DeScioli, et al. (2007). Audience effects on moralistic punishment. *Evolution and Human Behavior* 28(2): 75–84.

Kurzban, R., J. Tooby, et al. (2001). Can race be erased? Coalitional computation and social categorization. *Proceedings of the National Academy of Sciences* 98(26): 15387–15392.

Ladyna-Kots, N. (1935). Infant chimpanzee and human child. Museum Darwinianum, Moscow.

Lakoff, G., and M. Johnson (1980). *Metaphors we live by.* Chicago: University of Chicago Press.

Lakshminarayanan, V. R., and L. R. Santos (2008). Capuchin monkeys are sensitive to others' welfare. *Current Biology* 18(21): R999–R1000.

Layard, R. (2006). *Happiness: Lessons from a new science.* New York: Penguin Press.

Leitenberg, M. (2003). Deaths in wars and conflicts between 1945 and 2000. *Occasional Paper* (29).

Lerner, J. S., D. A. Small, et al. (2004). Heart strings and purse strings: Carryover effects of emotions on economic decisions. *Psychological Science* 15(5): 337–341.

Liberman, V., S. M. Samuels, et al. (2004). The name of the game: Predictive power of reputations versus situational labels in determining prisoner's dilemma game moves. *Personality and Social Psychology Bulletin* 30(9): 1175–1185.

Lieberman, M. D., R. Gaunt, et al. (2002). Reflection and reflexion: A social cognitive neuroscience approach to attributional inference. *Advances in Experimental Social Psychology* 34: 199–249.

Lind, M. (1999). Civil war by other means. *Foreign Affairs* 78: 123.

Lind, M. (2012, October 10). Slave states vs. free states, 2012. *Salon.* Retrieved February 3, 2013, from http://www.salon.com/2012/10/10/slave_states_vs_free_states_2012/.

List, J. A. (2007). On the interpretation of giving in dictator games. *Journal of Political Economy* 115(3): 482–493.

Loewenstein, G. (1996). Out of control: Visceral influences on behavior. *Organizational Behavior and Human Decision Processes* 65(3): 272–292.

Loewenstein, G., S. Issacharoff, et al. (1993). Self-serving assessments of fairness and pretrial bargaining. *The Journal of Legal Studies* 22(1): 135–159.

Lord, C. G., L. Ross, et al. (1979). Biased assimilation and attitude polarization: The effects of prior theories on subsequently considered evidence. *Journal of Personality and Social Psychology* 37(11): 2098.

MacDonald, A. W., J. D. Cohen, et al. (2000). Dissociating the role of the dorsolateral prefrontal and anterior cingulate cortex in cognitive control. *Science* 288(5472): 1835–1838.

MacIntyre, A. (1981). *After virtue.* Notre Dame, IN: University of Notre Dame Press.

Mackie, J. L. (1977). *Ethics: Inventing right and wrong.* Harmondsworth, UK, and New York: Penguin.

Macmillan, M. (2002). *An odd kind of fame: Stories of Phineas Gage.* Cambridge, MA: MIT Press.

Madison, L. (2012, October 23). Richard Mourdock: Even pregnancy from rape something "God intended." CBS News. Retrieved February 3, 2013, from http://www.cbsnews.com/8301-250_162-57538757/richard-mourdock-even-pregnancy-from-rape-something-god-intended/

Mahajan, N., M. A. Martinez, et al. (2011). The evolution of intergroup bias: Perceptions and attitudes in rhesus macaques. *Journal of Personality and Social Psychology* 100(3): 387.

Mahajan, N., and K. Wynn (2012). Origins of "Us" versus "Them": Prelinguistic infants prefer similar others. *Cognition* 124(2): 227–233.

Margulis, L. (1970). *Origin of eukaryotic cells: Evidence and research implications for a theory of the origin and evolution of microbial, plant, and animal cells on the Precambrian earth.* New Haven, CT: Yale University Press.

Mariner, J. (2001). No escape: Male rape in US prisons. Human Rights Watch.

Marlowe, F. W., J. C. Berbesque, et al. (2008). More "altruistic" punishment in larger societies. *Proceedings of the Royal Society B: Biological Sciences* 275(1634): 587–592.

Marsh, A. A., S. L. Crowe, et al. (2011). Serotonin transporter genotype (5-HTTLPR) predicts utilitarian moral judgments. *PLOS ONE* 6(10): e25148.

McClure, S. M., K. M. Ericson, et al. (2007). Time discounting for primary rewards. *Journal of Neuroscience* 27(21): 5796–5804.

McClure, S. M., D. I. Laibson, et al. (2004). Separate neural systems value immediate and delayed monetary rewards. *Science* 306(5695): 503–507.

McElreath, R., R. Boyd, et al. (2003). Shared norms and the evolution of ethnic markers. *Current Anthropology* 44(1): 122–130.

McGuire, J., R. Langdon, et al. (2009). A reanalysis of the personal/impersonal distinction in moral psychology research. *Journal of Experimental Social Psychology* 45(3): 577–580.

McMahan, J. (2009). *Killing in war.* New York: Oxford University Press.

Medak, P. (1986). Button, button, *The Twilight Zone.*

Mendez, M. F., E. Anderson, et al. (2005). An investigation of moral judgement in frontotemporal dementia. *Cognitive and Behavioral Neurology* 18(4): 193–197.

Mercier, H., and D. Sperber (2011). Why do humans reason? Arguments for an argumentative theory. *Behavioral and Brain Sciences* 34(2): 57.

Messick, D. M., and K. P. Sentis (1979). Fairness and preference. *Journal of Experimental Social Psychology* 15(4): 418–434.

Metcalfe, J., and W. Mischel (1999). A hot/cool-system analysis of delay of gratification: Dynamics of willpower. *Psychological Review* 106(1): 3–19.

Michor, F., Y. Iwasa, et al. (2004). Dynamics of cancer progression. *Nature Reviews Cancer* 4(3): 197–205.

Mikhail, J. (2000). Rawls' linguistic analogy: A study of the "Generative Grammar" model of moral theory described by John Rawls in *A Theory of Justice.* Cornell University, Dept. of Philosophy.

Mikhail, J. (2011). *Elements of moral cognition: Rawls' linguistic analogy and the cognitive science of moral and legal judgment.* New York: Cambridge University Press.

Milgram, S., L. Mann, et al. (1965). The lost letter technique: A tool for social research. *Public Opinion Quarterly* 29(3): 437–438.

Milinski, M., D. Semmann, et al. (2002). Reputation helps solve the "tragedy of the commons." *Nature* 415(6870): 424–426.

Mill, J. S. (1865). *On liberty.* London: Longman, Green, Longman, Roberts and Green.

Mill, J. S. (1895). The subjection of women. National American Woman Suffrage Association.

Mill, J. S. (1998). Utility of Religion. In *Three Essays on Religion.* Amherst, NY: Prometheus Books.

Mill, J. S., and J. Bentham (1987). *Utilitarianism and other essays.* Harmondsworth, UK: Penguin.

Millbank, D., and C. Deane (2003, September 6). Hussein link to 9/11 lingers in many minds. *Washington Post.* Retrieved October 29, 2011, from http://www.washingtonpost.com/ac2/wp-dyn/A32862-2003Sep5? language=printer.

Miller, E. K., and J. D. Cohen (2001). An integrative theory of prefrontal cortex function. *Annual Review of Neuroscience* 24(1): 167–202.

Miller, G. F., and P. M. Todd (1998). Mate choice turns cognitive. *Trends in Cognitive Sciences* 2(5): 190–198.

Mills, C. M., and F. C. Keil (2005). The development of cynicism. *Psychological Science* 16(5): 385–390.

Mitani, J. C., D. P. Watts, et al. (2010). Lethal intergroup aggression leads to territorial expansion in wild chimpanzees. *Current Biology* 20(12): R507–R508.

Mooney, C. (2012). *The Republican brain: The science of why they deny science—and reality.* Hoboken, NJ: Wiley.

Moore, A. B., B. A. Clark, et al. (2008). Who shalt not kill? Individual differences in working memory capacity, executive control, and moral judgment. *Psychological Science* 19(6): 549–557.

Moore, G. E. (1903/1993). *Principia ethica.* Cambridge, UK: Cambridge University Press.

Moretto, G., E. Ladavas, et al. (2010). A psychophysiological investigation of moral judgment after ventromedial prefrontal damage. *Journal of Cognitive Neuroscience* 22(8): 1888–1899.

Morrison, I., D. Lloyd, et al. (2004). Vicarious responses to pain in anterior cingulate cortex: Is empathy a multisensory issue? *Cognitive, Affective & Behavioral Neuroscience* 4(2): 270–278.

Musen, J. D. (2010). The moral psychology of obligations to help those in need. Undergraduate thesis, Department of Psychology, Harvard University.

Nagel, J., and M. R. Waldmann (2012). Deconfounding distance effects in judgments of moral obligation. *Journal of Experimental Psychology: Learning, Memory, and Cognition* 39(1).

Nagel, T. (1979). *The possibility of altruism.* Princeton, NJ: Princeton University Press.

Navarrete, C. D., M. M. McDonald, et al. (2012). Virtual morality: Emotion and action in a simulated three-dimensional "trolley problem." *Emotion* 12(2): 364.

New York Times (1976, February 15). "Welfare queen" becomes issue in Reagan campaign.

Nichols, S. (2002). How psychopaths threaten moral rationalism: Is it irrational to be amoral? *The Monist* 85(2): 285–303.

Nichols, S., and J. Knobe (2007). Moral responsibility and determinism: The cognitive science of folk intutions. *Nous* 41: 663–685.

Nichols, S., and R. Mallon (2006). Moral dilemmas and moral rules. *Cognition* 100(3): 530–542.

Nietzsche, F. (1882/1974). *The Gay Science.* New York: Random House.

Nisbett, R. E., and D. Cohen (1996). *Culture of honor: The psychology of violence in the South.* Boulder, CO: Westview Press.

Nisbett, R. E., K. Peng, et al. (2001). Culture and systems of thought: Holistic versus analytic cognition. *Psychological Review* 108(2): 291.

Nisbett, R. E., and T. D. Wilson (1977). Telling more than we can know: Verbal reports on mental processes. *Psychological Review* 84(3): 231.

Norenzayan, A., and A. F. Shariff (2008). The origin and evolution of religious prosociality. *Science* 322(5898): 58–62.

Norton, M. I., and D. Ariely (2011). Building a better America—one wealth quintile at a time. *Perspectives on Psychological Science* 6(1): 9–12.

Nowak, M., and K. Sigmund (1993). A strategy of win-stay, lose-shift that outperforms tit-for-tat in the Prisoner's Dilemma game. *Nature* 364(6432): 56–58.

Nowak, M. A. (2006). Five rules for the evolution of cooperation. *Science* 314(5805): 1560–1563.

Nowak, M. A., and K. Sigmund (1992). Tit for tat in heterogeneous populations. *Nature* 355(6357): 250–253.

Nowak, M. A., and K. Sigmund (1998). Evolution of indirect reciprocity by image scoring. *Nature* 393(6685): 573–577.

Nowak, M. A., and K. Sigmund (2005). Evolution of indirect reciprocity. *Nature* 437(7063): 1291–1298.

Nowak, M. A., C. E. Tarnita, et al. (2010). The evolution of eusociality. *Nature* 466(7310): 1057–1062.

Nozick, R. (1974). *Anarchy, state, and utopia*. New York: Basic Books.

O'Neill, P., and L. Petrinovich (1998). A preliminary cross-cultural study of moral intuitions. *Evolution and Human Behavior* 19(6): 349–367.

Obama, B. (2006). Speech at the Call to Renewal's Building a Covenent for a New America conference. Retrieved February 3, 2013, from http://www.nytimes.com/2006/06/28/us/politics/2006obamaspeech.html?pagewanted=all

Ochsner, K. N., S. A. Bunge, et al. (2002). Rethinking feelings: An fMRI study of the cognitive regulation of emotion. *Journal of Cognitive Neuroscience* 14(8): 1215–1229.

Olsson, A., and E. A. Phelps (2004). Learned fear of "unseen" faces after Pavlovian, observational, and instructed fear. *Psychological Science* 15(12): 822–828.

Olsson, A., and E. A. Phelps (2007). Social learning of fear. *Nature neuroscience* 10(9): 1095–1102.

Onishi, K. H., and R. Baillargeon (2005). Do 15-month-old infants understand false beliefs? *Science* 308(5719): 255–258.

Packer, C. (1977). Reciprocal altruism in Papio anubis. *Nature* 265: 441–443.

Padoa-Schioppa, C. (2011). Neurobiology of economic choice: A good-based model. *Annual Review of Neuroscience* 34: 333.

Paharia, N., K. S. Kassam, J. D. Greene, M. H. Bazerman (2009). Dirty work, clean hands: The moral psychology of indirect agency. *Organizational Behavior and Human Decision Processes* 109: 134–141.

Parfit, D. (1984). *Reasons and persons*. Oxford, UK: Clarendon Press.

Paxton, J. M., T. Bruni, and J. D. Greene (under review). Are "counter-intuitive" deontological judgments really counter-intuitive? An empirical reply to Kahane et al. (2012).

Paxton, J. M., and J. D. Greene (2010). Moral reasoning: Hints and allegations. *Topics in Cognitive Science* 2(3): 511–527.

Paxton, J. M., L. Ungar, and J. D. Greene (2011). Reflection and reasoning in moral judgment. *Cognitive Science* 36(1) 163-177.

Pedersen, C. A., J. A. Ascher, et al. (1982). Oxytocin induces maternal behavior in virgin female rats. *Science* 216: 648–650.

Perkins, A. M., A. M. Leonard, et al. (2012). A dose of ruthlessness: Interpersonal moral judgment is hardened by the anti-anxiety drug lorazepam. *Journal of Experimental Psychology* 999. doi:10.1037/a0030256.

Petrinovich, L., P. O'Neill, et al. (1993). An empirical study of moral intuitions: Toward an evolutionary ethics. *Journal of Personality and Social Psychology* 64(3): 467.

Pinillos, N. Á., N. Smith, et al. (2011). Philosophy's new challenge: Experiments and intentional action. *Mind & Language* 26(1): 115–139.

Pinker, S. (1997). *How the mind works*. New York: WW Norton & Company.

Pinker, S. (2002). *The blank slate: The modern denial of human nature*. New York: Viking.

Pinker, S. (2007). *The stuff of thought: Language as a window into human nature*. New York: Viking.

Pinker, S. (2008). Crazy love. *Time* 171(4): 82.

Pinker, S. (2011). *The better angels of our nature: Why violence has declined*. New York: Viking

Pizarro, D. A., and P. Bloom (2003). The intelligence of the moral intuitions: Comment on Haidt (2001). *Psychological Review* 110(1): 193–196; discussion, 197–198.

Plato (1987). *The Republic*. London: Penguin Classics.

Plutchik, R. (1980). *Emotion, a psychoevolutionary synthesis*. New York: Harper & Row.

Politisite (September 13, 2011). CNN Tea Party debate transcript part 3. Retrieved February 3, 2012, from http://www.politisite.com/2011/09/13/cnn-tea-party-debate-transcript-part-3-cnnteaparty/#.USAY2-jbb_I.

Posner, M. I., and C. R. R. Snyder (1975). *Attention and cognitive control: Information processing and cognition*, ed. R. L. Solso, pp. 55–85. Hillsdale, NJ: Erlbaum.

Poundstone, W. (1992). *Prisoner's Dilemma: John von Neumann, game theory, and the puzzle of the bomb*. New York: Doubleday.

Powell, J. (2012, November 15). Why climate deniers have no scientific credibility—in one pie chart. *DeSmogBlog*. Retrieved February 3, 2013, from http://www.desmogblog.com/2012/11/15/why-climate-deniers-have-no-credibility-science-one-pie-chart.

Premack, D., and A. J. Premack (1994). Levels of causal understanding in chimpanzees and children. *Cognition* 50(1): 347–362.

Putnam, R. D. (2001). *Bowling alone: The collapse and revival of American community*. New York: Simon & Schuster.

Putnam, R. D., D. E. Campbell (2010). *American grace: How religion divides and unites us*. New York: Simon & Schuster.

Quine, W. V. (1951). Main trends in recent philosophy: Two dogmas of empiricism. *The Philosophical Review* (60) 20–43.

Rand, D. G., S. Arbesman, et al. (2011). Dynamic social networks promote cooperation in experiments with humans. *Proceedings of the National Academy of Sciences* 108(48): 19193–19198.

Rand, D. G., A. Dreber, et al. (2009). Positive interactions promote public cooperation. *Science* 325(5945): 1272–1275.

Rand, D. G., J. D. Greene, et al. (2012). Spontaneous giving and calculated greed. *Nature* 489(7416): 427–430.

Rand, D. G., H. Ohtsuki, et al. (2009). Direct reciprocity with costly punishment: Generous tit-for-tat prevails. *Journal of Theoretical Biology* 256(1): 45.

Rangel, A., C. Camerer, et al. (2008). A framework for studying the neurobiology of value-based decision making. *Nature Reviews Neuroscience* 9(7): 545–556.

Ransohoff, K. (2011). Patients on the trolley track: The moral cognition of medical practitioners and public health professionals. Undergraduate thesis, Department of Psychology, Harvard University.

Rathmann, P. (1994). *Goodnight Gorilla.* New York: Putnam.

Rawls, J. (1971). *A theory of justice.* Cambridge, MA: Harvard University Press.

Rawls, J. (1999). *A Theory of Justice,* rev. ed. Cambridge, MA: Harvard University Press.

Reuters (2008, September 10). No consensus on who was behind Sept 11: Global poll. Retrieved October 29, 2011, from http://www.reuters.com/article/2008/09/10/us-sept11-qaeda-poll-idUSN1035876620080910.

Richeson, J. A., and J. N. Shelton (2003). When prejudice does not pay effects of interracial contact on executive function. *Psychological Science* 14(3): 287–290.

Rodrigues, S. M., L. R. Saslow, et al. (2009). Oxytocin receptor genetic variation relates to empathy and stress reactivity in humans. *Proceedings of the National Academy of Sciences* 106(50): 21437–21441.

Roes, F. L., and M. Raymond (2003). Belief in moralizing gods. *Evolution and Human Behavior* 24(2): 126–135.

Royzman, E. B., and J. Baron (2002). The preference for indirect harm. *Social Justice Research* 15(2): 165–184.

Rozenblit, L., and F. Keil (2002). The misunderstood limits of folk science: An illusion of explanatory depth. *Cognitive Science* 26(5): 521–562.

Rozin, P., L. Lowery, et al. (1999). The CAD triad hypothesis: A mapping between three moral emotions (contempt, anger, disgust) and three moral codes (community, autonomy, divinity). *Journal of Personality and Social Psychology* 76(4): 574.

Ruse, M. (1999). *The Darwinian revolution: Science red in tooth and claw.* Chicago: University of Chicago Press.

Ruse, M., and E. O. Wilson (1986). Moral philosophy as applied science. *Philosophy* 61(236): 173–192.

The Rush Limbaugh Show (2011, September 22). Retrieved February 3, 2013, from http://www.rushlimbaugh.com/daily/2011/09/22/quotes the big voice on the right.

Russell, S. J., and P. Norvig (2010). *Artificial intelligence: A modern approach.* Upper Saddle River, NJ: Prentice Hall.

Sachs, J. (2006). *The end of poverty: Economic possibilities for our time.* New York: Penguin Group USA.

Sarlo, M., L. Lotto, et al. (2012). Temporal dynamics of cognitive-emotional interplay in moral decision-making. *Journal of Cognitive Neuroscience* 24(4): 1018–1029.

Saver, J. L., and A. R. Damasio (1991). Preserved access and processing of social knowledge in a patient with acquired sociopathy due to ventromedial frontal damage. *Neuropsychologia* 29(12): 1241–1249.

Schachter, S., and J. Singer (1962). Cognitive, social, and physiological determinants of emotional state. *Psychological Review* 69(5): 379.

Schaich Borg, J., C. Hynes, et al. (2006). Consequences, action, and intention as factors in moral judgments: An fMRI investigation. *Journal of Cognitive Neuroscience* 18(5): 803–817.

Schelling, T. C. (1968). The life you save may be your own. In S. B. Chase, ed., *Problems in public expenditure analysis.* Washington, DC: Brookings Institute.

Schlesinger Jr, A. (1971). The necessary amorality of foreign affairs. *Harper's Magazine* 72: 72–77.

Seligman, M. (2002). *Authentic happiness: Using the new positive psychology to realize your potential for lasting fulfillment.* New York: Free Press.

Semin, G. R., and A. Manstead (1982). The social implications of embarrassment displays and restitution behaviour. *European Journal of Social Psychology* 12(4): 367–377.

Seyfarth, R. M., and D. L. Cheney (1984). Grooming, alliances and reciprocal altruism in vervet monkeys. *Nature* 308(5959): 3.

Seyfarth, R. M., and D. L. Cheney (2012). The evolutionary origins of friendship. *Annual Review of Psychology* 63: 153–177.

Shenhav, A., and J. D. Greene (2010). Moral judgments recruit domain-general valuation mechanisms to integrate representations of probability and magnitude. *Neuron* 67(4): 667–677.

Shenhav, A., and J. D. Greene (in prep.). Utilitarian calculations, emotional assessments, and integrative moral judgments: Dissociating neural systems underlying moral judgment.

Shenhav, A., D. G. Rand, et al. (2012). Divine intuition: Cognitive style influences belief in God. *Journal of Experimental Psychology: General* 141(3): 423.

Shergill, S. S., P. M. Bays, et al. (2003). Two eyes for an eye: the neuroscience of force escalation. *Science* 301(5630): 187.

Shiffrin, R. M., and W. Schneider (1977). Controlled and automatic information processing: II. Perceptual learning, automatic attending, and a general theory. *Psychological Review* 84: 127–190.

Shiv, B., and A. Fedorikhin (1999). Heart and mind in conflict: The interplay of affect and cognition in consumer decision making. *Journal of Consumer Research* 26(3): 278–292.

Sidanius, J., F. Paratto (2001). *Social Dominance.* New York: Cambridge University Press.

Sidgwick, H. (1907). *The methods of ethics.* Indianapolis, IN: Hackett Publishing Company Incorporated.

Singer, P. (1972). Famine, affluence and morality. *Philosophy and Public Affairs* 1: 229–243.

Singer, P. (1979). *Practical ethics.* Cambridge, UK: Cambridge University Press.

Singer, P. (1981). *The expanding circle: Ethics and sociobiology.* New York: Farrar Straus & Giroux.

Singer, P. (1994). *Rethinking life and death.* New York: St. Martin's Press.

Singer, P. (2004). *One world: The ethics of globalization.* New Haven, CT: Yale University Press.

Singer, P. (2005). Ethics and intuitions. *The Journal of Ethics* 9(3): 331–352.

Singer, P. (2009). *The life you can save: Acting now to end world poverty.* New York: Random House.

Singer, P., and H. Kuhse (1999). *Bioethics: An Anthology.* Malden, MA: Blackwell Publishers.

Singer, T., B. Seymour, et al. (2004). Empathy for pain involves the affective but not sensory components of pain. *Science* 303(5661): 1157–1162.

Singer, T., R. Snozzi, et al. (2008). Effects of oxytocin and prosocial behavior on brain responses to direct and vicariously experienced pain. *Emotion* 8(6): 781.

Sinnott-Armstrong, W. (2006). Moral intuitionism meets empirical psychology. In T. Horgan and M. Timmons, eds., *Metaethics after Moore,* pp. 339–365. New York: Oxford University Press.

Sinnott-Armstrong, W. (2008). Abstract + concrete = paradox. In J. Knobe and S. Nichols, eds., *Experimental philosophy,* pp. 209–230. New York: Oxford University Press.

Sinnott-Armstrong, W. (2009). *Morality without God?* New York: Oxford University Press.

Sloane, S., R. Baillargeon, et al. (2012). Do infants have a sense of fairness? *Psychological Science* 23(2): 196–204.

Sloman, S. (1996). The empirical case for two systems of reasoning. *Psychological Bulletin* 119(1): 3–22.

Sloman, S., Fernbach, P. M. (2012). I'm right! (For some reason). *New York Times.* Retrieved November 8, 2012, from http://www.nytimes.com/2012/10/21/opinion/sunday/why-partisans-cant-explain-their-views.html?_r=0.

Slovic, P. (2007). If I look at the mass I will never act: Psychic numbing and genocide. *Judgment and Decision Making* 2: 79–95.

Small, D. A., and G. Loewenstein (2003). Helping a victim or helping the victim: Altruism and identifiability. *Journal of Risk and Uncertainty* 26(1): 5–16.

Small, D. M., and G. Loewenstein (2005). The devil you know: The effects of identifiability on punitiveness. *Journal of Behavioral Decision Making* 18(5): 311–318.

Smart, J. J. C., and B. Williams (1973). *Utilitarianism: For and against.* Cambridge, UK: Cambridge University Press.

Smith, A. (1759/1976). *The Theory of Moral Sentiments.* Indianapolis: Liberty Classics.

Smith, J. M. (1964). Group selection and kin selection. *Nature* 201: 1145–1147.

Smith, M. (1994). *The moral problem.* Oxford, UK, and Cambridge, MA: Blackwell.

Snopes.com (2012, November 7). Letter to Dr. Laura. Retrieved February 3, 2013, from http://www.snopes.com/politics/religion/drlaura.asp.

Sober, E., and D. S. Wilson (1999). *Unto others: The evolution and psychology of unselfish behavior.* Cambridge, MA: Harvard University Press.

Stanovich, K. E., and R. F. West (2000). Individual differences in reasoning: Implications for the rationality debate? *Behavioral and Brain Sciences* 23(5): 645–665.

Stephens-Davidowitz, S. (2012). The effects of racial animus on a black presidential candidate: Using google search data to find what surveys miss. Available at http://www.people.fas.harvard.edu/-sstephen/papers/RacialAnimusAndVotingSethStephensDavidowitz.pdf.

Stevenson, B., and J. Wolfers (2008). Economic growth and subjective well-being: Reassessing the Easterlin paradox. National Bureau of Economic Research.

Stevenson, R. L. (1891/2009). *In the South Seas.* Rockville, MD: Arc Manor.

Stich, S. (2006). Is morality an elegant machine or a kludge. *Journal of Cognition and Culture* 6(1–2): 181–189.

Stoll, B. J., N. I. Hansen, et al. (2010). Neonatal outcomes of extremely preterm infants from the NICHD Neonatal Research Network. *Pediatrics* 126(3): 443–456.

Strohminger, N., R. L. Lewis, et al. (2011). Divergent effects of different positive emotions on moral judgment. *Cognition* 119(2): 295–300.

Stroop, J. R. (1935). Studies of interference in serial verbal reactions. *Journal of Experimental Psychology: General* 121(1): 15.

Stuss, D. T., M. P. Alexander, et al. (1978). An extraordinary form of confabulation. *Neurology* 28(11): 1166–1172.

Sunstein, C. R. (2005). Moral heuristics. *Behavioral and Brain Sciences* 28(4): 531–542; discussion, 542–573.

Susskind, J. M., D. H. Lee, et al. (2008). Expressing fear enhances sensory acquisition. *Nature Neuroscience* 11(7): 843–850.

Suter, R. S., and R. Hertwig (2011). Time and moral judgment. *Cognition* 119(3): 454–458.

Swami, P. (2009, June 15). GOP hits Obama for silence on Iran protests. CBS News.

Tajfel, H. (1970). Experiments in intergroup discrimination. *Scientific American* 223(5): 96–102.

Tajfel, H. (1982). Social psychology of intergroup relations. *Annual Review of Psychology* 33(1): 1–39.

Tajfel, H., and J. C. Turner (1979). An integrative theory of intergroup conflict. *The Social Psychology of Intergroup Relations* 33: 47.

Talmy, L. (1988). Force dynamics in language and cognition. *Cognitive Science* 12(1): 49–100.

Tennyson, A., and M. A. Edey (1938). *The poems and plays of Alfred Lord Tennyson.* New York: Modern Library.

Tesser, A., L. Martin, et al (1995). The impact of thought on attitude extremity and attitude-behavior consistency. In *Attitude strength: Antecedents and consequences,* ed. R. E. Petty and J.A. Krosnick, 73-92. Mahwah, NJ: Lawrence Erlbaum.

Thomas, B. C., K. E. Croft, et al. (2011). Harming kin to save strangers: Further evidence for abnormally utilitarian moral judgments after ventromedial prefrontal damage. *Journal of Cognitive Neuroscience* 23(9): 2186–2196.

Thompson, L., and G. Loewenstein (1992). Egocentric interpretations of fairness and interpersonal conflict. *Organizational Behavior and Human Decision Processes* 51(2): 176–197.

Thomson, J. (1985). The trolley problem. *Yale Law Journal* 94(6): 1395–1415.

Thomson, J. (2008). Turning the trolley. *Philosophy and Public Affairs* 36(4): 359–374.

Thomson, J. J. (1976). Killing, letting die, and the trolley problem. *The Monist* 59(2): 204–217.

Thomson, J. J. (1990). *The realm of rights.* Cambridge, MA: Harvard University Press.

Tienabeso, S. (2012, January 25). Warren Buffett and his secretary talk taxes. ABC News.

Tonry, M. (2004). *Thinking about crime: Sense and sensibility in American penal culture.* New York: Oxford University Press.

Tooley, M. (2008). The problem of evil. *Stanford Encyclopedia of Philosophy.*

Trémolière, B., W. D. Neys, et al. (2012). Mortality salience and morality: Thinking about death makes people less utilitarian. *Cognition.*

Trivers, R. (1971). The evolution of reciprocal altruism. *Quarterly Review of Biology* 46: 35–57.

Trivers, R. (1972). Parental investment and sexual selection. In B. Campbell, ed., *Sexual selection and the descent of man, 1871–1971,* pp. 136–179. Chicago: Aldine.

Trivers, R. (1985). *Social evolution.* Menlo Park, CA: Benjamin/Cummins Publishing Co.

Unger, P. K. (1996). *Living high and letting die: Our illusion of innocence.* New York: Oxford University Press.

Union of Concerned Scientists (2008). Each country's share of CO_2 emissions. Retrieved November 7, 2011, from http://www.ucsusa.org/global warming/science and impacts/science/each-countrys-share-of-co2.html.

United Nations (2011). Human development report 2011. Retrieved February 3, 2013, from http://hdr.undp.org/en/media/HDR 2011 EN Complete.pdf.

United Nations Office of Drugs and Crime (2011). Global study on homicide. Retrieved February 3, 2013, from http://www.unodc.org/documents/data-and-analysis/statistics/Homicide/Globa study on homicide 2011 web.pdf.

US Energy Information Administration (2009). Emissions of greenhouse gases in the United States in 2008. USDOE Office of Integrated Analysis and Forecasting.

US General Accounting Office (1990). Death penalty sentencing: Research indicates pattern of racial disparities.

US House of Representatives (2008). Final vote results for roll call 681. Retrieved February 3, 2013, from http://clerk.house.gov/evs/2008/roll681.xml.

US Senate (2008). US Senate roll call votes 110th Congress, 2nd session, on passage of the bill (HR 1424 as amended). Retrieved Feburary 3, 2013, from http://www.senate.gov/legislative/LIS/roll call lists/roll call vote cfm.cfm?congress=110&session=2&vote=00213.

Valdesolo, P., and D. DeSteno (2006). Manipulations of emotional context shape moral judgment. *Psychological Science* 17(6): 476–477.

Valdesolo, P., and D. DeSteno (2007). Moral hypocrisy: Social groups and the flexibility of virtue. *Psychological Science* 18(8): 689–690.

Valdesolo, P., and D. DeSteno (in press). Moral hypocrisy: The flexibility of virtue. *Psychological Science.*

Vallone, R. P., L. Ross, et al. (1985). The hostile media phenomenon: Biased perception and perceptions of media bias in coverage of the Beirut massacre. *Journal of Personality and Social Psychology* 49(3): 577.

van Yperen, N. W., K. van den Bos, et al. (2005). Performance-based pay is fair, particularly when I perform better: Differential fairness perceptions of allocators and recipients. *European Journal of Social Psychology* 35(6): 741–754.

Variety (1989). TV Reviews—Network: Everybody's baby. 3335(7): May 31.

Von Neumann, J., and O. Morgenstern (1944). *Theory of games and economic behavior.* Princeton, NJ: Princeton University Press.

Wade-Benzoni, K. A., A. E. Tenbrunsel, and M. H. Bazerman (1996). Egocentric interpretations of fairness in asymmetric, environmental social dilemmas: Explaining harvesting behavior and the role of communication. *Organizational Behavior and Human Decision Processes* 67(2): 111–126.

Waldmann, M. R., and J. H. Dieterich (2007). Throwing a bomb on a person versus throwing a person on a bomb: Intervention myopia in moral intuitions." *Psychological Science* 18(3): 247–253.

Walster, E., E. Berscheid, et al. (1973). New directions in equity research. *Journal of Personality and Social Psychology* 25(2): 151.

Warneken, F., B. Hare, et al. (2007). Spontaneous altruism by chimpanzees and young children. *PLOS Biology* 5(7): e184.

Warneken, F., and M. Tomasello (2006). Altruistic helping in human infants and young chimpanzees. *Science* 311(5765): 1301–1303.

Warneken, F., and M. Tomasello (2009). Varieties of altruism in children and chimpanzees. *Trends in Cognitive Sciences* 13(9): 397.

Wert, S. R., and P. Salovey (2004). Introduction to the special issue on gossip. *Review of General Psychology* 8(2): 76.

West, T. G., and G. S. West (1984). *Four texts on Socrates.* Ithaca, NY: Cornell University Press.

Whalen, P. J., J. Kagan, et al. (2004). Human amygdala responsivity to masked fearful eye whites. *Science* 306(5704): 2061–2061.

Wiggins, D. (1987). *Needs, values, and truth: Essays in the philosophy of value.* Oxford, UK: Blackwell.

Wike, R. (2009, September 21). Obama addresses more popular U.N. Pew Research Global Attitudes Project.

Wilson, D. S. (2003). *Darwin's cathedral: Evolution, religion, and the nature of society.* Chicago: University of Chicago Press.

Wilson, T. D. (2002). *Strangers to ourselves: Discovering the adaptive unconscious.* Cambridge, MA: Harvard University Press.

Winkelmann, L., and R. Winkelmann (2003). Why are the unemployed so unhappy? Evidence from panel data. *Economica* 65(257): 1–15.

Winner, C. (2004). *Everything bug: What kids really want to know about insects and spiders.* Minocqua, WI: Northword Press.

Wittgenstein, L. (1922/1995). *The Tractatus Locigo-Philosophicus,* trans. C. K. Ogden. London: Routledge and Kegan Paul.

Wolff, P. (2007). Representing causation. *Journal of Experimental Psychology: General* 136(1): 82.

Woodward, A. L., and J. A. Sommerville (2000). Twelve-month-old infants interpret action in context. *Psychological Science* 11(1): 73–77.

Woodward, J., and J. Allman (2007). Moral intuition: Its neural substrates and normative significance. *Journal of Physiology–Paris* 101: 179–202.

World Bank (2012, February 29). World Bank sees progress against extreme poverty, but flags vulnerabilities. Retrieved February 3, 2013, from http://www.worldbank.org/en/news/press-release/2012/02/29/world-bank-sees-progress-against-extreme-poverty-but-flags-vulnerabilities.

Wright, R. (1994). *The moral animal: Why we are, the way we are: The new science of evolutionary psychology.* New York: Vintage.

Wright, R. (2000). *NonZero: The logic of human destiny.* New York: Pantheon.

Image Credits

2.1   Magic Corner: Courtesy of author.

2.2   Simulate Violence: From Crushman, F., K. Gray, A. Gaffey, and W. B. Mendes (2012). Simulating murder: the aversion to harmful acrtion. *Emotion* 12(1): 2–7; reprinted with permission from APA.

2.3   Eyespot Diagram: From Haley, K. and D. Fessler (2005). Nobady's watching? *Evolution and Human Behavior*, 26(3): 245; reprinted with permission from Elsevier.

2.4   Shapes: From Hamlin, J. K.; K. Wynm, and P. Bloom(2007). Social evaluation by preverbal infants. *Nature*: 450(7169): 557–559; reprinted with permission from Macmillan Publishers Ltd

2.5   Decision Tijme: Courtesy of the author.

3.1   Cities in Public Goods Game: Image courtesy of the author, with data from: Herrmann, B., C. Thöni, and S. Gachter(2008). Antisocial punishment across societies. *Science* 319(5868): 1362–1367.

3.2   Cooperating Levels: From Herrman, B., C Thöni, and S. Gachter, (2008). Antisocial punishment across societies. *Science* 319(5868): 1362–1367. Reprinted with permission from AAAS. Image adapted by permission of Herrman, Thoni, and Gachter.

3.3   Scientific Literacy: From Kahan, D. M., et al (2012). The polarising impact of science literacy and numeracy on perceived climate change risks. *Nature Climate Change* 2(10): 732–735; reprinted with permission form Macmillan Publishers Ltd.

3.4   Force Applied: From Shergill, S. S., P. M. Bays, C. D. Firth, and D. M. Wolport (2003). Two eyes for an eye: the neuroscience of force escalation. *Science* 301(5630): 187–187; reprinted with permission from AAAS.

4.1   Footnridge: Courtesy of author.

4.2   Switch: Courtesy of author.

4.3   Dual-process morality emotions: This work uses emoticons licensed under the Creative commons Attribution 3.0 Unported License. To view a copy of this license, visit http://creativecommons.org /licenses/by/3.0/or send a letter to Creative Commons, 444 Castro Street, Suite 900. Mountain View. California. 94041, USA.

5.1   Three brains: *Top row*: From McClure, S. M., D. I. Laibson, G. Loewenstein, and J. D. Chohen (2004). Separate neutral Systems valued immediate and delayed monetary rewards. *Science* 306(5695): 503–507. Reprinted with permission from AAAS. Adapted with permission of authors *Middle row*: From Ochsner, K. N., S. A Bunge, J. J. Gross, and J. D. E. Gabrieli (2002). Rethinking feeling: An fMRI study of the cognitive regulation of emotion. *Journal of Cognitive Neuroscience* 14(8): 1215–1229; © 2002 by the Massachusetts Institute of Technology. *Bottom row*: From Cunningham, W. A., M. K. Johnson, C. L. Raye, J. C. Gathenby, J. C. Gore, and M. R. Banaji (2004). *Psychological Science* 15(12): 806–813; © 2004 by Cunningham, Johnson, Raye, Gatenby, Gore, & Banaji. Reprinted by Permission of SAGE Publications.

5.2   Fearful Eyes: From Whalem, P. J., J. Kagan, R. G. Cook, et al. (2204). Human amygdala reponsivity to masked fearful eye white. Science 306(5704): 2061–2061; Reprinted with permission from AAAS. From *Good Night, Gorilla* by Peggy Rathmann, copyright © 1194 Peggy Rathmann. Used by permission of G. P. Putnam's Sons, a division of Penguin Group (USA) LLC.

9.1   Remote footbridge: Courtesy of author.

9.2   Footbridge switch: Courtesy of author.

9.3   Footbridge pole: Courtesy of author.

9.4   Obstacle collide: Courtesy of author.

9.5   Loop case: courtesy of author.

9.6   Collision alarm: courtesy of author.

9.7   Action plans: Courtesy of author.

9.8   Casual chain footbridge: Courtesy of author.

9.9   Causal chain loop: Courtesy of author.

9.10   Diagram loop: Courtesy of author.

9.11   Spatial loop: Courtesy of author.

9.12   Diagram dual-process emoticons: This work uses emoticons licensed under the Creative Commons Attribution 3.0 Unported License. To view a copy of this license, visit http://creativecommons.org/licenses/by/3.0/ or send a letter to Creative Commons, 444 Castro Street, Suite 900, Mountain View, California, 94041, USA.

Index